Lewis R. Heim

Lewis R. Heim

Mechanical Genius of Modern Industry

by

Robert V. Jacobs

THE
CONNECTICUT
PRESS
Cheshire, CT

First Edition

Cataloging-in-Publication Data 2019935649

Name: Jacobs, Robert V.
Title: Lewis R. Heim: Mechanical Genius of Modern Industry / by Robert V. Jacobs.
Description: Softcover edition | Cheshire, CT: The Connecticut Press, 2019 / Illustrated, includes annotations, appendices, and index, 456 pp.

Identifiers: LCCN 2019935649 | ISBN 978-0-99-77907-2-6
BISAC: BIO003000 BIOGRAPHY & AUTOBIOGRAPHY / Business | HIS036060 HISTORY / United States / 20th Century | TEC046000 TECHNOLOGY & ENGINEERING / Steel Industry / Machinery / Centerless Cylindrical Grinder / Heim Rod End

Printed in the United States of America

*Dedicated to the memory of my sister, Lynda,
who provided the foundation for starting this book
and the inspiration to finish it.*

*Lynda Carol Jacobs
1960–2014*

Preface

I am the great-grandson of Lewis R. Heim through his daughter, Florence, and her daughter, Elizabeth Anne. Since Mr. Heim lived a long time, I had the privilege of meeting my great-grandfather even though I was very young and retain only a vague remembrance. I was taken to his bearings plant in Fairfield, Connecticut, many times and given various bearings to play with while the adults talked. When I was young, even though my parents told me of my great-grandfather and his inventions of the centerless grinder and the rod end bearing, I did not understand the significance of his accomplishments until I wrote this biography.

I started doing research in 2010 using a set of patents issued to Mr. Heim that were obtained in 2001 by my sister, Lynda, from the United States Patent and Trademark Office *(USPTO)* in Washington, DC. Lynda presented the collection to our grandmother *(Florence)* on her 100th birthday in October 2001. For patents issued prior to 1976, the USPTO did not publish a master list by the inventor's name. Unless you knew the patent number, you had to conduct a manual search from microfilm or patent record books. My sister spent many days combing through microfilm and found 59 patents granted to Lewis Heim. As I conducted research on this biography, I discovered many additional patents for a total of 92, including the original patents for the first two-wheel Centerless Cylindrical Grinder.

My great-grandfather's inventions and achievements benefited many industries, with their greatest impact on the manufacture of small cylindrical components and mechanical bearings used in automobiles, aircraft, and general machinery. I wrote this book to ensure that an accurate history of his accomplishments resides in one place.

R. V. J.
Windsor, CT
2019

Preface

Contents

Contents

Acknowledgments

Family Members

Lynda C. Jacobs: Lewis Heim's great-granddaughter via Elizabeth A. Van Etten Jacobs and Florence H. Van Etten

George L. Jacobs: husband to Heim's granddaughter, Elizabeth A. Van Etten Jacobs

Charles W. Heim: Heim's grandson via Charles R. Heim

Carl H. Van Etten: Heim's grandson via Florence H. Van Etten

Bonnie Jean H. Perkins: Granddaughter of Alfred H. Heim via Richard B. Heim

Institutions and Companies

Smithsonian Institution: National Museum of American History, Washington, DC. Catalogs, advertisements, articles, and photos of the Ball and Roller Bearing Company, the Heim Centerless Grinder, and the Heim Company

    Repository for Heim Centerless Grinder #209

    Hall of Tools Records on Grinding

United States National Archives and Records Administration:
    National Archives, New York , NY
        Court documents on Ball & Roller Bearing vs. The F. C. Sanford
        Manufacturing Co., 1920–1924
    National Archives, Boston, MA
        Court documents: The Heim Grinder Co. vs. The Fafnir Bearing
        Co., 1924–1926
    National Archives: College Park, MD
        Patent Assignments

United States Department of the Interior, National Park Service
    National Register of Historic Places
        Registration form for the Ball and Roller Bearing Company

Danbury Museum and Historical Society
    Danbury hatting history

Danbury, CT Town Clerk's Office
    Danbury land records

Danbury Public Library
    City directories and various articles

Bridgeport, CT Public Library
    Various articles

American Precision Museum, Windsor, VT
    Repository of Heim Centerless Grinder No. 158.
    APM Library: *American Machinist* magazine

Worcester Polytechnic Institute, Worcester, MA.
    Gordon Library: *Machinery* magazine
    Articles, advertisements, and photographs

The Ball and Roller Bearing Company, New Milford, CT
    David Nohe *(CEO)* and Michael Smith (sales manager)
    Date: April 11, 2013: History, catalogs, and photographs

The Heim Company, Fairfield, CT
    Thomas Crainer, vice president and general manager
    Mike Nastasia, manufacturing engineer
    Andy Henn, manufacturing engineer
    November 30, 2016: History, catalogs, and photographs

Flugmuseum Messerschmitt, Manching, Germany
    Photograph: Slotted plain bearing obtained from a WWII Messerschmitt Bf 109

Illustrations

The following list covers illustrations placed within the chapters. Illustrations placed in the appendices are listed in the table of contents for each appendix. Numbers in italics and parentheses indicate patent numbers.

Chapter 7: Centerless Grinder Competitors

Chapter 14: Accomplishments

Introduction

Lewis Rasmus Heim *(September 19, 1874–March 29, 1964)* was an American machinist, engineer, and manufacturer who invented the Centerless Cylindrical Grinder and the Rod End Bearing known as the "Heim Joint. He was also one of the pioneers of modern antifriction and spherical bearings.

Lewis Heim had an inherent ability to visualize complex mechanisms and mechanical processes that resulted in the creation of novel machines, manufacturing methods, tools, and mechanical bearings. He was granted his first patent in 1896 at the age of 22 and his last patent in 1968, four years after his death at age 89. His inventions ranged from machines to manufacture hats and automate the ironing of fold collars, to precision grinders and swaging devices for bearing components plus plain and antifriction bearings used in submarines, tanks, aircraft, trucks, automobiles, and industrial machinery. With every invention Heim sought to improve product designs, increase production rates, and reduce costs to compete in highly competitive markets.

As the inventor of the centerless grinder, which was critical to the growth and success of the automobile and aircraft industries, Heim saw his invention revolutionize machine shop practices around the world through the high-speed production of cylindrical metal components used in the internal combustion engine, mechanical bearings, and machines. His invention continues in wide use over 100 years after its conception.

As the originator of the Heim Joint, his rod end bearings were installed in nearly every American warplane produced at the end of World War II. He became internationally known as an industrialist and designer of machinery and at the time of his retirement, he was president of the Heim Company of Fairfield, Connecticut. During his long career, which spanned more than seven decades, he was granted 92 patents for his inventions, many of which are still in use today.

In addition to his life story, this biography provides a detailed summary of Mr. Heim's inventions and contributions to the advancement of machine tools and mechanical bearings, including detailed technical discussions and explanations of their importance. In support of these discussions, I have included summaries of the changing technologies, events, and people that influenced Heim's life and career. Materials used for the biography were assembled from historical publications, product catalogs, public records, legal documents, patents, and trade articles, as well as interviews of industry executives and descendants.

In the early 2000s, the Internet became a virtual encyclopedia containing vast amounts of information that previously had been difficult to obtain. I discovered many websites containing statements or stories on the origins of the centerless grinder and the rod end bearing, with much of the information either inaccurate or erroneously credited to others for Mr. Heim's work. Given the lack of historical information on Mr. Heim's accomplishments, what started out as a small project to publish a website page evolved into this biography.

Chapter One

The Late 1800s

The Industrial Revolution

Lewis Heim lived almost nine decades, and his life and career were shaped by the rapidly changing times of the late nineteenth and early twentieth centuries. Influences included the Industrial Revolution, farm life, automobiles, and two world wars. The Industrial Revolution is divided into two periods and involved the transition from hand production methods to those of machines, new chemical and iron production processes, water and steam power, and the development of machine tools. It originated in Britain around 1760 and spread throughout Western Europe and North America, and marks a major turning point in history with a sustained improvement of living standards.

The First Industrial Revolution saw the introduction of new technologies involving textiles, steam engines, and iron making and ended sometime between 1820 and 1840. After a long transition period that included the American Civil War, the Second Industrial Revolution began around 1870 and ended around the First World War with mass production and the assembly line. It was characterized by rapid industrial development in Western Europe, the United States, and Japan and brought forth new technologies, especially electricity, the internal combustion engine, new materials, and communication technologies. Whereas the First Industrial Revolution was centered on low-cost iron, steam technologies, and textile production, the Second Industrial Revolution evolved around low-cost steel, railroads, oil, chemicals, and electricity. These ingredients led to the creation of the internal combustion engine, electric motors, heavy machinery, machine tools, and modern mechanical bearings.

Mechanical bearings are generally overlooked in history, but were an important development in the latter stages of the Industrial Revolution and a critical component of modern machines. The function of a bearing is to provide support while allowing motion *(linear or free rotation)* between moving parts. Mechanical bearings include the class of rolling bearings *(ball and roller type)* and plain bearings such as spherical types in which one surface slides over the other. Most spherical bearings use metal-on-metal sliding surfaces but can include other materials such as ceramics and plastics. Without bearings, modern industry *(machines and transportation)* would not exist, leaving the world stuck in the First Industrial Revolution.

Modern bearings need to support heavy, moving loads and, in many cases, high-speed movement. The two enemies of bearings are friction and wear, which lead to poor performance and component failure. These are addressed through high-strength materials, precision manufacturing, lubrication, and cooling. The strength and wear resistance of steel and its alloys are far superior to iron and softer metals. Precision is required to produce bearing components manufactured to exact dimensions so that the parts fit together and move with the least amount of resistance. Lubricants such as grease and oil are needed to minimize friction, wear, and heat generation. Lubricants are one of the many chemicals derived from petroleum through various refining processes.

Iron was the primary metal used for strength before 1860. Steel, which is made from iron and small amounts of carbon, and hence called carbon steel, is at least two times stronger than iron. And alloy steels, containing nickel, chrome, and other elements, are many times stronger than basic carbon steel. In addition, steel and its alloys, due to superior physical properties, are more resistant to cracking and failure under long-term dynamic loads.

Steel had been known and manufactured well before 1860, but was expensive compared to iron. In 1856, Sir Henry Bessemer, an English engineer and businessman, patented the first inexpensive industrial process for the mass production of steel from molten pig iron. The Bessemer process revolutionized steel manufacture by increasing the scale and speed of production, reducing labor requirements, and decreasing overall production costs. After patenting his process in 1856, steel became an inexpensive replacement for iron in many products and eventually became widely used in machines, structures, and applications requiring high-strength materials, such as precision machine tools and bearings.

Petroleum *(crude oil)* had been used since ancient times but was in raw form and not readily available. In 1847, the first process to distill kerosene from petroleum was invented by James Young of Scotland. Several years later, Young and his business partners built the first commercial refinery to produce kerosene, various types of oils, and paraffin wax. Kerosene provided a cheaper alternative to whale oil as a fuel for lighting and demand grew quickly worldwide. In America, the demand for petroleum products set off the search for supplies, leading to the first commercially drilled well by Edwin Drake near Titusville, Pennsylvania, in 1859. Later, refineries started producing various products from petroleum, including gasoline, lubricants, and other chemicals. One of the most famous of the refiners was the Standard Oil Company founded by John D. Rockefeller. Prior to its formation in 1870, Rockefeller and his partners operated a refinery in Cleveland that was one of the first to produce many types of petroleum-based chemicals, including kerosene, jellies, waxes, and lubricants for machines.

The abundance of the new fuels and chemicals plus the new steel materials led to the development of the internal combustion engine and the birth of the automobile industry in the late 1800s. As the auto industry grew, demand increased not only for fuel *(gasoline and diesel)*, but also for the greases and oils to lubricate bearings and other moving parts of engines used in automobiles and industrial machines.

Electricity was another key component of the Second Industrial Revolution that was a product of experiments in magnetism and electricity made by Michael Faraday. In 1821, Faraday developed a rotating device using electromagnetism that was the first direct current electric motor. However, it was not until the late 1800s that the world would see the greatest progress as electricity turned from a scientific curiosity into an essential tool for modern life and became a driving force of the Second Industrial Revolution.

In the 1880s, as the demand for electricity accelerated, two types of large power plants were built. The first type were those using steam to rotate a turbine that was connected to an electric generator. These plants burned carbon-based chemical fuels *(wood, coal, oil, and natural gas)* to evaporate water and generate high pressure and temperature steam. The other type was the hydroelectric plant that used water pressurized from a change in elevation to rotate turbines connected to electric generators. The low-cost electricity generated by these two types of plants was used to provide lighting and power to homes, cities, and factories. In factories, electric motors replaced steam and water to drive assembly lines, industrial machinery, and many other processes.

However, as important as electricity was in lighting, electric motors, and other powered devices, there was one application that was essential to closing out the Second Industrial Revolution. Without this electricity-based technology, the automobile industry and mass production would have taken much longer to develop, if at all. This technology allowed for the creation of materials critical to the production of precision machine tools and the identical and interchangeable parts of internal combustion engines, automobiles, and aircraft. This technology, which will be discussed later, was commercialized at the turn of the century and produced the final component to the family of machine tools necessary for commercialization of assembly line and mass-production technologies.

German Immigration

The population of America grew rapidly after colonization in the early 1600s. By 1700 approximately 250,000 people, mostly Europeans, had settled in the eastern shore areas between Virginia and Massachusetts[1] By 1800, the population increased to 5.3 million located along the Atlantic Ocean and extending westward to the Mississippi River. Due to high birth and immigration rates, by 1900, America's population rose to seventy-six million, spread out over the land area of the lower forty-eight states[2]

Germans were a large part of the newly created and growing America. German immigration to America began before 1700 but the major waves of immigrants took place after 1850. Many had immigrated to Pennsylvania seeking religious freedom and others arrived through a redemption system where they were required to work for their sponsor for a defined period to repay the cost of the voyage.[3]

The Industrial Revolution was late to arrive in Germany. Whereas it started in Britain around 1760 and moved to America around 1780, it did not arrive in Germany until after 1850. Germany experienced many changes during this time, with one being reforms to promote peasant land ownership. As these reforms continued, many rural peasants became landless and indigent, forcing them to leave Germany for America. The large waves of German migration began around mid-century, in which the 1850s saw around one million Germans arrive, mostly small farmers and their families. The 1870s saw more than 700,000 Germans immigrate, with another 1.4 million in the 1880s.[4,5] After arriving in America, many continued west to the north-central states of Wisconsin, Minnesota, Michigan, North Dakota, Nebraska, and Iowa. However, some stayed in the east and settled in Pennsylvania and New York, all of which already had large German populations.

Changes on the Farm

In the early years of America, agriculture was the primary industry. In 1800 almost 75 percent of working people earned their living operating small farms or plantations. And even though the First Industrial Revolution was well under way, manufacturing at that time was in its infancy, employing less than three percent of the labor force. By 1900, the Second Industrial Revolution expanded manufacturing's role in the economy to 20 percent while agriculture shrank to 40 percent. But due to the large increase in population between 1800 and 1900, the total number of people working in agriculture *(farming and ranching)* rose tremendously from around 1.4 million to more than 11 million.[6]

Fig. 1-1: Charles George and Frederika Kleifoth Heim, Danbury, Connecticut: ca. 1900.

The Second Industrial Revolution had a tremendous impact on agriculture and farm life. In the latter half of the 1800s, animal-driven farm machines were giving way to steam-powered machines and later, machines driven by the internal combustion engine. Farmers benefited from increased mechanization, greater transportation options, and improved cultivation methods. Nonetheless, the financial situations of many farming families became precarious. Record crop yields resulted in lower prices while production costs increased, a combination that threw many farmers into deep debt.

Adding to these burdens, farmers endured a long period of depressed prices for grain, cotton, and other commodities during the 1870s and 1880s. Then came the Panic of 1893, another economic depression involving shaky railroad financing and overbuilding, bank failures, and farm closings.

These events encouraged many farmers in the 1880s and 1890s to move to the cities, where earning a living was easier. The attraction to the city was powerful, for large numbers of farmers succumbed to the temptations of urban promises and packed their bags. Jobs, higher wages, and such technological wonders as electricity and the telephone gradually took their toll on the rural population.

From the Farm to the City

There are very few records for Lewis Heim and his family during his childhood. His parents were Charles George and Frederika *(Kleifoth)* Heim, both of whom emigrated from western Germany *(Figure 1-1)*. Charles Heim was born in 1840 in the Hesse region near Frankfurt and in 1870 made the voyage across the Atlantic and settled in East Fishkill, near Poughkeepsie, in Dutchess County, New York. Dutchess County, located in southeastern New York, is bordered by the Hudson River and upper western Connecticut. The Dutch settled the area in the late 1600s, and from around 1715 to 1730, most incoming settlers were Germans.[7] Frederika *(Frances)* Kleifoth-Heim, born in 1844, was from the Hamburg area and immigrated shortly thereafter and joined Charles to start their lives in America.

Charles and Frances Heim worked a dairy farm in the town of East Fishkill in the southern region of Dutchess County, New York. While working the farm, Charles and Frances had five children: two girls and three boys. The two girls were born first and last with the oldest, Wilhelmina, born on December 21,1873, and the youngest, Laura Louise, born on March 23, 1885. In the middle were the three boys, Lewis Rasmus, born September 19, 1874, Alfred Henry, born May 23, 1877, and William Charles, born December 3, 1878.

As farmwork during those times was mainly manual labor, Lewis and his siblings spent most of their time assisting their parents in the daily chores. Most rural children at that time did not receive a formal education and Lewis was no exception, with his education consisting of attending a "little red schoolhouse" in the area, up through the eighth grade.[8]

There is little doubt that life on the farm exposed Lewis to the rigors of farming and the labor-intensive machines used during that time. Whereas Lewis never liked farmwork, he developed a passion for machines and things mechanical. The access to machines at an early age most likely gave Heim the opportunity to experiment with improvements and new concepts, and hone his skills as a machinist and inventor. With virtually no opportunities to pursue his passion on the farm, in 1890, at the age of 16, Heim moved to Danbury, Connecticut, to seek work.[9]

Chapter Two

Danbury and Hatting Machines

Hat City

Danbury is located in central western Connecticut, about 70 miles northeast of New York City and 30 miles southeast of East Fishkill, where Heim was born. First settled in 1685, most were farmers but the town quickly transitioned into a center for making hats. The first hatmaker of record in Danbury was Zadoc Benedict, who founded a firm in 1780 that turned out three hats a day. In 1787, Danbury manufacturers Oliver Burr and Company imported an English hatter to train apprentices who became expert hatters and opened their own shops.[1]

Around 1800, Danbury became a center for the manufacture of hats, and hatting became Danbury's main industry. At first, all hats were handmade with simple tools but, over time, machines replaced hand tools and hats were produced at a faster rate. As machines took over some of the labor, many of the smaller hat shops closed because they could not compete with the larger, speedier factories. In 1860, there were 12 large hat factories employing 2,000 workers and making around 1.5 million hats annually. By 1887, 30 hat companies were producing around 5 million hats a year and employed around 4,000 hatters, with many hundreds more employed in satellite companies supplying machines, goods, and services to the major hat-making companies. Around 1892, the hatting industry shifted from making hats ready to sell to producing unfinished hat bodies that were sold to smaller hat shops for finishing. By 1904, Danbury was producing around 24 percent of America's finished hats and 75 percent of the unfinished "rough" hat bodies. The turn of the century was the period of peak hat production in Danbury that led to its nicknames "The Hat Capital of the World," "Hat City," and "Danbury Crowns Them All."[2]

Danbury was not the only city making hats. There were many other cities with large concentrations of hatting-related companies, including New York, Philadelphia, Boston, Chicago, and Cincinnati. Plus there were many more people employed by the hat industry in the other Connecticut towns, including its next-door neighbor, Bethel, and South Norwalk, located on Long Island Sound. New York was also a major center for wholesale, commercial, and retail sales, with many hundreds of shops from which to purchase all types of men's and women's hats.

Hats made in the late 1800s were generally made from felt, straw, or leather. With felt hats there were two wholly separate processes involved in their manufacture. The first process was manufacture of the felt and what was called the "rough" hat body. The second process was making the finished hat complete with its final shape, color, stiffness, brim, lining, bands, or other features. Some companies made both the felt and the finished hat but there were many companies that specialized in one or the other. In the first process, the manufacture of felt, there were many operations involved. A short summary is provided as follows:[3]

Felt is a non-woven fabric created by felting animal fur or wool fibers, and at that time was the most common material used to make hats. Felting, or more appropriately "wet felting," is the process of causing small fibers to bind together *(interlock)* when treated with hot water and rubbing. The fibers contain "scales" that are directional *(similar to pine cones)* and "interlock" with each other when rubbed together and lubricated with hot water.

In the late 1800s, fur felt was widely used in hats, with fur producing a much softer feel than wool. Fur felt hats were made from rabbit and hare, with higher-priced hats made from the less plentiful beaver, nutria, and muskrat. With rabbit and hare, fur for felting is obtained from the back, flanks, and belly, with the best fur obtained from the small area making up the center of the back. With beaver, nutria, and muskrat, only the short and fine underfur was used since fur on the rest of the animal produced poor-quality felt.

Felting is believed to have originated in Asia where nomadic tribes discovered the tendency of fibers to mat together when exposed to warm and wet conditions. Over the years, Western cultures adopted and improved the processes for felting by adding machines and chemicals. By the late 1800s an enhanced process for wet felting called "carroting" was the preferred method for producing high-quality felt. In the carroting process, cleaned and sheared animal skins were treated with a solution of dilute mercuric nitrate that opened up the "scales" on the fur fibers to enhance the interlocking action and thus the felt quality. After treating, the skins were dried in an oven that caused the thin fur to turn a shade of orange similar to carrots, which led to the term "carroting." But the solution of mercuric nitrate was very toxic and generated vapors that resulted in widespread cases of mercury poisoning that led to slurred speech, uncontrolled shaking, and stumbling. This led to the coining of phrases "Mad as a Hatter" and "The Danbury Shakes." Due to its severe health hazard, the use of mercuric nitrate in felt production was eventually banned.

After drying, the skins containing the carrotted fur were sent to the "cutting" room, where the fur was processed through machines equipped with rollers and rotary knives that sheared the skin from the fur. The cut fur, now skin-free, was then sorted, graded, and blown in a machine to remove any remaining dirt and hair. The blown fur was then weighed and sent to a "coning" machine that made a fur mat called an embryo hat. In this machine, enough fur for one hat is dropped from a shelf above a perforated copper cone. A fan with its intake under the cone creates a vacuum that attracts the falling fur, causing it to settle evenly across the cone surface. After formation of the fur mat, wet burlap is wrapped around it, followed by a metal cover. The wrapped and covered fur mat is then immersed in hot water to initiate felting of the fur. After "wetting" with hot water, the cone-shaped fur mat is carefully removed from the copper cone, placed on a table, and rolled by hand between wet cloths to start the shrinking or felting of the fur. The whole process is called "starting" and results in a fur cone about one-half its original size. To complete the felting process, the tender fur cone is put through several machines that promote shrinking through multiple wetting and rolling operations.

After this process, the fur cone is a fully felted, unfinished rough hat body, which is then thoroughly dried and shaved to remove the fuzzy surface. In the final operation, the rough hat is "stiffened" by applying a solution of shellac to add firmness to the felt. The stiffened rough hat body is now ready for manufacture into its final hat design. The above description was taken from an article titled "How a Felt Hat Is Made," published in a series of seven articles in *The American Hatter (January–September 1921)*. And since Heim worked in a Danbury hat factory, the following pictures *(Figs. 2-1 and 2-2)* taken from the article provide an idea of the working conditions he experienced.

In the 1880s, with the advancement of the Second Industrial Revolution, many changes took place in Danbury. By 1883, more than 140 phones were operating in the area. In 1887, horse-drawn trolley service in the form of the Danbury and Bethel Street Railway began service and operated a fifteen-mile network of track. Also in 1887, the first electric light was switched on. As hatting grew in Danbury, so did the railroads. Not only did they bring in large machines and coal for fuel, but they also provided a way to transport hats to other cities for sale. The Danbury and Norwalk Railroad was the first railroad to service Danbury and its hatting industry, starting in 1852. It was absorbed by the Housatonic Railroad in 1887 and then taken over by the New York,

1. Cleaning and shearing animal pelts.

2. Carroting animal pelts with mercuric nitrate.

3. Cutting fur from carroted pelts (along wall).

4. Sorting cut fur.

Fig. 2-1: Felt-Making Process for Hats from The American Hatter, 1921.

5. Blowing, cleaning & grading fur.

6. Coning fur into embryo hats.

7. Wetting and multi-rolling rough hat bodies.

8. Applying shellac to stiffen rough hat bodies.

Fig. 2-2: Felt-Making Process for Hats from American Hatter Magazine, 1921.

New Haven and Hartford Railroad in 1892. By 1890, Danbury was a city of almost 20,000 people when Lewis Heim left his parents' farm in East Fishkill for the big city in Connecticut.[4]

At the time of Heim's arrival, some of the larger hat-making companies included E. A. Mallory & Sons, D. E. Loewe & Co., Davenport and Von Gal, E. F. Davis & Co., the Lee Hat Manufacturing Co., Merritt and Knox, H. McLachlan & Co., and Simon & Delohery, with the latter two making only unfinished rough hat bodies. Some popular hats at that time made from fur felt included the Fedora *(Figure 2-3)* and the Black Top Hat. There were also several companies in Danbury that manufactured machines used by hat manufacturers. Five of the larger companies included Doran Brothers, the New Machine Company, Charles H. Reid, the Hat Curling Company, and the Turner Machine Company *(Figure 2-4)*.

When Heim arrived in Danbury to look for work in the fall of 1890 he had just turned sixteen years old. Heim took up residence at 33 Town Hill Avenue and found his first job as a "car driver," which is the description listed in the city directory. Since automobiles were just beginning to appear, the term car driver refers to the operator of a horse-drawn street or trolley car. In 1891 he moved to Railroad Avenue and found work at the local railroad company as a "fan maker." Outside of any experience when he was growing up on the farm, this is the first indication that Heim worked on machines.

During 1892, Heim's older sister, Wilhelmina *(Minna),* around 19 years old, moved to Danbury and found employment as a nurse at the Danbury Hospital. Sometime between 1892 and 1894, Heim joined a local hat manufacturer where he took a job as one of the many "hatters" in

Fig. 2-3: Fedora Hat Made from Fur Felt: ca. Early 1900s.

Danbury, making and repairing hats. Heim was not listed in the Danbury City Directory in 1893 or 1894, so there is no record of the year he left the railroad company. After taking his new job as a hatter, he and his sister Minna moved from Railroad Avenue to 27 North Street. On the same street at numbers 22 and 24 was one of Danbury's hat manufacturers, T. Brothwell and Company. Heim may have worked for Brothwell and moved to North Street to be near his employer but also could have moved there to accommodate more people. For in 1894, the rest of the Heim family, consisting of his parents, two brothers, and younger sister left the farm in East Fishkill and moved in with Lewis and Minna at 27 North Street. Charles, the father, took a job delivering milk. Around the same time Heim's two younger brothers also started their work careers in Danbury as the middle brother, Alfred, turned 17 and the youngest, William, turned 16. Sometime in 1894 or 1895, both brothers followed Lewis into the hatting industry and took jobs as hatters in local factories.

Hat-Making Machines

In 1895, while working as a hatter, Heim started to use his innate skills in machine design to develop improvements in hat-making machines. Machines at the turn of the twentieth century were strictly mechanical devices and most were driven by belt systems common to machine shops of the era. Most machines consisted of intricate combinations of shafts, gears, springs, levers, pulleys, and journal bearings all made from wood and metal materials. Hat making at that time was a labor-intensive process with machines being developed and improved as a result of competition to speed production, increase quality, and lower manufacturing costs.

As previously described, one of the processes that felt hats undergo during manufacture is stiffening. After the hat is shaped, a stiffening compound is applied to give hat bodies, brims, and crowns rigidity. Stiffening compounds came in a variety of formulations, with most made from the sap of pine trees *(gum)* and shellac. The gum and shellac were dissolved in an alkali solution or solvent, brushed on and/or pressed into the felt, then dried.

In January 1896, at the age of 21, Lewis filed his first patent application for a hat-stiffening machine. This invention provided a "method" and "device" for holding

Fig. 2-4: Turner Machine Company Business Card: 1894.

hats during application of the stiffening compound and was an "improvement" on existing machines. The "method" portion of the invention consisted of a process for applying the stiffening compound evenly and thoroughly while the "device" portion of the patent related to the mechanism for implementing the method.

Heim's invention consisted of a machine that firmly held the hat in place and rotated it during application of the stiffening compound *(Figure 2-5)*. The machine freed the operator from having to hold or manipulate the hat during the process, resulting in improved application of the compound and faster production. The primary improvement was a hat-block consisting of a wooden cone fitted with an expandable rubber tip and expandable metal fingers that were equipped with small spurs. The hat block was attached to a hollow shaft supported horizontally by journal bearings. At the opposite end of the hat block, the hollow shaft was connected by pipe to a source of compressed air. Pulleys mounted on the hollow shaft were connected to the factory belt system as the source of power for rotation. Mounted to one side of the machine was the brush used to apply the stiffening compound. The brush was connected to a rigid, hollow shaft and mounted on sliding arms. The hollow shaft was connected using a flexible hose to an elevated reservoir holding the stiffening compound.

In operation, the operator would place a hat on the cone-shaped hat block, open

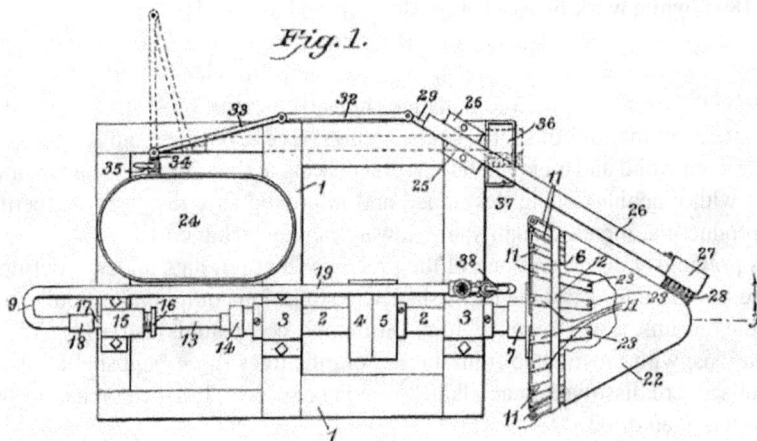

Fig. 2-5: Hat Stiffening Machine—Top View. Patent 573,876, filed Jan. 8, 1896.

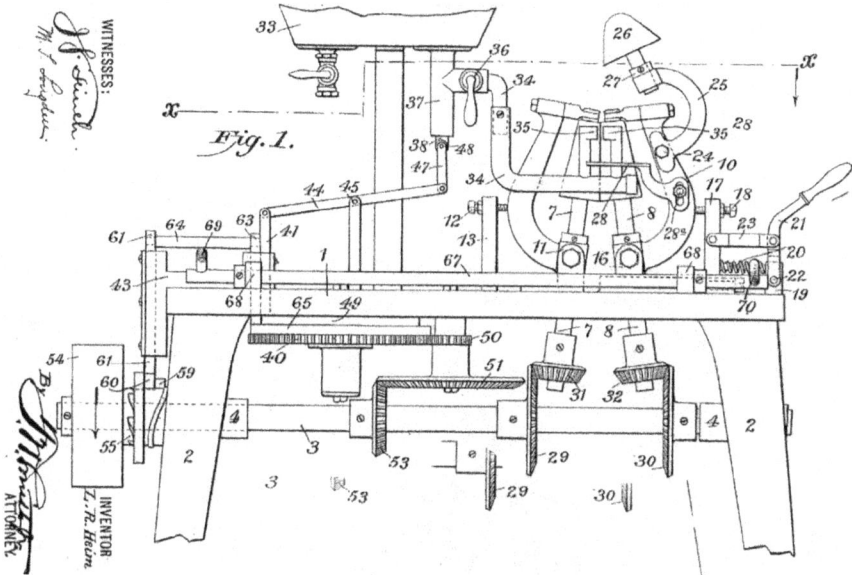

Fig. 2-6: Machine for Stiffening Brims of Hats. Patent 580,396 Filed Aug. 14, 1896.

the valve on the compressed air line, and inflate the hat block to hold the hat. The operator would then actuate the pulley system to rotate the hat block. With the hat firmly secured and rotating, the operator would then manipulate the brush across the outside of the hat to evenly apply stiffening compound. Even though Heim's first hat-making machine appears fairly simple, the invention saved time, improved quality, and lowered the cost of stiffening hats. There were two witnesses to the signing of Heim's first patent application. One was his brother Alfred and the other was John Simon Roth, Heim's future father-in-law. Roth, being older and an experienced hatter, may have been a mentor to young Lewis. The patent, number 573,876, was granted later that year on December 29.

As mentioned before, the hatting industry was always under constant pressure to reduce costs and improve product quality. Undoubtedly, Heim was exposed to these pressures, as well as to the repetitive and tedious jobs hatters performed that led to conception of his invention. It is not known where Heim developed his invention, but most likely it was developed on his own time outside the company, maybe at his home or some other place nearby where he had the space and tools to work on machines. Given that he was a witness to the invention, Roth may have provided guidance or other aid such as a garage to work on his ideas.

Patent rights historically belong to the company that pays an employee and funds invention development. If such were the case in the 1890s, then Heim would have been required to assign the patent to his employer and so state in the patent. Given that he retained ownership, Heim most likely developed his first and subsequent hat-making inventions on his own time and cost.

Apparently, hat companies liked what Heim invented and used it to make their hats. That success led Heim to develop other machines to improve the hat-making

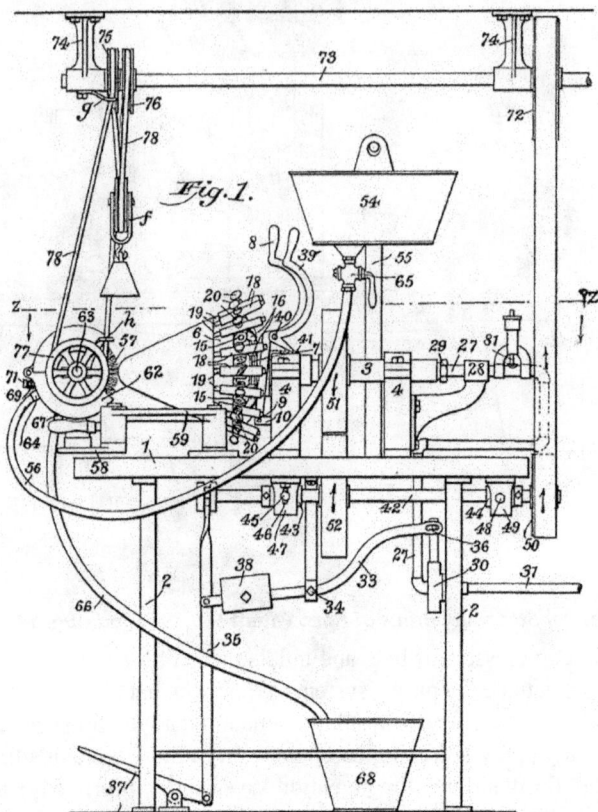

Fig. 1.

WITNESSES:

INVENTOR
L.R.Heim

Fig. 2-7: Machine for Stiffening Crowns of Hat Bodies.

Patent 581,554, filed Aug. 14, 1896.

process. Later that same year *(1896)*, Heim developed two new machines. One machine was designed for stiffening the brims of hats *(580,396)* and the other for stiffening the crowns of hat bodies *(581,554)*.

Heim's second hat-making machine *(580,396)* was designed specifically for automatically applying stiffening compound to the brims of hats and was more complex than his first machine. The brim-stiffening machine *(Figure 2-6)* contained a cone support on which the hat was placed and positioned such that the brims were in a vertical plane between two conical rollers. Stiffening compound was applied simultaneously to both sides of the hat brim and then pressed into the material when the brim rotated between the two rollers.

Heim designed this machine with several novel features to automate application of the stiffening compound and improve the quality of the finished product. One was the use of a clutch to engage and disengage the mechanical drive controlling rotation of the two conical rollers. A second feature was the use of a gear and cam system to

14

Fig. 2-8: Lewis R. Heim and Anna E. Roth: ca. 1898.

regulate and automatically shut off the flow of stiffening compound so that the machine could continue the rolling process for a timed period before automatically stopping the machine by disengaging the clutch. As described by the Turner Machine Company in its 1901 advertisement *(Figure 2-11)*, the invention provided for a controlled and timed process that reduced the cost of labor while achieving greater uniformity of compound application.

Heim's third invention *(581,554)* to stiffen the crowns of hats was a modification of his first invention *(573,876)* to stiffen the bodies of hats. It used a similar cone-shaped device for holding and rotating hats, plus an applicator brush assembly designed to swing to one side to facilitate mounting and removing hats on the cone-shaped hat block *(Figure 2-7)*. It also included other features such as a foot treadle that performed two functions: one was to actuate the compressed air valve for the hat block and the other, by further depressing the treadle, actuated friction pulleys for shaft and hat-block rotation. All together, these features combined to automate and quicken the entire process.

Also during 1896, the entire Heim family moved from their North Street address to 80 Franklin Street in the southwest section of town. They all lived at this address until 1898 when 24-year-old Lewis married 21-year-old Anna Elizabeth Roth of New York City *(Figure 2-8)*. Lewis and Anna moved from Franklin Street to a house at 45 Stevens Street.

In 1899, Heim developed his most complex machine for the hat-making industry. This was a machine to iron hats and contained several novel features to not only automate ironing of the entire hat surface, but to provide flexibility in the duration of contact with the various hat materials.

In patent 645,593, Heim developed a machine that used a rotating main shaft con-

15

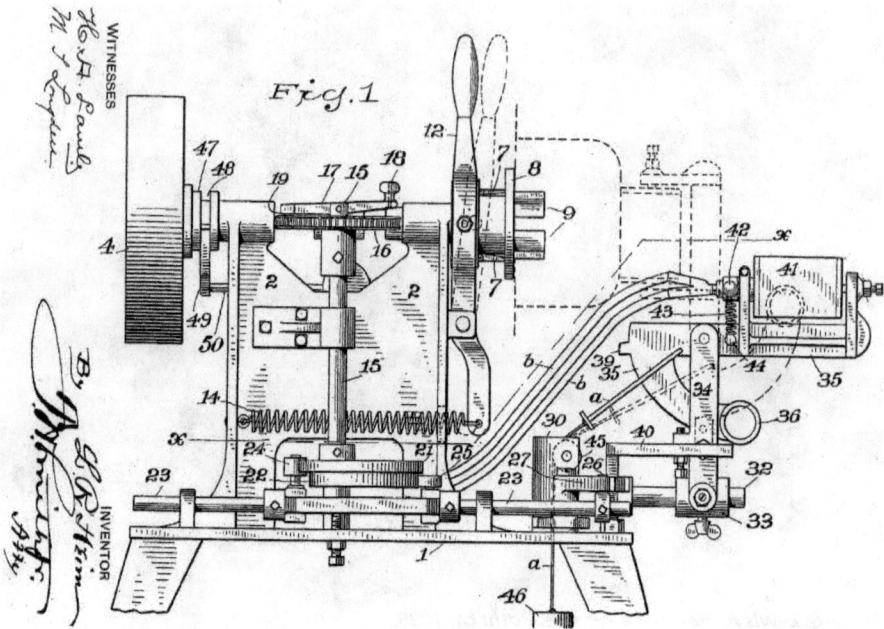

Fig. 2-9: Hat Ironing Machine. Patent 645,593 Filed Sept. 16, 1899.

taining a clutch at one end and a hat-holding chuck at the other end. Installed near the middle of this shaft was a worm gear system that operated an offset push rod designed to move a pivoting iron assembly into various positions to iron hat surfaces such as the crown, brim, and band. Within the push rod assembly, Heim incorporated a system using two cams that controlled the position and duration of iron contact to make the ironing process automatic and less likely to cause burning of hat surfaces. Two cams were required to achieve the movement and timing needs of ironing the various hat surfaces. Given the many hat types, apparently no machine at the time could be designed to iron all types automatically. So, Heim incorporated a means to allow the operator to independently move the iron to a desired position for additional ironing, as required by the hat design. As shown in Figure 2-9 and the other figures in the patent, Heim's hat-ironing machine was very innovative for a young machinist.

Other machines designed by Heim included a hat-clearing machine *(645,592)* for trenching out excess stiffening compound and a machine for curling the brims of hats. Apparently, marring of the brim underside and hat crown was an acute problem with brim-curling machines. In Heim's hat-curling machine *(695,393)*, he redesigned the existing machine to incorporate new features that allowed the operator to properly feed and remove hats without marring its various surfaces. Existing curling machines were designed to place a hat brim between two wheels, with one wheel larger than the other. The smaller wheel was designed with a convex periphery that fit into the concave periphery of the larger wheel. The small wheel was mounted on a spindle assembly that would swing away from the larger wheel that was fixed in place. The operator would place the hat brim into the space in front of the large wheel and close the small wheel against the large wheel for curling.

In Heim's hat-curling machine, he reversed the fixed and movable wheel assemblies to allow more space for feeding and removing the hat. To prevent damage to the hat, he added supports and guides to protect the hat crown and brim underside, and added other features to improve adjustability to accommodate a wide range of hat designs while keeping the machine simple and easy to use.

Heim's hat-making inventions consisted of original designs and redesigns of existing machines, each with novel and mechanically complex features. With each hat-making invention, Heim strove to provide a machine with greater automation and flexible operation while delivering improved product quality at lower cost. By April 1902, at the age of 27, Heim had been granted a to-

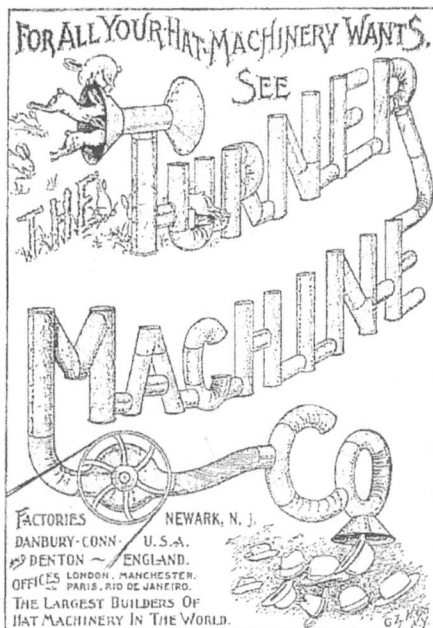

HATTERS' MACHINERY

IRON and BRASS FOUNDRY.

General Jobbing and Repairs.

FOR ALL YOUR HAT-MACHINERY WANTS,
SEE THE TURNER MACHINE CO

FACTORIES
DANBURY-CONN.   NEWARK, N. J.
DENTON ~ ENGLAND.   U.S.A.
OFFICES LONDON, MANCHESTER.
PARIS, RIO DE JANEIRO.
THE LARGEST BUILDERS OF
HAT MACHINERY IN THE WORLD.

Our Machine Shop is one of the
Best Equipped in New England,
And we are in a position to undertake any kind of work. We run a special department to fill all your requirements in the Repair Work and Jobbing Line.
Importers of and Dealers in Hatters' Supplies,

Fig. 2-10: Turner Machine Company Advertisement: 1905 Danbury City Directory.

tal of seven patents for his family of complementary hat-making machines. This was quite an accomplishment for a young man still in his twenties and with only a grade-school education.

The machines invented by Heim were built and sold by the Turner Machine Company, which was a large multinational manufacturer with factories in Danbury, New Jersey, and England and advertised itself as "The Largest Builders of Hat Machinery in the World" *(Figure 2-10)*. The advertisement, which was published in the 1905 Danbury City Directory, is interesting in that it shows rabbits entering Turner Machine and hats falling out. In February 1901, Turner Machine published an advertisement in which it featured Heim's machine for automatically stiffening the brims of stiff or soft hats, stating that it saved costs in labor and materials *(Figure 2-11)*.

Sometime in 1901 or early 1902 Turner Machine bought all but one of Heim's hatting machine patents with the seventh, the patent for a hat-curling machine *(695,393)*, sold to Fernando Olmstead. The price paid by Turner Machine to Heim is not known but

17

Fig. 2-11: Turner Machine Company Advertisement Promoting the Heim Automatic Brim-Stiffening Machine. The American Hatter (Feb. 1901).

18

at the time it must have appeared to be a fair price. However, Heim later discovered that his patents were worth far more than he received. In a newspaper article published in 1944, Heim revealed that since he was young and inexperienced, he had sold his patents for the proverbial song and missed reaping early riches from his innovations.[5]

Turner Machine continued to sell Heim's inventions to hat manufacturers for a significant period with one, the felt-stiffening machine, used by the hatting industry well into the 1960s.[6] Heim must have been very encouraged by his success making hatting machinery because he soon left his job at the hat company to start his own business.

Regarding Lewis's two brothers, Alfred and William, the Danbury city directories provided some information on the activities of William, but not much on Alfred. With Alfred, the directory showed him living with his parents at 80 Franklin Avenue but did not include his profession until 1900, when he was 22 years old and listed as a machinist.

But with William, the city directory contained more information on his addresses and profession. In 1899, he left the hatting trade to become a bicycle maker and posted two addresses, with one for his home address on Franklin Avenue and the second for his work address at 2 River Street. It also listed Lewis's home address as 2 River Street, but then showed Lewis back at his Stevens Street address in 1900. Relatives said that Lewis did many things for his family that included help in buying property and starting businesses. So, it is probable that Lewis either rented or purchased the River Street property and set up William *(21 years old)* to manufacture bicycles. In 1901, William relocated his bicycle manufacturing business to 29 Elm Street, where he remained until 1905.

It is interesting to note that even though Heim was inventing many types of hat-making machines, he retained his profession as a "hatter" in the Danbury City Directory until he changed it to "machinist" in 1901. Also that same year, in October, Lewis and Anna had their first child, a daughter they named Florence Elizabeth.

Chapter Three

The Collar Roller and
the Heim Machine Company

The Collar Roller

N ear the end of 1901 or early 1902, Heim became interested in machines that ironed laundered clothing. In 1903, Heim, along with a co-inventor, developed a special machine designed for the typical fashion worn at the turn of the century. Many people familiar with the times know that shirts were made with detachable collars and cuffs, which were held on the shirt with buttons or other types of clasps. Collars, due to constant rubbing against a person's neck, soiled easily, leaving a dirty ring around the inside that required laundering to remove.

The detachable collar was invented in 1827 by Hannah Lord Montague of Troy, New York. Detachable collars fastened either at the front or the back of a shirt with a collar button, a stud on a shank, or shaft, that slips through two small eyelets on a collar. Collars detached from the body of shirts for laundering, thus extending shirt life. They became popular money-saving items as a solution to "ring-around-the-collar" when clothing was custom-sewn, and their use quickly spread throughout the nation.

Troy became a center for shirt and collar manufacturing for America and by 1897 there were twenty-five manufacturers producing around 100 million collars and cuffs a year. In the early 1900s, there were two types of detachable collars. The first type was "stand-up" or "standing collars" that were of the straight cylinder type with no folds and generally worn to formal functions. The second type was "fold collars" that would turn down or fold over and would be worn to less formal functions and by office workers. Examples of stand-up and fold collars are shown in Figure 3-1 and can be seen worn on Heim and his brothers in a family photo in Appendix 1 *(A1-3-3)* taken in 1916.

Detachable collars were generally white in color and made of cotton and linen fabrics made stiff by dipping them in thin cooking starch. Some collars were bonded to stiff cardboard and others to celluloid *(film stock)*. Each collar type had many variations and generally ranged in height from 2 to 3 inches, with many inducing chafing of the neck with too much sudden motion of a person's head.

Detachable collars eventually became a status symbol of the growing office worker class, from which the term "white-collar worker" originated. The collars came in a variety of styles and were relatively inexpensive at around 35 cents apiece. But when sold at 4 to 5 dollars for a box of twenty-five, the cost per collar was much less. Detachable collars were sold nationwide, along with colorful collarless shirts in mail order catalogs from Sears-Roebuck, Montgomery Ward, and others. By 1900, most men in America covered their neck in the vise-like grip of the starched detachable collar. Even after shirts became mass manufactured, removable collars remained popular and were a common part of men's and some women's wardrobes into the 1930s.[1]

As with all clothing, detachable collars became soiled after use and required laundering. After washing, collars were stiffened with starch then ironed and shaped to the proper curve to fit around a person's neck. The process of ironing and shaping stand-up and fold collars was mainly a manual task that led many people to drop off soiled collars at their local steam laundry. In the early 1900s, Danbury supported many laundries such as the Danbury Troy Laundry at 5 Maple Avenue, City Steam Laundry on Main Street, and many other steam laundries run by Chinese owners.

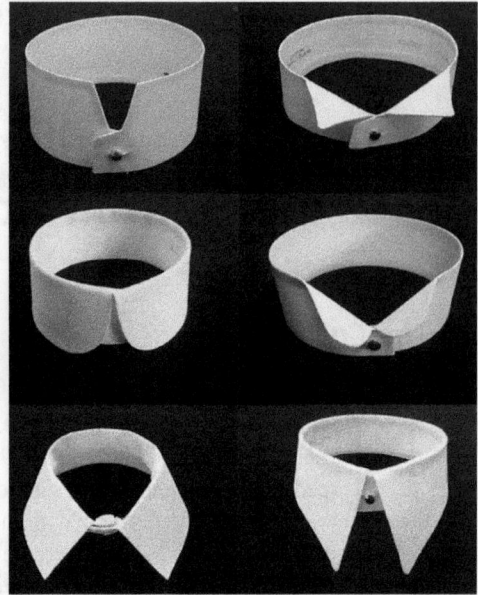

Fig. 3-1: Examples of Detachable Collars ca. 1900; Stand-Up in upper left. Fold Collars in upper right plus middle and bottom rows.

As with most businesses at that time, there were many new machines and processes being introduced to improve, automate, and lower the cost of laundering clothes. In 1902, Heim teamed up with Edwin Targett to design a new machine that automatically ironed and shaped detachable fold collars. On January 17, 1903, they filed a patent application with the title "Machine for Shaping Fold-Collars" that incorporated a special feature much appreciated by those wearing detachable collars. Their invention *(Figure 3-2)* not only automated the process of ironing and shaping fold collars, it also softened the razor edge created during ironing. The machine performed a four-step process to form a close folded edge, smoothly iron the collar without damage to the material, shape the collar, and dry out surplus moisture. The first step was to fold the collar and iron the fold in the collar. The collar was placed on a wedge-shaped component called a "standard" *(component #21 in Figure 3-2)* over which a heated shoe *(component #22)* was actuated down to first form the fold, and second to moisturize and iron the fold. It was the design of the standard and shoe that "softened" the sharp edge of the fold to alleviate chafing of a person's neck. The second step was to final iron and shape the collar to fit around a person's neck. After ironing the fold, a shaping roll pulled the collar off the standard and pressed it between two vertical rolls to final iron and dry out the collar while imparting a curved shape.

The fold collar ironing and shaping machine was derived from Heim's hat-ironing and brim-curling machines. It used many of the same concepts and features such as upright rollers with journal bearings, rotating shafts with worm gears, interlocked and synchronized gear sets, elbow levers and tension springs, and was driven by the

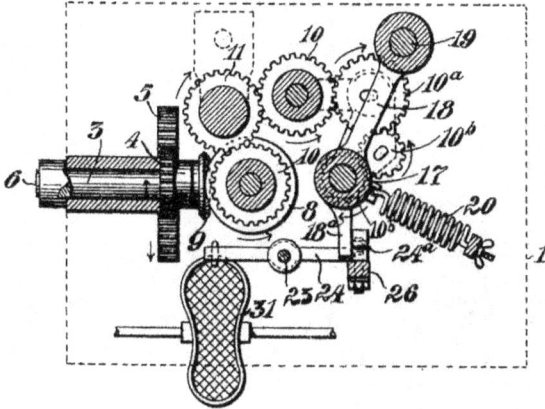

Fig. 3.

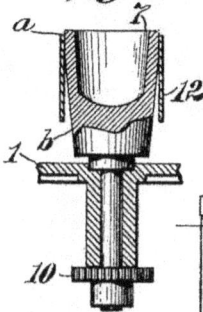

Fig. 5.

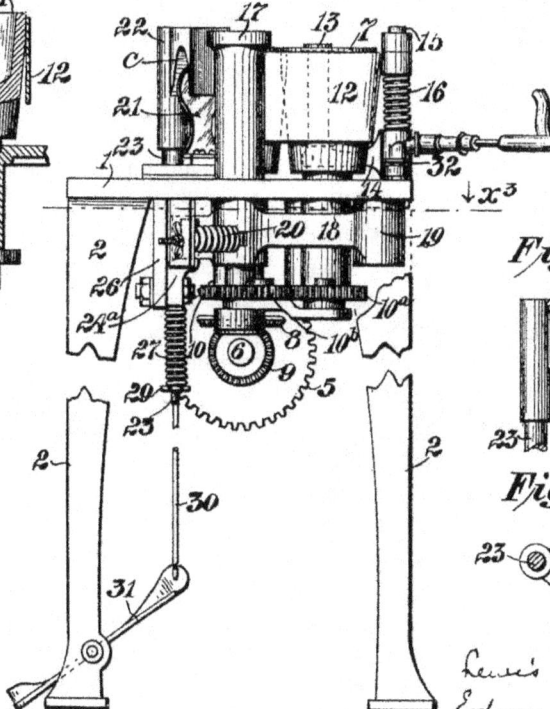

Fig. 4.

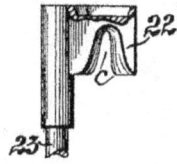

Fig. 6.

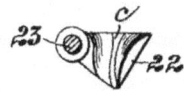

Fig. 7.

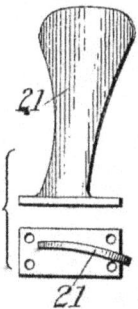

Fig. 8.

Fig. 3-2: Machine for Shaping Fold Collars. Patent 780,750 Filed Jan. 17, 1903.

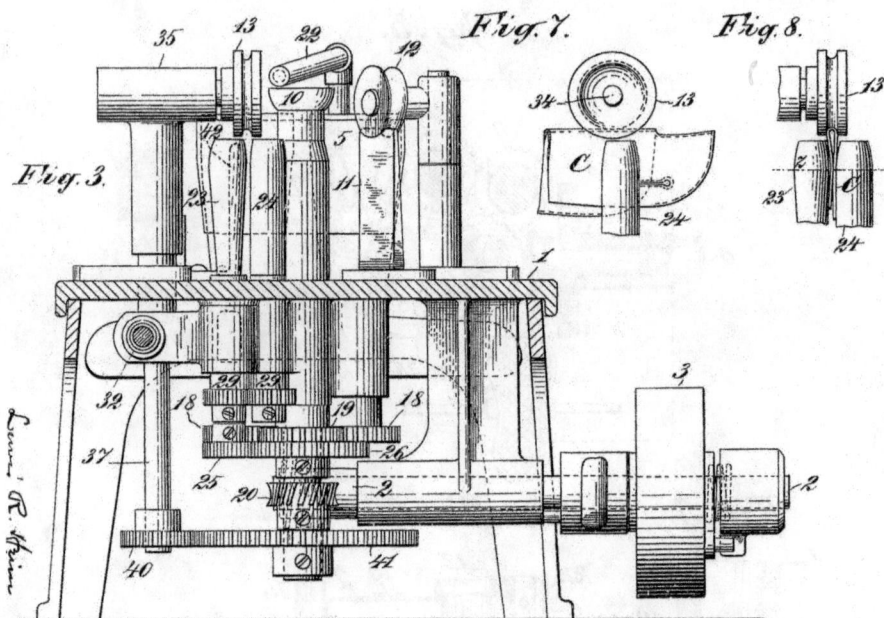

Fig. 3-3: Machine for Ironing Fold Collars. Patent 865,977 Filed June 18, 1906.

factory belt and pulley system. The machine became known as the "Heim Fold Collar Shaper" and the "Heim Collar Roller" and was granted letters patent *(780,750)* two years later on January 24, 1905.

It is not known how Heim became interested in fold-collar laundry machines. However, since there was a co-inventor on the patent, one possible explanation is that Targett approached Heim with ideas for an automated machine and asked for help in its design. Targett may have been involved in the laundry business and knew the processes used to launder detachable collars as well as the need for automation and lower costs. He also may have known Heim through the hatting industry, heard about Heim's hat machines, or was introduced through acquaintances. With his new machine for ironing collars, Targett later partnered with Henry M. Siemon to organize the steam laundry business of Targett and Siemon on Bridge Street in Danbury.[2]

Over the next three years *(1904–1906)*, Heim filed three more patent applications for improvements in his collar-shaping and ironing machine, with all carrying the same title of "Machine for Ironing Fold Collars." The last patent for this machine, No. 865,977, contained improvements that became the final design of this class of laundry machine. Heim incorporated a second ironing process in the form of a roller where the surface rotated faster than the speed of the collar passing beneath to burnish the fold *(component 13 in Figure 3-3)*. This burnishing process softened and polished the fold that removed the "Gay Nineties Razor Edge" from stiff fold collars. Heim marketed his collar roller machine as the "Heim Fold Collar Shaper" that can be seen in business envelopes sent to the Excelsior Carbon Manufacturing Company in 1912 and the Norton Company in 1915 *(see Appendix 2, A2-1)*. At that time, the Norton Company was a well-established manufacturer of grinding wheels and machines, so it is highly likely that the business between the two companies related to grinders and abrasive wheels Heim used in manufacturing his collar-ironing machines.

The Heim Machine Company

In 1903, Heim did not have a shop in which to manufacture his fold collar laundry machines. His younger brother, William, who quit his hatting job in 1898, operated a bicycle manufacturing shop at 29 Elm Street. It is not known where Heim built his machines but he may have used space in William's shop.

In 1904, Heim and his brother Alfred leased factory space in a former hatting machine shop at 20–22 Maple Avenue in Danbury and founded the Heim Machine Company to build the collar roller.[3] The factory consisted of a house and two-story machine shop originally built by Joseph Nutt, an English-born machinist. The shop was located on a one-third–acre site

JOSEPH NUTT,

Inventor and Manufacturer of

Hatters' Special Machinery

MACHINERY DESIGNED AND BUILT TO ORDER.

General Machine Jobbing, Etc.

JOSEPH NUTT & CO.,

MANUFACTURERS OF

Self-Conforming,

Spring Steel

Hat Wires,

Made entirely by Automatic Machinery of the very best quality of spring tempered steel. Are **Far Superior** to other makes of Wires.

Try a Sample Order and be Convinced.

PRICES GIVEN UPON APPLICATION.

Fig. 3-4: Joseph Nutt Advertisement. January 1892.

at the intersection of Maple Avenue and Crosby Street. The tracks of the former New York and New Haven Railroad pass diagonally in front of the buildings and gave the lot a triangular shape.

Joseph Nutt fabricated wires and racks for hats and operated in one of the many satellite industries supporting Danbury hat manufacturers. Nutt also called himself an inventor and manufacturer of "Hatters' Special Machinery" that were designed and built to order. His business, Joseph Nutt and Company, was advertised as "Manufacturers of Self-Conforming, Spring Steel Hat Wires" *(Figure 3-4)*. Nutt operated his machine shop and resided in the house with his family until his death in 1896. His son *(M. J. Nutt)* took over the business until 1900, when the property was sold to a local businessman, Frederick M. Thompson.

As described above, the collar roller had many parts requiring precision machining such as milling, drilling, shaping, and grinding. The Heim brothers may have used the machines and equipment in the former Nutt shop to build the collar roller but soon found they needed additional equipment, either to fill gaps in manufacturing or to meet the demand for orders. According to the Danbury Land Records, on March 31, 1905, the Heim Machine Company purchased a large selection of machines and equipment *(shown below)* from the Prentiss Tool and Supply Company located at 115 Liberty Street in New York City.[4]

One (1) - Lodge & Shipley cone-driven engine lathe w/taper attachment: $372.00
One (1) - Prentiss lever crank shaper: $290.00
One (1) - Barnes back geared power feed drill press: $130.00
One (1) - Millers Falls hack saw: $ 19.00
One (1) - 12 inch Blount grinder on stand with counter shaft: $ 24.00

Total: $835.00

All the companies listed above *(Lodge & Shipley, Cox & Prentiss, W. F. & John Barnes, Millers Falls, and J. G. Blount)* were, at that time, well-known tool or machine tool manufacturers. Lodge & Shipley were known for their high-quality and very large lathes. Cox & Prentiss, as well as Blount, were well-known manufacturers of grinding machines.

Shortly thereafter, on April 5, 1905, the Heim Machine Company purchased the Nutt property and "all tools and machinery used in the wire business and hat-rack

Fig. 3-5: Jones and Lamson Flat Turret Lathe: ca. 1900.

business" from Frederick M. Thompson. The shop they bought was only the first floor of the Maple Avenue building, with the second floor retained by Thompson. The purchase was concluded on April 15 when a Bond Deed was issued for a total price of $2,500, with $500 paid at purchase and the balance paid in installments.[5]

Later that year on October 5, the Heim brothers bought their single most expensive machine, a flat turret lathe *(Figure 3-5)* from the Jones and Lamson Machine Company of Springfield, Vermont. Jones and Lamson was a well-known maker of machine tools for making guns and sewing machines. The receipt posted in the Danbury Land Records indicated the price paid by Heim was $1,283 with terms of $100 cash down payment with the balance paid at $100 per month.[6]

A turret lathe is a machine tool used for repetitive production of duplicate cylindrical parts and is composed of a long horizontal bed with a rotating chuck at one end into which is inserted the cylindrical part to be machined. Opposite the face of the chuck is a turret mounted on a base that slides along the length of the machine bed. The turret contains several different cutting tools and rotates on the base to access and position the cutting tool for use on the work to be machined. The turret lathe was an excellent machine for fabricating cylindrical parts, including gears that required multiple cutting, boring, and threading operations. As seen in his patent illustrations, Heim used many highly machined cylindrical parts in his collar roller. As expensive as it was, the flat turret lathe was a very versatile machine that was not only essential but also helped lower the cost of producing Heim's fold collar–ironing machines.

The Danbury Troy Laundry

In 1906, Lewis Heim and his brother Alfred, along with a John R. Booth, bought the Danbury Troy Laundry, an existing business established in the late 1800s. The Danbury Troy laundry was originally located at 5 Maple Avenue at the corner of White Street but later, under its former owner, moved to 158 – 162 Main Street in the center of town. An advertisement *(Figure 3-6)* published in the 1893 Danbury City Directory states that it is the oldest steam laundry in the city and includes an illustration of clowns throwing detachable collars inside a large collar-like ring.

The Heim brothers bought the laundry not only as a livelihood but also as a laboratory to test and perfect laundry machines developed by Lewis. After purchasing the business, the partners incorporated on October 4, 1906, with a total funding of $18,000 in capital stock *(180 shares at $100 par value).*[7] John Booth became the manager of the laundry while the brothers continued working at the Heim Machine Company. They advertised the Danbury Troy Laundry as the best-equipped steam laundry in New England *(Figure 3-7)* and operated the collar-ironing and shaping machines manufactured at the Heim Machine Company.

In early 1907, Heim developed a second laundry machine that expanded his ironing process to the traditional flat *(stand-up)* collar, cuffs, and similar garments. The purpose of the invention was to improve the automation process for ironing and shaping many types of collars and cuffs in one machine. The machine was specifically designed to burnish *(smooth, soften, and polish)* both edges of the traditional stand-up collar as well as both edges of cuffs.

Fig. 3-6: Advertisement for the Danbury Troy Laundry: 1893.

Called an Automatic Edge Ironer for Collars and Cuffs *(patent 1,064,644)*, the machine differed from the original collar roller in that a table was constructed with horizontal rollers that automatically moved flat garments through each of three processes, one each for moistening, ironing, and shaping. The machine utilized a system of horizontal and vertical shafts that rotated to drive the feed rolls, the moisturizing disks, the ironing rolls, and the shaper rolls *(Figure 3-8)*. To ensure that all stages worked together, Heim designed a complex set of gears, pinions, belts, and chains to drive all shafts in synchronization. In order to start the process, all the operator had to do was place a collar or cuff on the table and slide it under the first set of feed rolls.

One unique feature of the invention was the use of "spreader guides" that were cam-like devices installed on the moistening disks and ironing rolls. They were connected under the table to the shafts rotating the disks and rolls and when actuated, would independently slide horizontally as required to accommodate the size and position of the garment. The spreader guides made the machine very flexible to accommodate different widths of collars and cuffs.

High-Grade

Laundry Work,

REQUIRES MODERN MACHINERY AND METHODS.

===*The*===

DANBURY TROY LAUNDRY CO.

is the best Equipped Steam Laundry Plant in New England.
If your laundry work is not satisfactory give us a trial.
A postal or telephone call will receive prompt attention.

TELEPHONE 560.

158, 160 and 162 Main Street,

JUST SOUTH OF CITY HALL.

Visitors Welcome at all Times.

Fig. 3-7: Advertisement for the Danbury Troy Laundry Operated by Alfred Heim: 1912.

The main feature, and the one for which Heim collar-ironing machines were known, was the mechanism for removing the rough, razor-like edges produced when ironing starched garments. The heated ironing rolls, as operated in his fold collar machine, were designed to rotate at a speed greater than the speed of the edge being ironed, thus creating a rubbing effect on the material and achieving the desired softer feel to the collar.

Between 1903 and 1911, Lewis Heim was granted a total of seven patents for his detachable collar-ironing and shaping machines. The machines manufactured by Heim Machine were marketed not only in the United States, but also internationally in Canada, Europe, China, and Australia.[8] The Heim Machine Company made and sold the collar-ironing and shaping machines for more than ten years until the fall of 1915 when the business was sold.

During the early years of the Heim Machine Company, Heim not only built up his manufacturing business, he did a lot of moving within Danbury. In 1898, Heim was living with his family at 80 Franklin Avenue. After getting married, Lewis and Anna moved about a half-mile south to 45 Stevens Street. In 1903, Lewis and Anna moved to another house on the same street at number 23. In the following year *(1904)*, they moved northeast to 9 Granville Avenue, which was much closer to the Heim Machine Company at 22 Maple Street. Then, one year later *(1905)*, Lewis and

29

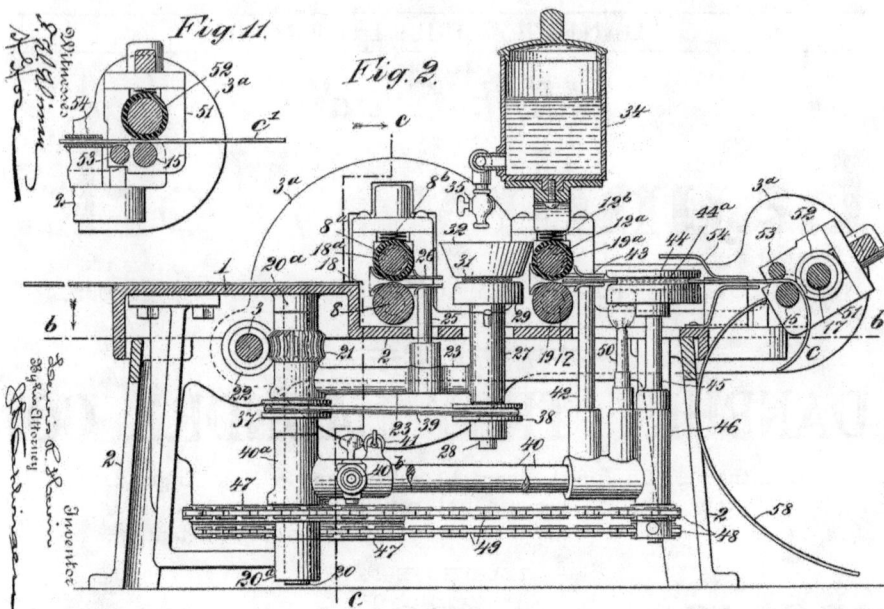

Fig. 3-8: Automatic Edge Ironer for Collars and Cuffs. Patent 1,064,644, filed March 28, 1907.

Anna moved a half-mile northwest to 63 Balmforth Avenue where they resided until 1924. Heim probably rented most of the properties he lived in until he bought the house on Balmforth Avenue. But the Danbury Land Records show that Heim dabbled in real estate, with many properties bought and sold during that time.

As previously discussed, Danbury was a major center for making hats that employed many people, including Lewis, Alfred, and William Heim. However, the Heim family found that there were better opportunities in other types of businesses, with William being the first to leave the hatting profession in 1899 to manufacture bicycles. Later, in 1905, with Lewis and Alfred busy making and selling laundry machines, William, at 27 years of age, left the bicycle trade and opened a music store. Located at 270 Main Street in Danbury, the Heim Music Store was established as a dealer in sheet music, musical instruments, and phonographs *(Figure 3-9)*. Apparently, William and the youngest sibling, Laura, were schooled in music and had a passion to pursue it as a career. Laura, who listed her occupation as a music teacher in the Danbury City Directory *(1908)*, helped William with his music business.

During the same period up through 1910, there were several other events affecting the Heim family. First was the marriage of Alfred Heim *(30)* to Clara Barnum *(25)* of Danbury on June 12, 1907. Seven months later, Charles George Heim *(father)* died at the age of 67 on January 8, 1908. Two months later, on March 12, son Charles Roth Heim was born to Lewis and Anna. And three years after getting married, on June 23, 1910, Alfred and Clara had their first and only child, Richard Barnum Heim.

Wm. C. Heim,

DEALER IN

Pianos
Musical
Merchandise
Phonographs
Sheet Music

IN FACT, EVERYTHING IN THE MUSIC LINE

270 Main Street,   ∴   **Danbury, Conn.**

Fig. 3-9: Advertisement for Heim's Music Store, 1912.

The Ball and Roller Bearing Company

In 1909, Alfred left the Heim Machine Company to work full time at the Danbury Troy Laundry, leaving Lewis as the sole proprietor.[9] Around the same time *(1909 or 1910)*, with the Heim Machine Company doing well, Lewis, now 35 years old *(Figure 3-10)*, turned his attention to making ball and roller bearings for the automobile industry. Heim would later name his business the Ball and Roller Bearing Company and made his bearings in the same building used to make his laundry machines. Within a short time, to accommodate the new business, Heim built the first of three wood frame additions on to the former Nutt machine shop.[10]

In 1912, John Booth and Lewis Heim, both original owners of the Danbury Troy Laundry, sold their shares to Alfred and his new partner, James G. Durkin, who lived on Garfield Avenue in the south side of Danbury.[11] After Alfred left the Heim Machine Company to focus on the Troy Laundry, the 1912 Danbury City Directory recorded the changes, as it was the first year listing Alfred as president

31

Fig. 3-10: Lewis R. Heim, Age 35, ca. 1909

and treasurer, and no longer affiliated with Heim Machine. The year 1912 was also the first year in which the Ball and Roller Bearing Company was listed in the city directory as a separate business from the Heim Machine Company.

Chapter 4

The Ball and Roller Bearing Company

Development of the Modern Bearings Industry

As discussed in the first chapter, the Industrial Revolution was a transition to new manufacturing processes in which machines replaced people to perform manual tasks. Famous machines of the 1800s include the cotton-spinning machines used in the textile industry; the steam engine used for power in locomotives, ships, and factories; mechanical reapers used on farms; and the sewing machine. Many types of new and very large machines were developed to operate the burgeoning steel, paper, and chemical industries. And in the late 1800s, one of the most famous machines, the internal combustion engine, was developed, which gave rise to the age of the automobile.

Machines are defined as tools containing moving parts that perform a specialized task. In order to facilitate operation, machines are designed with components called mechanical bearings that perform two specific functions. One is to constrain and promote relative motion between moving parts and the other is to reduce friction and wear to allow smooth, precise operation over the machine's lifetime.

During most of the Industrial Revolution up to the late 1800s, there was essentially only one type of mechanical bearing used in machines. This bearing is called a "journal" bearing and belongs to the family of "plain bearings" that includes the "jewel" bearing and the modern-day "spherical" bearing. Plain bearings consist of two surfaces rubbing against each other with one surface generally fixed to constrain a second, moving surface. Outside of jewel bearings, most plain bearings use a lubricant such as grease or oil to reduce friction and wear. Jewel bearings consist of natural jewels such as rubies, diamonds, and sapphires and were used primarily in mechanical watches, clocks, and later in sensitive measuring equipment such as the compass.

Virtually all machines developed during the 1800s used some type of journal bearing, with most being a shaft or wheel connected to a stationary support. Examples include the axle and wheel assembly used on wagons and the journal box that supported the rail car frame on the wheel axle assembly. Figure 4-1 shows a typical railcar journal bearing with the oil lubrication packing placed underneath and in contact with the rotating axle. If the oil leaked or dried out, which happened often, the bearing overheated and failed. This is where the term "hot box" originated, and it applies to other vehicles using journal boxes.

In the latter part of the 1800s, a second type of mechanical bearing was developed as a product of the Second Industrial Revolution. This was the rolling element or rolling bearing. Rolling element bearings, also known as ball and roller bearings, differ from plain bearings in that the load is carried by placing metal balls or rollers *(cylinders)* between two metal rings called races. The relative motion of the races causes the balls or rollers to roll with little resistance.

The concept of rolling bearings had been around for centuries. The Romans used wooden ball bearings to support a rotating table on ships dating back to the early part of the first century AD. Around 1500, Leonardo da

Fig. 4-1: Railcar Journal Bearing ca. 1906.

Vinci produced drawings of machines that incorporated ball bearings. However, rolling element bearings had to wait until the latter part of the 1800s before becoming practical to manufacture. The primary issues with rolling bearings are the requirements for manufacture. In order to work properly, the balls in ball bearings must be perfectly spherical and of identical diameter. The race *(inner and outer)*, that is the component that retains the rolling balls, must be perfectly circular and the grooves concentric. In roller bearings, the rolls must be perfectly round, the same diameter, and the same length while the race must be perfectly flat across the width and perfectly circular in diameter. These specifications require manufacturing within very precise dimensions or the bearing would not move smoothly with low friction and wear.

The machines, materials, and processes necessary to make such bearings were not available during most of the Industrial Revolution. It was only after 1865 that the key technologies required to make precision bearings slowly become commercially available. First there was high-strength steel, later lubricants made from refined petroleum and, by the mid-1880s, electricity. Electricity led to the development of artificial abrasives, which was the final technology critical to the mass production of precision bearings. Artificial abrasives were necessary in the making of the artificial grinding wheel; they first became available in 1896 and rapidly replaced emery as the primary grinding abrasive. After this, the ability to grind metal parts *(bearings, shafts, bolts, castings)* to precision dimensions at low prices allowed manufacturers of all types to produce precisely made interchangeable parts that became the foundation of mass-production technologies.

Note: Emery, also known as natural emery, is an abrasive mineral primarily consisting of the crystalline form of aluminum oxide [Al_2O_3]. In early grinding, emery was ground to a powder, then glued to a cloth [emery cloth] or a rigid material such as a round piece of wood or steel called an emery wheel.

Ball bearings were the first of the modern rolling element bearings to appear. Early ball bearing designs did not work well due to poor manufacturing processes that produced out-of-round balls and rough surfaces in both the balls and bearing races.

It was not until the 1860s that the first practical ball bearing was used in the bicycle. Jules Suriray of France patented a radial ball bearing in 1869 and then used it to win the world's first bicycle road race. However, widespread use of ball bearings had to wait another two decades for machines to produce high-quality hardened steel balls and races made to precision dimensions and in large quantities.

First to arrive were the ball-making machines. In 1883, Friedrich Fischer, a German engineer, invented a ball bearing grinder that produced metal balls to exact sizes in large volumes. In 1887, Henry Richardson of the United States invented a similar ball bearing grinder that produced hardened spherical balls suitable for use in bearings.

However, these machines were not conventional grinders. Neither machine used an abrasive to machine the balls. Instead, the machines used two circular plates aligned in parallel that contained grooves embedded in the surface. The balls to be ground were inserted between the plates. One plate would rotate while the other remained fixed. The pressure and motion within the grooves created by the rotating plate removed excess material and produced hardened spherical balls of the same diameter.

The term "hardening" of steel is mentioned many times in this book. By definition, hardening is a metallurgical and metalworking process used to increase the hardness of metal. Increasing material hardness increases its ability to carry weight without deforming, as well as increasing its service life, two important properties in mechanical bearings.

For purposes of this book there are two forms of hardening applicable to bearings. The first is "work hardening" that is used in the process of making the hardened spherical balls mentioned above. Work hardening is also called "strain hardening" or "cold working," and refers to the process in which the steel is strained past its yield point, causing the material to become more difficult to deform and thus hardened.

The second process for hardening steel is "martensitic hardening," more commonly known as quenching and tempering. In this process, steel is heated to very high temperatures then "quenched," generally in water, to rapidly change its crystalline structure, resulting in very hard steel called martensite. In this structure, the steel can be too hard and will have some brittle properties that are not suitable for use in bearings. To "toughen" this very hard steel, martensitic steel is "tempered" by heat-treating to some temperature below that of hardening, followed by slow cooling in air. Outside of the ball-making process described above, all references to hardened steel refer to the quench-and-temper processes.

Even though the ball-making process did not require the new grinding wheels made with an artificial abrasive, the race-making machines did. In early ball bearings, the races were made in the shape of cups and cones. As applications for ball bearings grew, they were designed with circular inner and outer races *(rings)* that retained the balls while allowing rotation.

In 1896, with the advent of the new grinding wheels and machines, ball bearing races made to tight tolerances of 0.00025 inches *(0.006 mm)* became possible.[1] Around 1896, the Diamond Machine Company of Providence, Rhode Island, and the Pratt and Whitney Company of Hartford, Connecticut, were two of the first companies to manufacture specialized grinding machines for ball bearing races. The Diamond Machine Company was established as a manufacturer of emery grinding wheels, then ex-

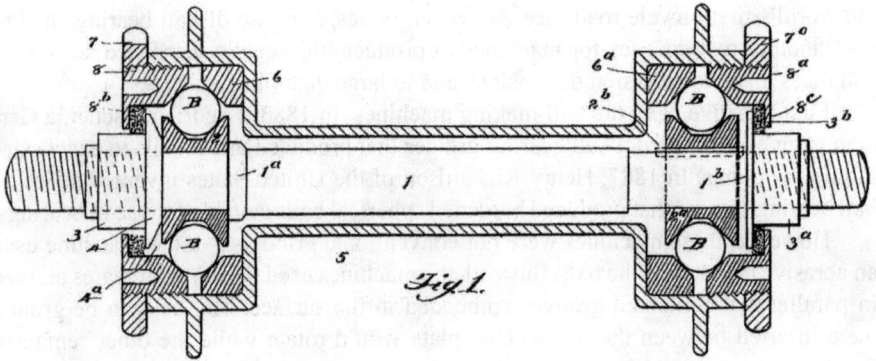

Fig. 4-2: Dodge Brothers Dirt-Resistant Ball Bearing for Bicycles. Patent 567,851 Filed July 20, 1895.

panded into universal grinding machines, gun-boring machines, and lathes. The Pratt and Whitney Company was founded in 1860 and in the early years made guns and gun-making machines for the Colt Patent Firearms Manufacturing Company. Pratt and Whitney established the concept of interchangeable parts and developed accurate measurement gages incorporated into their machines.

The first widespread application of ball bearings was in the bicycle industry. Bicycles were a popular mode of transportation in the late 1800s and a growing industry. Ball bearings used in bicycles were very small and experienced light loads and low speeds. The small, easily made, and low-cost components of bicycles allowed manufacturers of specialized grinding machines to experiment with different machining processes that resulted in ball bearings designed to minimize friction and wear. As the grinding processes improved, larger ball bearings designed for heavier loads and higher speeds became possible and were eventually used in roller bearings.

An example of an early ball bearing designed for bicycles is found in patent 567,851 granted to Horace and John Dodge in 1896. They first manufactured bicycles before founding the Dodge Brothers Company that manufactured precision automobile engines and chassis. The Dodge ball bearing *(Figure 4-2)* was unique in that it was the first dirt-resistant bearing designed with two complementary cone-shaped sections that formed a split outer race.[2]

Friedrich Fischer's invention of the ball-grinding machine in 1883 and the subsequent founding of his bearing manufacturing company, FAG *(Fischers Aktien-Gesellschaft)*, are generally considered the start of the modern rolling bearings industry.[3] In the United States in 1890, very few companies made rolling element bearings used in machines. Ball bearings, due to limits in manufacturing and capabilities, were primarily used in bicycles and small machines. As the 1890s progressed, new grinding technologies and the advent of the automobile industry set the stage for the modern bearings industry.

Automobiles required both journal bearings and rolling bearings. Journal bearings were primarily used in the internal combustion engine that had an ample reservoir of lubricant. However, in components such as the wheels and drivetrain, rolling-type bearings were required. Design specifications for ball and roller bearings, including loads, rotational speed, and operating temperatures varied widely depending on the application. Support requirements varied to constrain either radial *(weight)* or axial *(thrust)* movements, with some applications requiring both. Roller bearings have an advantage over ball bearings with their ability to operate under higher loads by spreading the forces over a greater surface area. For both bearing types, the most important requirements were that the bearings reduce friction and wear to an absolute minimum to ensure smooth, quiet operation, durability, and a long life. The ability to reduce friction using rolling surfaces is why rolling bearings are referred to as antifriction bearings.

In the early 1900s, as the automobile industry expanded, many supporting industries, including rolling bearing manufacturers, got their start. In addition to automobiles, trucks, and airplanes, many machines required antifriction rolling bearings in order to function. To meet the needs of individualized applications, new companies were either founded or expanded from their other products to manufacture various types of ball and roller bearings. A list of the more prominent companies in the early years of the industry is provided to illustrate the various types of rolling bearings developed.

Bearing Manufacturing Companies Founded between 1883 and 1912:

1. 1883: FAG *(Fischers Aktien-Gesellschaft)*
       Founder: Friedrich Fischer
       Location: Germany
       Inventor of the Ball-Grinding Machine
       Bearing Type: Ball

2. 1892: Hyatt Roller Bearing Company
       Founder: John Wesley Hyatt
       Location: Newark/Harrison, New Jersey
       Bearing Type: Spiral-Wound Roller

3. 1894: The Ball Bearing Company
       Founder: Winfield S. Rogers
       Location: Boston, Massachusetts
       Bearing Types: Ball and Roller.

4. 1898: The Standard Roller Bearing Company
       Founder: Samuel S. Eveland
       Location: Philadelphia, Pennsylvania
       Bearing Types: Radial Journal Roller, Roller Thrust, and Tapered Roller

5. 1899: Timken Roller Bearing Axle Company
       Founder: Henry Timken
       Location: St. Louis, Missouri, and Canton, Ohio

Bearing Type: Tapered Roller

6. 1907: SKF *(Svenska Kullagerfabriken)* Ball Bearing Company
   Founder: Sven Wingquist
   Location: Gothenburg, Sweden
   Bearing Type: Double-Row Self-Aligning Ball
   1909: First U.S. subsidiary in New York
   1915: First U.S. factory in Hartford, Connecticut

7. 1908: New Departure Manufacturing Company
   Founders: Albert and Edward Rockwell
   Location: Bristol, Connecticut
   Bearing Type: Double-Row Angular Contact Ball

8. 1910: The Ball and Roller Bearing Company
   Founder: Lewis R. Heim
   Location: Danbury, Connecticut
   Bearing Types: Ball Thrust, Roller Thrust, and Journal Roller

9. 1911: Fafnir Bearing Company
   Founder: Howard S. Hart
   Location: New Britain, Connecticut
   Bearing Type: Ball

10. 1912: The Torrington Company *(Bearing Manufacturing)*
   Founder: None. Acquired small ball bearings business in 1912.
   Location: Torrington, Connecticut
   Bearing Types: Ball and Needle Roller

All the above companies were major contributors to the evolution of the modern rolling bearings industry. Most had unique bearing designs that found large markets in transportation and industrial machinery. A synopsis of each company is given below.

FAG Bearing Company

Friedrich Fischer invented the first grinder for making the balls for ball bearings in 1883. His machine allowed steel balls to be ground absolutely round and to the same diameter and manufactured in large volumes. Fischer's ball-grinding machine laid the foundation for the entire rolling bearing industry. In 1896, Fischer built a large factory in Schweinfurt, Germany, that manufactured 10 million balls per week.

Fischer died of a stroke in 1899 and his company fell into financial trouble. George Schafer took over, and by 1909, FAG produced more than 40 percent of the world's ball bearings, with many exported to the United States. In 1911, Ray Harroun, whose car engine was equipped exclusively with FAG ball bearings, won the inaugural race of the Indianapolis 500.[4]

Hyatt Roller Bearing Company

In 1892, John Hyatt *(1837–1920)* developed the first commercially used roller bearing. Hyatt was known as a prolific inventor whose first patent, issued in 1861, was a grinding machine for knives and other sharp-edged tools. Hyatt invented the first widely used roller bearing using spiral-wound rollers. At the time, grinding machines could not achieve the precise dimensions required to use solid-cylinder rollers, so Hyatt devised a method to fabricate rollers from coiled strips of flat spring steel wound into cylinders using a roller-making machine that he also developed. Hyatt's roller design had many benefits over solid-cylinder rollers available at that time. Spiral-wound rollers were more flexible, did not heat up as much from friction, and were more durable due to less wear. Hyatt's roller bearings were used in the axles and transmissions *(Figure 4-3)* of automobiles well into the 1920s.

In the same year of his invention, Hyatt founded the Hyatt Roller Bearing Company in New Jersey and by 1895 was selling around $2,000 of bearings per month. Even though his bearings were selling well, Hyatt was not a businessman and the company became deep in debt. In the same year, Hyatt hired a 20-year-old draftsman by the name of Alfred P. Sloane Jr. who recently graduated from the Massachusetts Institute of Technology. Even though Sloane realized the potential of Hyatt's roller bearing, the low pay forced Sloane to leave in 1897 for a better-paying position at a new company developing an electric refrigerator.

In 1899, with Hyatt still in financial difficulty, Sloane's father *(A. P. Sloane, Sr.)* bought Hyatt and hired back Alfred to make the company profitable. Within six months as the manager of production, Sloane made a profit. In 1901, at the age of 26, Sloane became president of Hyatt and positioned the company for rapid growth and expansion by selling to the major automobile companies.

Sloane learned a big lesson early in his career on the importance of precision manufacturing when Hyatt's first order shipped to the Cadillac Automobile Company was refused. Cadillac was formed in October 1902 from the remnants of the Henry Ford Company and was managed by Henry Leland, a former engineer at Brown and Sharpe. Leland rejected Hyatt's bearings since the manufacturing tolerances did not meet Cadillac's specification of +/- 0.001 inches *(0.025 mm)*. During a meeting between Leland and Sloane at Cadillac in Detroit, Leland apparently impressed upon Sloane the importance of uniform precision in manufacturing automobile parts for reliability and interchangeability. Sloane accepted Leland's advice and implemented standards for ensuring precision manufacturing at Hyatt. Under Sloane's leadership, Hyatt became a major supplier of bearings used in axles and transmissions and sold not only to Cadillac, but also to the Ford Motor Company and the major suppliers to the automobile industry. In 1916, General Motors, in a strategy to own key suppliers, purchased Hyatt Roller Bearing, New Departure Manufacturing, and other companies and formed a parts and accessories division called the United Motors Company with Alfred Sloane as president. Sloane remained with General Motors for the next 40 years. He became president of GM in 1923, retired as chairman of the board in 1956, and then formed one of the world's most prestigious management schools at his alma mater, MIT.

Fig. 4-3: Hyatt Roller Bearing Advertisement Showing Spiral-Wound Rollers. The Saturday Evening Post, *June 5, 1915.*

Fig. 4-4: Bantam Ball Bearing Company Advertisement in Machinery, August 1922.

The Ball Bearing Company

The Ball Bearing Company was founded by Winfield S. Rogers in Boston in 1894. At that time and through the turn of the century, ball and roller bearings were just becoming available to replace journal bearings.[5] But due to poor manufacturing processes, mainly grinding technology, rolling bearings made in large quantities to precise dimensions were not available. Grinding was mostly conducted on centered machines using steel grinding wheels coated with emery. Solid grinding wheels composed of man-made abrasives were just introduced to the market and were years away from large-volume availability. Given the poor quality of manufacture, rolling *(antifriction)* bearings were generally shunned in favor of journal bearings. However, high-quality rolling bearings were in demand and companies able to produce such bearings survived and grew. The Ball Bearing Company of Boston was one such company credited with pioneering design and manufacturing processes to improve rolling bearing quality and, in 1899, was the first to publish a catalog for ball bearings in the United States.[6]

In 1901, the Ball Bearing Company sold both its Boston and Keane, New Hampshire, plants to the Standard Roller Bearing Company of Philadelphia. Rogers left the Boston company around 1903, moved to Connecticut, and founded a new company in Bantam, about 40 miles northeast of Danbury. Rogers was, in addition to a businessman, an avid inventor. Inventions patented by Rogers include rolling bearings *(patents 639,775 and 1,182,796)*, a signal lantern *(229,054)*, a ratchet drill *(401,375)*, a puzzle padlock *(1,013,566)*, plus other items.

Rogers first named his company the Anti-Friction Bearing Company and later changed its name to the Bantam Ball Bearing Company *(Figure 4-4)*, which was commonly referred to as the Bantam Company and continued the Boston product line of manufacturing both ball and roller bearings. Around 1915 or 1916, Bantam discontinued manufacture of roller bearings to focus on ball bearings. In July of 1928, the entire operation in Bantam was moved to a new plant in South Bend, Indiana, at which time its name was changed back to the Ball Bearing Company. Over the next twelve years, the Ball Bearing Company increased production through facilities expansion and increases to its workforce, from 55 to 375. In 1935, the company was acquired by the Torrington Company as part of its plan to fuel long-term growth. Under Torrington, the company name was changed to the Bantam Bearings Company to reflect its expanded range of bearing types.

The Standard Roller Bearing Company

In 1898, Samuel S. Eveland founded the Standard Roller Bearing Company in Philadelphia, Pennsylvania, to manufacture journal roller and roll thrust bearings.[7] Eveland was both a businessman and an inventor. Eveland invented and patented both antifriction rolling bearings and machine tools used in the manufacture of bearings. Some of the bearings he invented are listed below, with Figure 4-5 illustrating his roller bearing designed for end thrust loads:

Roller Bearing: Patent 650,383, filed Sept. 22, 1899.
Antifriction End Thrust Device: Patent 676,939, filed Nov. 21, 1899.
Antifriction Bearing: Patent 687,954, filed Dec. 10, 1900.
Antifriction End Thrust Device: Patent 718,111, filed Nov. 30, 1901.
Ball Bearing for Vertical Shafts: Patent 935,648, filed Dec. 6, 1907.

In 1901, Eveland bought the Ball Bearing Company and merged it with Standard Roller Bearing and six other bearing and bearing equipment manufacturers. As Standard Roller Bearing grew, it manufactured a wide range of annular ball, thrust, and tapered roller bearings. The company made many contributions to the advancement of antifriction bearings, with one of the most notable the development of 52100 steel, a high-carbon, low-alloy steel containing chromium that greatly improved bearing life.[8]

Around 1918 or 1919, Eveland sold Standard Roller Bearing to the Marlin-Rockwell Company *(MRC)* and later the name was changed to the Standard Steel Bearing Company. MRC merged with Gurney Ball Bearing in 1924 and purchased Strom Ball Bearing in 1925. In 1964, Marlin-Rockwell merged with TRW and was renamed TRW Bearings in 1979. In 1986, TRW sold the Bearings Division to Swedish firm SKF Group, an international conglomerate composed of nearly 200 companies operating in many countries.[9]

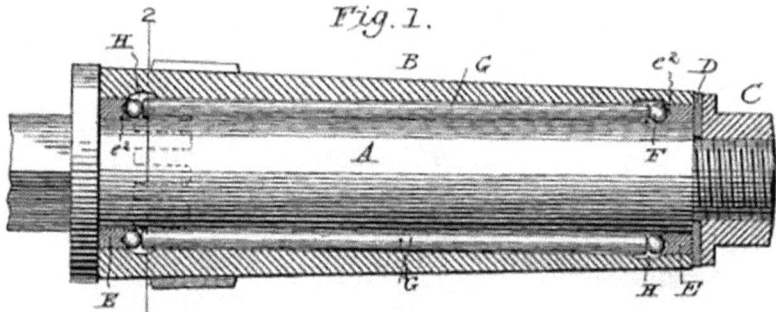

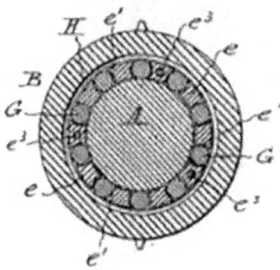

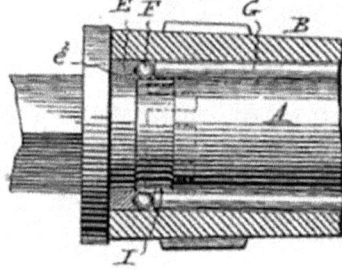

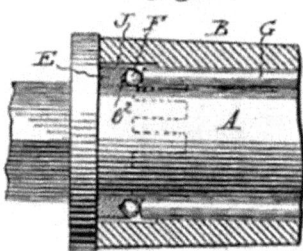

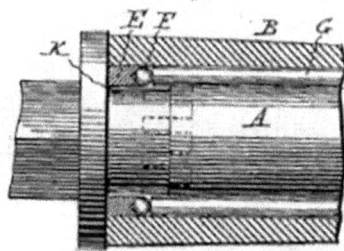

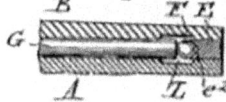

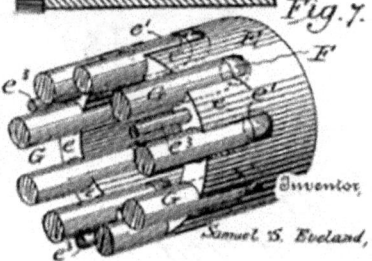

Fig. 4-5: Eveland Roller Bearing Designed for End Thrust Loads. Patent 650,383, filed Sept. 22, 1899.

Fig. 1

Fig. 2

Fig. 3

Fig. 4

Fig. 5

Fig. 6

Fig. 7

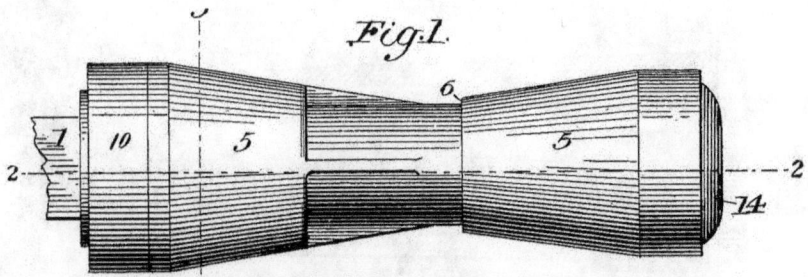

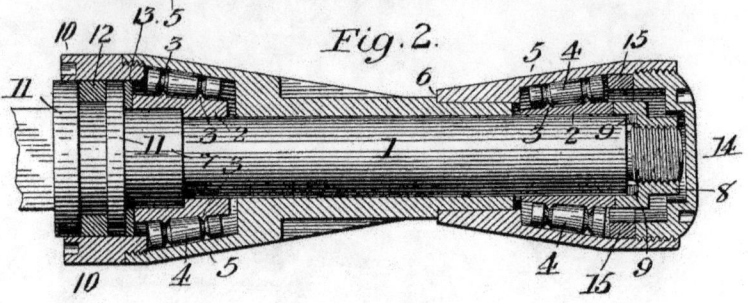

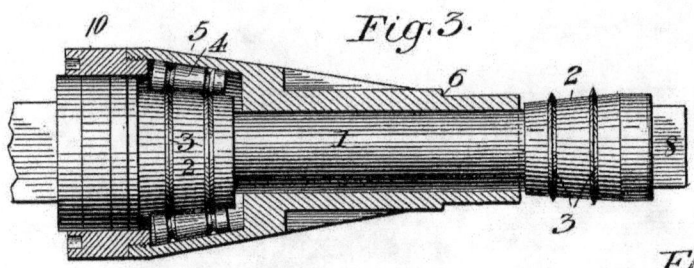

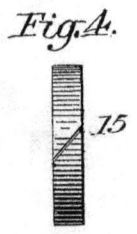

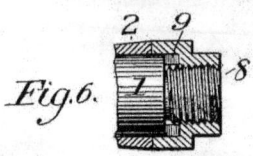

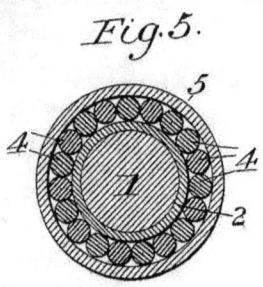

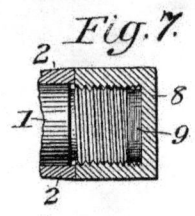

Witnesses:
David Attelson.
T. Percs Carr

Inventors
Henry Timken and
Reginald Heinzelman,

Fig. 4-6: Timken Tapered Roller Bearing. Patent 606,635, filed Aug. 27, 1897.

Timken Roller Bearing Axle Company

In 1897, Henry Timken, a carriage maker, developed a bearing designed with tapered rollers to allow freight wagons to make sharp turns. Patented in 1898, Timken's bearing *(Figure 4-6)* was designed for use in wheel axles to replace friction *(journal)* bearings that required lubricants to function. Tapered roller bearings were a breakthrough for the automobile industry because they could handle both weight *(radial forces)* and cornering *(thrust forces)* more reliably and without a reservoir of lubricant that is required for journal bearings. In 1899, as Sloane was hired to fix Hyatt Roller Bearing, Henry Timken of St. Louis, Missouri, founded a company to sell his new tapered roller bearing.

Two years later, Timken relocated to Canton, Ohio, in order to serve the burgeoning automobile industry, with major manufacturing centers in Detroit and Cleveland. Timken eventually expanded its bearing product lines to serve many other industries and today is a global manufacturer of bearings, high-performance steels, and engineered components.

SKF (Svenska Kullagerfabriken) Ball Bearing Company

In 1906, Sven Wingquist, a Swedish engineer, invented the double-row self-aligning ball bearing. This bearing *(Figure 4-7)* had an inner ring machined with two rows of races and a common sphered raceway machined into the outer ring. With this design, the bearing is self-aligning and insensitive to shaft deflections and misalignment while rotating. Wingquist subsequently founded SKF in Sweden in 1907 and in 1909 established a subsidiary in New York. Six years later in 1915, SKF built its first United States factory in Hartford, Connecticut.

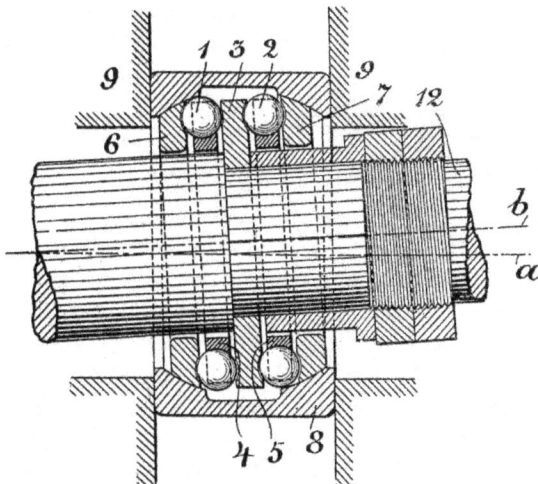

Fig. 4-7: Wingquist/SKF Double-Row Self-Adjusting (Aligning) Bearing. Patent 980,582, filed Nov. 14, 1908.

New Departure Manufacturing Company

New Departure was originally a manufacturer of doorbells in Bristol, Connecticut. Founded in 1888 by brothers Edward and Albert Rockwell, they branched out into the bicycle industry in 1898 to make coaster brakes and wheel hubs. In 1904, the Rockwell brothers ventured into the automobile business, producing a car with a 15-horsepower engine. This led to the production of the Rockwell Taxi Cab in 1907 that was known for its distinctive orange-yellow color and later used by the Yellow Taxicab Company in New York City.

In 1906, Alfred Rockwell invented the first angular-contact ball bearing *(Figure 4-8)* designed to accept thrust *(axial)* loads. New Departure sold its bearings for use in automobile front wheels and machine tools. After 1910, bearings became the principle product of New Departure and in 1916 the company was purchased by General Motors and became a division under its parts supplier, United Motors Corporation.

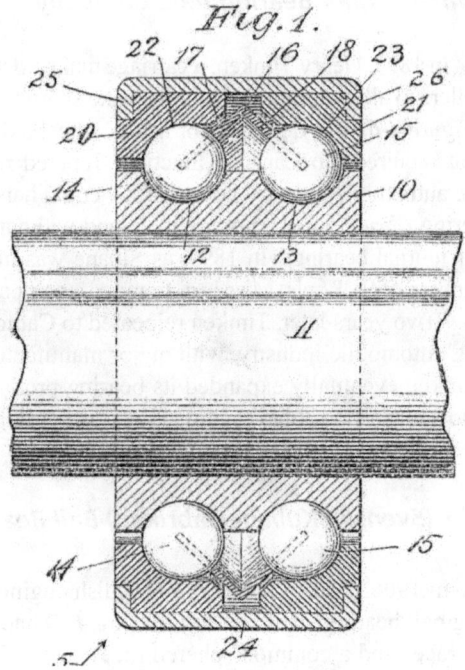

Fig. 4-8: Rockwell/New Departure Double-Row Angular Contact Ball Bearing. Patent 921,464, filed May 10, 1906.

The Ball and Roller Bearing Company

In 1910, the Ball and Roller Bearing Company *(B&RB)* was founded by Lewis R. Heim in Danbury, Connecticut. The B&RB shared the same building as the Heim Machine Company, which was founded in 1904 to make collar ironing and other machinery.[10] From the beginning, the B&RB manufactured ball thrust, roller thrust, and journal roller bearings mostly for machinery and vehicles. The B&RB became a prime supplier to the United States military during the First World War, with bearings supplied on army tanks and to the navy for use in pumps and submarine periscopes. Outside of the military, the B&RB was a major supplier of journal roller bearings to the paper and textile industries.[11] In later years, the B&RB expanded to make radial ball and roller bearings, piston pins, valve lifters, valve lifter rolls, steel bushings, and small hardened and ground cylindrical rolls.

Typical bearings produced by the B&RB are shown in Figures 4-9, 4-10, and 4-11. A description of the bearings along with specifications can be seen in Catalog

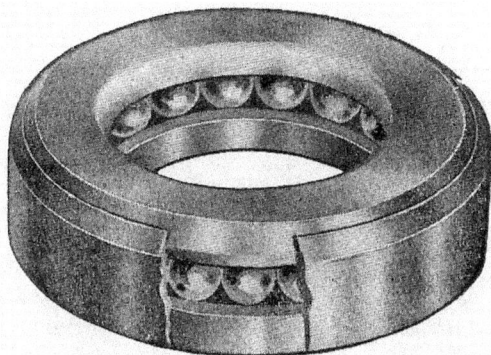

Fig. 4-9: B&RB Ball Thrust Bearing.

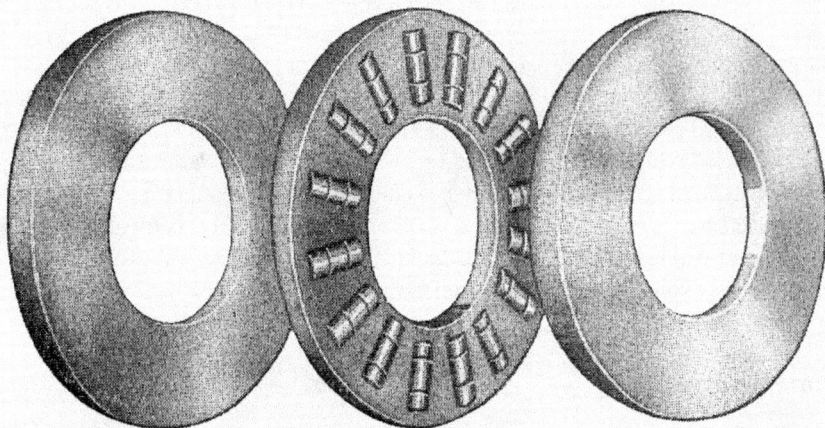

Fig. 4-10: B&RB Roller Thrust Bearing.

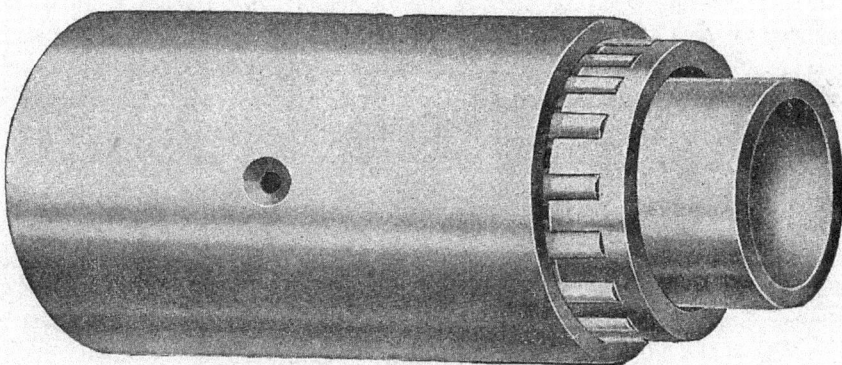

Fig. 4-11: B&RB Journal Roller Bearing.

#11 located in Appendix 2 *(A2-3)*. Even though the B&RB was officially founded in 1910, Heim, in many publications, such as Catalog #11, used the 1904 founding date of the Heim Machine Company. Also found in Appendix 2 are typical advertisements for B&RB bearings placed in trade magazines *(A2-2)*.

Fafnir Bearing Company

Howard S. Hart was an engineer, inventor, and businessman who, along with Norman Cooley, founded the Hart and Cooley Manufacturing Company in New Britain, Connecticut, in 1901. Hart and Cooley manufactured stamped steel registers for ventilation systems that replaced registers made from traditional heavy cast iron.

On November 5, 1910, two inventions were filed for a lower-cost ball bearing fitted with a new type of ball cage used to hold and align the steel balls. One application was filed by employees Edward Goodwin and Edward House, and the second was filed by Howard Hart. These inventions became patents 981,551 and 988,336, both granted in 1911. Hart's ball cage design is shown in Figure 4-12. Around the same time, another inventor, William Hasselkus, who was a citizen of Germany working at Hart and Cooley, developed an improved machine for grinding steel balls used in ball bearings that was granted patent 989,524 in April 1911.

With these inventions, in 1911, Howard Hart founded the Fafnir Bearing Company in New Britain, Connecticut, to manufacture ball bearings that competed against higher-priced bearings produced in England and Germany. The company grew over the years and expanded production to include many variations of ball bearings, including angular contact, ceramic, and pillow block types. Fafnir was acquired by the Torrington Company in 1985.

Torrington Company

The Torrington Company originated as the Excelsior Needle Company that was organized in Wolcottville *(Torrington)*, Connecticut, in 1866 to produce sewing machine needles. Its business grew over the years, with expansion into knitting machine latch needles and heavy hook needles used to make shoes. Excelsior Needle later expanded to serve the growing bicycle industry by forming the Torrington Swaging Company to manufacture spokes for bicycle wheels. In 1917, both companies were merged into a single corporation called the Torrington Company of Connecticut.

In 1912, prior to the merger, Torrington acquired a small ball bearing manufacturing business that evolved into Torrington's primary business in the 1920s. After the start of the Depression in 1929, bearing sales dropped off significantly, forcing Torrington to search for new business. Around 1933, one of its research engineers, Edmund Brown, developed a new bearing type with long, slender rollers that resembled needles. Similar in design to a journal roller bearing, Brown's bearing was designed for constraining shafts, with the ability to use the shaft as the inner race. Figure 4-13, taken from patent 2,038,474, illustrates Brown's roller bearing using thin, needle-like rolls. The Torrington "needle" roller bearing became the bulk of Torrington's bearing production for many years.

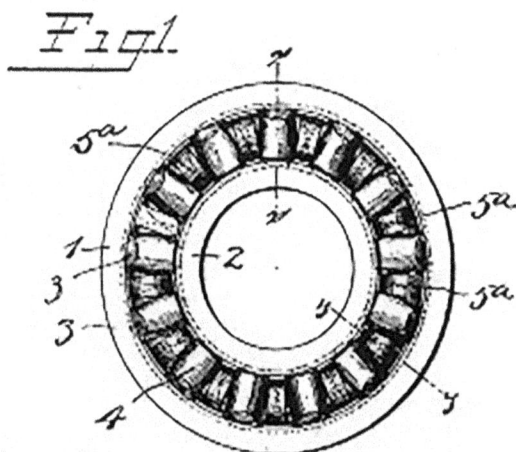

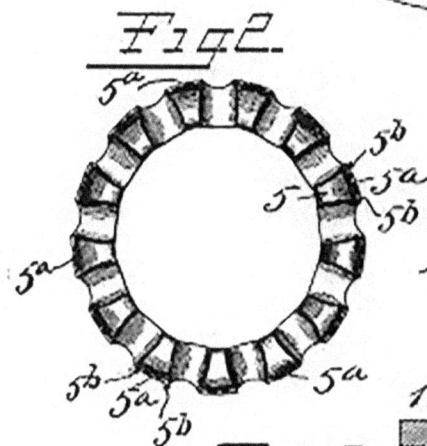

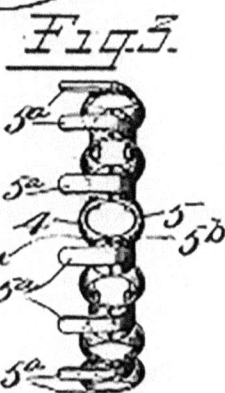

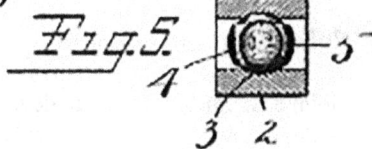

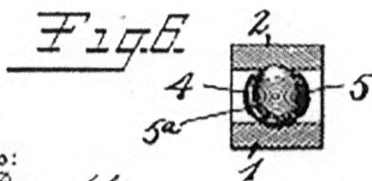

Fig. 4-12: Hart Ball Cage for Ball Bearings. Patent 988,336, filed Nov. 5, 1910.

In 1935, Torrington acquired the Bantam Ball Bearing Company, which helped fuel the company's size and growth. Ingersoll-Rand bought Torrington in 1968 and allowed Torrington to remain as an independent subsidiary. Torrington purchased Fafnir in 1985 to become the largest bearing manufacturer in North America. In 2003, Ingersoll-Rand sold Torrington to the Timken Company, which in turn sold Torrington to the Japanese corporation JTEKT in 2009.

The above list of bearing manufacturers is not intended to be all-inclusive, as there were many other companies that started during the same period. Examples include the Auburn Ball Bearing Company that was established in 1893 in Auburn,

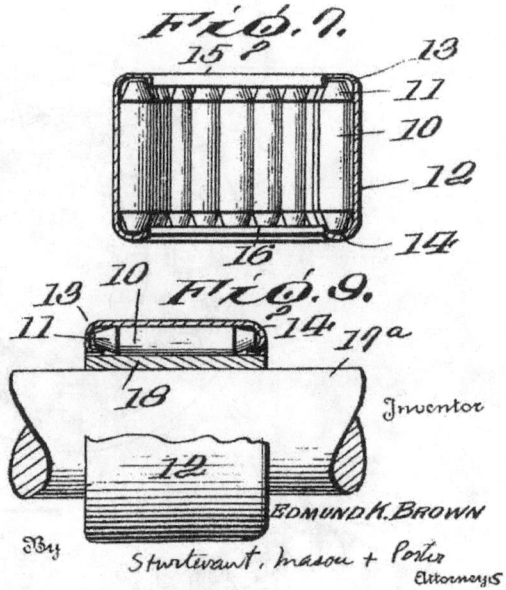

Fig. 4-13: Torrington Needle Bearing. Patent 2,038,474, filed Nov. 7, 1933.

New York, and sold bearings with a nearly frictionless V-groove design. American Ball Bearing Company started in 1911 and made rolling bearings for heavy industrial equipment. Other companies established in the early days include the Gwilliam Company, the U.S. Ball Bearing Manufacturing Company, and the Bearings Company of America.

Incorporation of the Ball and Roller Bearing Company

In 1909, Alfred Heim left the Heim Machine Company to devote his full attention to operating the Danbury Troy Laundry Company. This left Lewis as the sole owner of the Heim Machine Company that made his collar-ironing machines. At that time, Heim began producing ball and roller bearings of his own design. The primary reason Heim became interested in bearing manufacture was due to the lack of suitable rolling bearings for his detachable collar laundry machines.[12] A second reason was that Heim saw an opportunity to enter the burgeoning market for precision bearings used in all types of machines, a far larger market than his laundry machines. Over the next few years, Heim developed his bearing designs and sold them to various industries. During this period, Heim attracted the attention of investors who offered to help fund his new company.

On May 19, 1914, the Ball and Roller Bearing Company incorporated to manufacture *"ball and roller bearings and other kinds of bearings, laundry machinery and all other kinds of machinery."* The Danbury incorporation records show that Heim took on two partners, John Henry Roth and William C. Barrett, to provide financing and management to operate the factory. Roth, who was Heim's brother-in-law, became secretary while Heim retained the titles of president and treasurer. Barrett remained solely an investor until 1928, when he became vice president and an active officer in the operation of the B&RB.

At incorporation, 1,000 shares of capital stock were issued at $100 par value for a total of $100,000.[13] Heim contributed the assets of his growing bearings company, valued at $10,000, and received 100 shares *(10 percent)* in the B&RB. The investors Roth and Barrett contributed the balance of $90,000 in capital and received 900 shares. The capital infusion allowed the B&RB to expand production just as the First World War was starting in Europe, which spurred demand for all types of industrial bearings.

Contrary to the description provided in the Danbury incorporation records, the Heim Machine Company was not part of the B&RB. Heim continued to own and operate his laundry machine company as sole proprietor in the same building as the Ball and Roller Bearing Company up through 1915.[14] The Heim Machine Company made machines while the B&RB made bearings. Machines made in the Heim Machine Company included not only laundry machines, but also machines used to manufacture his bearings.

After 1910, many sectors of the economy, such as the automobile industry, began to experience explosive growth. Virtually all companies manufacturing machines required some type of mechanical bearing to function smoothly and operate well over long time periods with minimal maintenance. Rolling-type, antifriction bearings rapidly replaced journal bearings in many applications. As demand grew for rolling bearings, so did the competitive pressures to produce high-quality products at lower prices. Competition forced all companies to develop new manufacturing and production methods to improve quality and lower their costs. Manufacturers of rolling bearings relied heavily on grinding and sought new machines and processes to produce precision-made bearing components at low cost and high production rates. In many cases, the bearing manufacturers had to develop their own machine tools in order to compete. These pressures led Lewis Heim to develop one of the most sought-after machine tools for manufacturing precision cylindrical parts: the Centerless Cylindrical Grinder.

Chapter 5

A Short History of Grinding Machines

Machine Tools

Grinding machines belong to the family of machines referred to as machine tools. They are essential in the modern world because they make the precision and interchangeable components to machines that perform functions much more efficiently than humans can, and some of which cannot be done by humans at all. Without grinding processes and machines, industries requiring precision machinery producing at mass-production quantities would not exist. This importance is captured in the book titled *History of the Grinding Machine* written by Robert Woodbury of MIT in 1959. As a precursor to the event of Heim's development of the centerless grinder, a short history of grinding machines, influential inventors, and major developments is provided with information taken from Woodbury and other sources.

From the late 1700s to the early 1900s, the period known as the Industrial Revolution, the United States experienced tremendous changes in manufacturing, transportation, and agriculture, mostly due to the invention of new technologies, especially large and precision-made machines. The period started with the development of the steam engine used to power factories, locomotives, and ships, and ended with the invention of the internal combustion engine for automobiles, trucks, and airplanes, along with electricity production for lighting, electric motors, and communications. The industries spawned from these technologies provided people with a large assortment of products that made life easier and more enjoyable, and greatly improved living standards. One of the lesser-known but critical industries was the manufacture of machine tools. Machine tools are the machines that make the machines used in all other industries and served as the foundation for the Industrial Revolution.

One of the first machine tools that started the Industrial Revolution was the cylinder boring machine developed by Englishman John Wilkinson *(1728–1808)* in the early 1770s. James Watt, a Scottish engineer and inventor who developed the first reliable steam engine around 1775, tried for many years to obtain accurately bored cylinders to accommodate tightly fitting pistons. The engine cylinders were made of iron that, due to the manufacturing processes, were out-of-round, causing steam leakage around the piston and poor engine performance. To alleviate the leakage problem, Watt used the novel metal-boring machine developed by Wilkinson in 1774. Wilkinson's cylinder-boring machine *(Figure 5-1)* was derived from his original machine for boring holes in solid iron used to make cannons and gun barrels in armories. This machine used a round cutting tool that was supported at one end and

Fig. 5-1: Wilkinson Cylinder-Boring and -Grinding Machine: ca. 1775.

pushed forward while the workpiece *(cannon or gun)* rotated.

The cylinder-boring machine used to make Watt's steam engines was modified from the original Wilkinson boring machine and designed specifically to machine large- diameter iron cylinders. The engine cylinder *(workpiece)* to be bored was open at both ends and fixed to a saddle support. A shaft supported at both ends and passing through the cylinder was fitted with a round gear-like cutting tool. As the cutting tool was inserted into one end of the cylinder, it was rotated while being pushed through the cylinder to mill *(cut)* the oblong wall into one that was round and straight. What made this machine unique was that it also doubled as a grinding machine. By replacing the cutting tool with a piston covered in emery, the boring machine also performed grinding to achieve dimensional requirements within 0.02 inches *(0.5 mm)*[1] The ability to grind the cylinder wall after milling was one of the first attempts to produce identical and interchangeable parts critical to mass production.

Wilkinson's boring machine is considered to be one of the first practical machine tools. To be classified as a machine tool, the device must contain a cutting tool, a means to support and constrain the work, plus a means to guide the movement of the cutting tool or the work against the other. In many machines, the cutting tool is mounted on a platform or tool holder that is able to slide or otherwise position itself relative to the work. In machine tools such as lathes, millers, and grinders, the tool holder is usually connected to a leadscrew that converts turning motion into linear motion for positioning relative to the work.

Since 1800, many types of machine tools have been developed for machining metals using various methods for removing material and shaping the work. All machine tools shape a workpiece by removing *(cutting)* material through shearing. To be effective, efficient, and long lasting, cutting tools *(bits or blades)*, must be harder than the material being sheared. Machines such as the lathe, planer, and boring machines use a cutting tool with a single point to remove material. A drill press uses a cutting bit containing two points of contact. Saws and milling machines generally use a rotating tool bit containing multiple cutting points.

The last type of machine tool, the grinder, is very different from the other cutting tools. The grinder uses a very hard rock-like wheel rotating at high speed that contacts the work with hundreds or thousands of very small cutting points that remove very tiny pieces of the work. The process in which the grinding wheel literally shears off microscopic pieces of material is called "abrasive cutting" and allows for dimensional accuracy better than one ten-thousandth *(0.0001)* of an inch *(0.00254 mm)* with modern machines. This ability makes the grinder the only machine tool capable of achieving exact dimensions and a fine finish on very hard materials. These capabilities are critical in order to manufacture smoothly operating, long-lasting, and tight-fitting parts such as shafts, gears, and plain and rolling element bearings.

As the 1800s progressed, machine tools evolved to shape work made from much harder materials, such as steel and alloyed metals. The beds of the machines were made of iron or steel with heavy and rigid construction to minimize vibration and deformation while providing solid support for the cutting tools and workpiece. Components of these machines such as the carriage, headstock, tailstock, leadscrew, and spindles were made to tightly fitting, precise dimensions to minimize play and maximize dimensional accuracy of the machined work.

Machines, such as the lathe, the milling machine, the boring machine, the drill press, and the planer, evolved quickly to incorporate these features. However, the grinding machine, due to the nature of its cutting tool, the grinding wheel, lagged behind in development. Since grinding was the only machine tool capable of achieving very high tolerances, its slow development hindered development of machines requiring precision-made interchangeable parts. It would take more than 100 years from the development of Wilkinson's boring and grinding machine for grinding to fully develop into a machine tool capable of precision grinding of very hard, high-strength metals. Starting around 1900, performing this process at low cost and high production rates became a requirement of the new transportation industry founded on the internal combustion engine.

The Universal Grinder

As the First Industrial Revolution transitioned into the Second Industrial Revolution after 1840, there were many inventors of precision machines for cutting and grinding metal. Of all the inventors making significant contributions in the advancement of grinding technology, the two most recognized are Joseph R. Brown and Charles H. Norton. Joseph Brown *(1810–1876)* learned about machines through his father, who manufactured watches and clocks in Providence, Rhode Island. Around 1850, Brown set up his own business making calibration devices and machine tools that in 1853 became the Brown and Sharpe Manufacturing Company and was generally referred to as Brown and Sharpe. Brown's partner, Lucien Sharpe, was an inventor himself but eventually focused on the business side of the operation, letting Brown concentrate on developing new machines.

The Brown and Sharpe Company developed, manufactured, and sold mathematical calibration and measuring devices as well as machine tools used to make precision components. Their machine tools included turret lathes, and automatic screw-making, milling, and grinding machines. Calibration and measuring devices included the automatic linear dividing engine, standard wire gauges, vernier calipers, and shop micrometers. Calibration was essential on all machine tools, especially grinding machines, since these were the devices that set the position of the cutting tool relative to the work. Without them, the machines could not be properly set up for accurate operation.

One of Brown's most famous inventions, and his last before he died, was the Universal Grinding Machine *(patent 187,770)*. His first commercial machine was completed in 1874 and it became one of the first grinding machine tools used for mass production. The machine *(Figure 5-2)* had the profile of a lathe and was designed for surface grinding of hardened metal cylindrical work that was held at both ends by spindles that rotated using overhead belts. The grinding wheel was mounted on one side of the machine to a sliding support table for positioning along the length of the work. In addition to sliding, the support table was designed to swivel, allowing the grinding wheel to perform both flat and tapered grinding.

Brown knew his universal grinding machine would be a success and immediately offered it for sale, as it appeared in the Brown and Sharpe product catalog issued in January 1875. Due to his death in July 1876, Brown never had the chance to improve his invention, but Brown and Sharpe continued to sell, improve, and develop new models. Even with all the improvements and changes to the universal grinding machine, the basic design invented by Joseph Brown remains in modern machines.

Henry M. Leland, who later founded both the Cadillac and Lincoln automobile companies, worked at Brown and Sharpe as a foreman during development of the universal grinder. In a manuscript kept at Brown and Sharpe, Leland summarized the importance of Brown's grinding machine as follows:

> *What I consider Mr. Brown's greatest achievement was the Universal Grinding Machine. In developing and designing this machine he stepped out on entirely new ground and developed a machine which has enabled us to harden our work first then grind it with the utmost accuracy, at the same time protecting the ways, the surface on which the platen travels, from emery and grit, also the improvement of revolving the work from dead centers thus eliminating the error of live spindles and live centers. If all these machines should suddenly be taken away, it is hard to imagine what the results would be. It would be impossible to make any more hardened work for the best parts of our machinery and tools, that would be round, true and accurate in every detail to the closest possible limits. This in my judgment is one of the most remarkable inventions and too much cannot be said in its praise, or in ac-*

knowledgment of Mr. Brown's perseverance, wonderful initiative and genius. I know of none who deserves a higher place or who has done so much for the modern high standards of American manufacture of interchangeable parts as Joseph Brown.[2]

Leland's acknowledgement highlights the most important characteristics required by manufacturers of grinding machine tools. The first is to be able to grind steel after it has been hardened which, when combined with tempering, gives steel additional toughness that reduces wear over time. The second is the ability to produce a truly round and dimensionally accurate product. This ability is why the grinder is irreplaceable in the production of interchangeable metal parts.

Charles H. Norton *(1851–1942)* was well known as an inventor and manufacturer of machine tools and grinding wheels critical to the automobile industry. Norton was first exposed to grinding in 1865 at the age of fourteen while working at the Seth Thomas Clock Works in Plymouth, Connecticut. Companies manufacturing clocks, watches, and sewing machines generally manufactured small parts made to precision dimensions. The grinding machine was a necessity to make these small parts.

With twenty years of machine tool experience, Norton in 1886 joined the Brown and Sharpe Manufacturing Company. By that time, Brown and Sharpe grinding machines were being built for both light grinding and heavy surface grinding, primarily of steam engine parts. Up to the 1880s, grinding machines generally used wheels up to 12 inches in diameter and 0.5 inch thick.[3] Larger machines designed for heavy surface grinding used wheels made from emery, with diameters up to 18 inches and 1-inch thick. Grinding using such large wheels required larger machines and improved designs in order to achieve re-

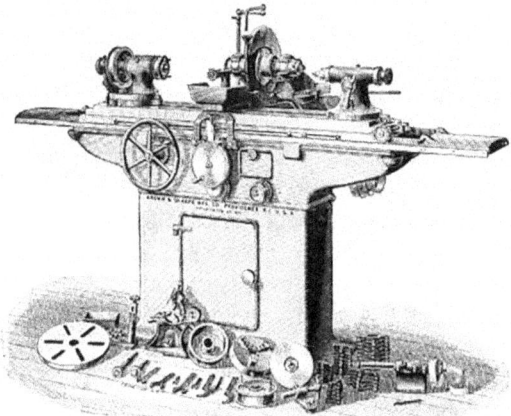

Fig. 240.—No. 2 Universal Grinding Machine: Front View.

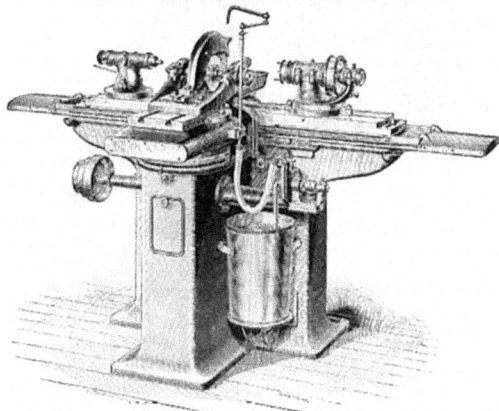

Fig. 241.—No. 2 Universal Grinding Machine: Rear View.
Brown & Sharpe Mfg. Co.

Fig. 5-2: Brown and Sharpe 1896 Universal Grinding Machine.

quired grinding tolerances.

After joining Brown and Sharpe, Norton was assigned to help improve Brown's universal grinding machine, which still contained defects eleven years after the first machines were sold. In addition to changes that addressed the defects, Norton is credited with the development of a device for performing internal grinding of cylindrical work. Internal precision grinding was just as necessary as external grinding with the manufacture of cannons and steam engine cylinders. The device was designed as an attachment to the universal grinder and generated significant sales for Brown and Sharpe.

While at Brown and Sharpe, Norton met and worked with Henry Leland and apparently impressed Leland enough that in 1890 Leland brought Norton to Detroit when he formed machine toolmaker Leland, Falconer and Norton. Norton was the chief designer as well as a partner in this company that made the universal miter trimmer and wet emery grinder.

Six years later, in 1896, Norton returned to Brown and Sharpe to design improved cylindrical grinders with wide wheels for heavy production grinding. However, politics within Brown and Sharpe against his ideas forced Norton to leave in 1898 and join the Norton Emery Wheel Company in Worcester, Massachusetts. The Norton Emery Wheel Company, founded by Frank B. Norton *(no relation to Charles Norton)* made artificial grinding wheels made with emery that were in great demand by industry.

At Norton Emery Wheel, Norton was given more freedom to put his ideas to work and developed large production grinding machines mainly for the nascent automobile industry. Larger machines were no different than smaller machines in the requirement for precision grinding. However, large machines with increased dimensions induced greater vibration and deflection, making high precision more difficult to achieve. Norton's experience with precision machine tools led him to design grinding machines that were massive compared to existing machines, with increased rigidity that reduced vibration during operation.

One of Norton's first machines, designed in 1900, was 8 feet long and used grinding wheels 24 inches in diameter by 2 inches wide. The machine, which later received an industry award for design, was larger, heavier, and more rigid than competing machines, and allowed operation over a range of wheel speeds while incorporating many automatic features. His first machine was sold to a New York company manufacturing printing presses, which operated it for thirty years. Norton designed and manufactured other grinding machines, many with micrometer movements allowing grinding of parts to a range of precision sizes. The automobile industry purchased many of Norton's grinding machines for grinding crankshafts, pins, camshafts, and other components made from metal. In 1906, with the company selling grinding machines in addition to artificial grinding wheels, the name was shortened to the Norton Company.[4]

After 1896, other manufacturers of grinding machines began competing against Brown and Sharpe and the Norton Company. Two of the more notable companies were the Landis Tool Company of Waynesboro, Pennsylvania, and the Heald Machine Company of Worcester, Massachusetts. Both of these companies primarily served the growing automobile industry and designed most of their machines for manufacturing automobile engine and drive train parts.

The Landis Tool Company was formed in 1897 out of a partnership between two brothers, Franklin and Abraham Landis. Landis Tool sold a large variety of grinding and screw-making machines. Their primary machine was a universal grinder similar to that made by Brown and Sharpe. The Landis cylindrical grinding machine was developed in 1883 and patented *(425,230)* in 1890. A device called a screw-cutting head that addressed problems with threading bolts was patented *(409,208)* in 1889. Grinding machines made by Landis, in addition to its universal grinder, included internal grinders and grinders for crankshafts, camshafts, rolls, and ball bearing races. Landis machines were widely used by the automobile industry for production grinding, with one very popular machine used to grind crankshafts. This machine, developed in 1905, eliminated torsion in the crankshaft during grinding and provided means for truing the grinding wheel without removing the crankshaft.

The Heald Machine Company was originally a foundry and machine shop formed by Stephen Heald in the 1850s. Heald's grandson James took over the company in 1903 and moved it to Worcester, where he made surface grinding machines designed specifically for automobile and gas engine parts. Heald Machine, which was located across the street from the Norton Company, developed and sold external and internal grinding machines.[5]

Two of Heald's most popular machines included a piston ring *(external surface)* grinder and an internal surface grinder for the cylinder walls of engine blocks. The automobile industry had difficulty manufacturing the cylinders of engines due to the thin wall design. The cylinder walls would deflect during manufacture, leaving ripples that caused the bypassing of gas and poor engine performance, which was basically the same problem addressed by Wilkinson's boring machine. In order to produce cylinder walls that were straight, round, and smooth, Heald developed a process that combined rough boring with finish-grinding, using a specially designed internal grinding machine *(Figure 5-3)* in which a high-speed grinding wheel was given an orbital movement around a fixed center with very light applied pressure to produce straight cylinder walls with a precision of 0.0005 inches.[6] Heald described his machine in a 1915 advertisement as follows:

The Heald Cylinder Grinding Machine has been especially designed for internal grinding on a great variety of machine parts and is particularly adapted for handling work which is of such shape that rotating in the usual manner is inconvenient or impossible.

This method of finishing work possesses many advantages over the ordinary way of smooth boring and reaming, and is especially valuable in gas and gasoline engines where the cylinder walls are thin and will spring away from the cutting tool easily.

Rigid in construction and built by skillful workmen from the highest materials, this machine is the ideal tool for internal grinding where extreme accuracy and uniformity is desired.[7]

The Artificial Grinding Wheel

Fig. 5-3: Heald Cylinder Grinding Machine for Internal Combustion Engines. **The Iron Age Aug. 3, 1905.**

After 1860, as mentioned earlier, advancement in design and use of the grinding machine lagged behind other machine tools. The reason the grinder lagged behind was not due to the machine, but to the design of the grinding wheel. The materials that formed grinding wheels could not grind the new steel materials at required production rates.

At the beginning of the Industrial Revolution, grinding machines used grinding wheels made from natural materials. These materials were: 1. emery *(aluminum oxide, iron oxide, and silica)*, 2. sandstone *(quartz and feldspar)*, and 3. diamond dust. The wheels used at that time included solid stone *(i.e., sandstone)* or wheels made from soft metal or wood and impregnated at the surface with an abrasive material such as emery, sand, or glass. Grinding wheels of this type were suitable for soft metal and light grinding of steel. However, they were wholly inadequate for heavy grinding of hardened steel materials produced in the late 1800s. What was required was a wheel that could withstand high centrifugal forces without surface degradation under high-volume, mass- production conditions.

Demands for better grinding wheels forced machinists to abandon wheels made with 100 percent natural abrasives and develop artificial wheels composed of natural abrasives mixed with a bonding agent to form a solid, rock-like material. Around 1825, the first artificial wheels appeared using natural abrasives mixed with gum resin. Soon after, other bonding materials such as vulcanized rubber, cement, and shellac were used to make solid grinding wheels. Advances in composition and capabilities of artificial wheels made from natural abrasives continued into the late 1800s. Grinding wheels made with emery in a bonding agent such as clay eventually became the most-used grinding material. These wheels were heated to very high temperatures, creating a vitrified bond that became known as the Emery Wheel.

However, even with all the improvements, artificial wheels still lacked the necessary characteristics required for grinding hard steels to fine tolerances under mass-production conditions. The perfect grinding wheel required an abrasive much harder than steel, that could be dispersed uniformly throughout the wheel, able to withstand centrifugal forces at high operating speeds, and be widely available at low cost. Since natural abrasives were not adequate, what was needed was an artificial grinding wheel made from artificial abrasives and a suitable bonding agent. This is where electricity, a key technology of the Industrial Revolution, provided the final ingredient, allowing the development of the perfect grinding wheel. Electricity led to the development of the electric furnace for the reduction of iron ores and aluminum oxide *(corundum)*. The specific feature of electric furnaces was the ability to generate very high temperatures, allowing for the melting and fusing of metal ores to create very hard compounds.

Just before 1900, two artificial abrasives were developed with hardness characteristics far above steel and steel alloys. The first was discovered and developed in 1891 by American chemist Edward G. Acheson *(1856–1931)*. Using an induction furnace in experiments in the reduction of iron ore, Acheson lined the furnace with a mixture of carbon and clay. Clay is a mixture of fine-grained hydrous silicates *(silicon oxides)* with varying amounts of aluminum, iron, magnesium, and alkali metals. The furnace, generating temperatures above 3100° F *(1700° C)* fused carbon to silicon, producing crystals of silicon carbide with a hardness rating more than 2.5 times that of hard steel. Acheson called his new abrasive Carborundum™ and in 1895 founded the Carborundum Company to produce the material. Since the Acheson process required large amounts of electricity, he located the plant at Niagara Falls, New York, which had ample and cheap electricity from the hydroelectric plant built earlier in 1881. Carborundum grinding wheels began to appear commercially in 1896 and because of the hardness of silicon carbide, the wheels quickly replaced emery in grinding metalwork.

The second artificial abrasive, artificial corundum, is a substance made from fused (melted) aluminum oxide *(Al$_2$O$_3$)*. Also known as Alundum™, artificial corundum is made in an electric arc furnace by passing an electric current between carbon electrodes. The electric arc generated in the furnace heats the material directly and generates temperatures in excess of 3700° F *(2037° C)*. The process for making Alundum was developed by Charles B. Jacobs of the Ampere Electro-Chemical Company of New Jersey in 1897. Since the electric arc furnace required large amounts of electricity, Jacobs moved his test facility to Rumford, Maine, in 1899, where he continued development work. Similar to New York, Maine had many hydroelectric plants with ample supplies of electricity. Alundum *(fused alumina)* is not as hard as Carborundum *(silicon carbide)* but is very close, and far superior to the emery-based artificial wheels. Alundum and Carborundum became the two most-used artificial abrasives in grinding wheels after 1900. The following table shows the hardness of the new grinding materials relative to steel.[8]

Relative Hardness: Knoop Scale

Hard Steel:	740
Tungsten Carbide:	1880
Corundum and Aluminum Oxide:	2000
Silicon Carbide:	2500
Boron Carbide:	2800
Diamond:	7000 +

In 1901, the Norton Emery Wheel Company bought the rights to make artificial corundum and gave it the trade name Alundum. Norton then built a plant to make Alundum near Acheson's Carborundum plant in Niagara Falls. Niagara Falls became the home to many other electrochemical companies, including the Pittsburgh Reduction Company *(Alcoa)* that smelted aluminum, and the Acetylene Light and Power Company *(Union Carbide)* and used electric arc furnaces to produce calcium carbide.

With the development and commercialization of grinding wheels made with artificial abrasives, the final component was in place to precision manufacture parts made from hardened steels at mass production volumes. The automobile industry now had all the tools it needed to mass-produce at low cost, interchangeable parts for internal combustion engines. Grinding technology would also be applied to the airplane, mechanical bearings, and many other industries to facilitate the next phase of industrialization.

By 1910, grinding machines using grinding wheels composed of man-made abrasives were widely available in many sizes, shapes, and grades.[9] As a result, costs to manufacture precision metal parts quickly dropped to help drive down the cost of automobiles and industrial machines. The automobile industry learned very quickly that precision manufacture of automobile parts was critical to success. Over time, automobile companies created engineering departments dedicated to grinding technology and its application to manufacturing and mass production. This lesson applied to all companies making precision, interchangeable metal parts, specifically manufacturers of rolling bearings.

Chapter 6

The Centerless Grinder

Ring Wheel Centerless Grinding Machines

By the early twentieth century, three methods of grinding were in use. These were internal, external, and end grinding. Cylindrical work *(internal and external)* was typically ground on machines that resembled lathes in construction and appearance that held and rotated workpieces at one or both ends. Many of the machines used were Brown and Sharpe universal grinding machines or similar machines made by other manufacturers. Performing grinding on machines using "centered" or "fixed" supports was labor intensive and time consuming for setup, machining, and removal. There was no relatively inexpensive way of rapidly grinding cylindrical objects, particularly the rolls that were used in roller bearings.

After founding the Ball and Roller Bearing Company in 1910, Heim began manufacturing various types of rolling or antifriction bearings, with roller bearings being sold at much higher volumes than ball bearings. Both precision dimensions and surface finish were very important factors to the successful, long-term performance of bearings. From the outset, Heim needed to manufacture the components of his bearings to tight tolerances and tried to produce roller bearing rolls to within a quarter-thousandth of an inch *(0.00025")*. But this was not always possible with existing grinding machines, forcing Heim to quote to customers maximum variations of one-half-thousandth of an inch *(0.0005")*. Even at 0.0005 inches *(0.0127 mm)*, Heim, as well as other manufacturers, found that bearings experienced shorter life spans.[1]

Bearing rollers were made of various grades of steel and cut from bar stock by screw machines or lathes. After cutting, the rolls were heat-treated to harden the steel that increased the load-carrying ability and life of the bearing. Once hardened, the rolls were ground to size on a lathe-type grinder equipped with fixed supports to hold the rolls. In 1910, Heim used two types of fixed support systems for grinding rolls for his bearings. The first was "centered" supports that required holes to be drilled into the ends of the roll into which pointed spindles were inserted. The other was "chuck" supports that securely held the work on the periphery at one or both ends. The grinding wheel for these machines was typically mounted on a slide running along one side of the machine and moved into place to perform grinding as done on a universal grinder. Because of the cost and time required for grinding rolls, Heim soon concluded there had to be a better way to manufacture roller-bearing rolls.

Heim worked on various ideas for a grinding machine that did not require fixed supports and could produce precision-made rolls at higher production rates. In the latter part of 1910, Heim designed his first centerless grinder consisting of a single grinding wheel mounted on a ring wheel chuck. In pursuit of his idea, around November 2,

1910, Heim contacted the Gardner Machine Company of Beloit, Wisconsin, regarding their line of ring wheel chucks. The Gardner Machine Company made metalworking machinery including single- and double-wheel disk grinders and polishing machines and carried an assortment of ring wheel chucks that supported grinding rings *(Figure 6-1)*. However, the attention required to growing Heim's nascent business delayed development of his idea. In the spring of 1913, as demand for roller bearings grew, Heim

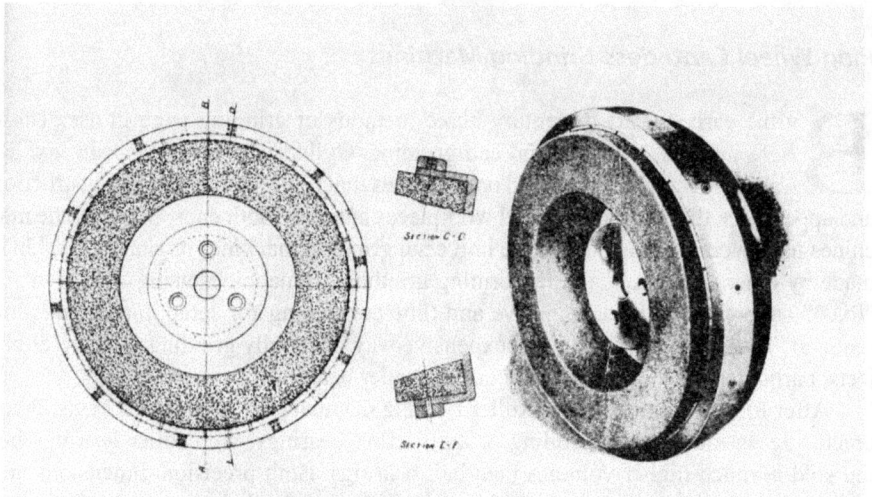

"Perfection" Ring Wheel Chucks

GARDNER "Perfection" Ring Wheel Chucks are made of steel throughout. The chuck body or shell is a one piece steel casting, machined all over. The outside retaining ring is a special steel forging machined all over. Rim of shell is partially slotted into twenty sections. Alternate sections are permanently riveted to retaining ring. Hollow set screws act on the remaining sections gripping the abrasive ring solidly and evenly. The shell is recessed to receive spindle collar, reducing overhang to a minimum. Two sets of backing plugs are furnished with each "Perfection" Chuck. When the abrasive ring becomes worn down to face of chuck, it is moved out one inch at a time and backed up by the plugs. By this means the abrasive ring can be used down to one inch in thickness which is the limit for safety.

"Perfection" Chucks are simple and quick acting; they are as light as is consistent with safety; they are interchangeable with steel disc wheels and can be used on various types of vertical surface, edge, face and knife grinders.

SPECIFICATIONS

SIZE	OUTSIDE DIAMETER OF RING WHEEL	EXTREME DIMENSIONS OF CHUCK	WEIGHT OF CHUCK WITHOUT RING
12	12 in.	13½x2½	33 pounds
14	14 in.	15 5-8x3	44 pounds
16	16 in.	17 5-8x4	67 pounds
18	18 in. ⎫ These sizes		
24	24 in. ⎬ not carried		
30	30 in. ⎭ in stock		

GARDNER MACHINE COMPANY
(The Disc Grinding Authorities)
BELOIT, WISCONSIN, U. S. A.

Fig. 6-1: Ring Wheel Chuck Sold by the Gardner Machine Company, 1910.

built his first concept of a "centerless" grinding machine. The machine consisted of a single-ring wheel chuck holding an abrasive wheel positioned against a flat steel plate. The plate contained a groove machined along the vertical length that guided the rolls sliding through the grinding zone. Since there were no spindles or chucks to hold the work, the grooved plate acted as a "centerless" support for the work.

To make his machine, Heim purchased one 14-inch ring wheel chuck and abrasive *(grinding)* wheel from the Gardner Machine Company. As seen in Figure 6-2, the ring wheel chuck and grinding wheel were mounted on a belt-driven shaft supported by two bearings. The grinding wheel was beveled around 30 degrees to the vertical from the inner to outer diameters. The centerless work support, often referred to as a "Carrier" or "Work Fixture" by Heim, was tilted at an angle equal to the bevel in the ring wheel to create parallel surfaces in the grinding zone. In operation, rolls were dropped into the groove at the top of the carrier that allowed gravity to pull the rolls downward past the grinding wheel. Contact of the grinding wheel against the roll would induce both roll rotation and friction of the roll against the carrier groove, which produced enough braking action of roll rotation to allow grinding to be performed.

During testing of the machine, Heim found that movement of the rolls through the grinding zone could be affected by positioning the carrier to either side of the vertical wheel centerline. Heim also experimented with the feed rates through the grind-

Fig. 6-2: Rebuilt Heim Single-Ring Wheel Centerless Grinder as designed in May 1913.
Picture shows single-ring wheel chuck without the grinding wheel attached.

ing zone and even though the carrier system was designed to feed rolls via gravity, Heim found that he could affect the speed and direction of the rolls by offsetting *(left or right)* the carrier relative to the center of the grinding wheel. By positioning the work to one side or the other of the vertical centerline, the rotating wheel surface induced both horizontal and vertical forces on the work. If placed against a downward moving wheel surface, the work feed rate accelerated downward. If placed against an upward-rotating wheel, the feed rate either slowed in its downward movement or, if moved far enough, reversed upward. It was on this machine, the single-ring wheel centerless grinder, that Heim first tested the natural drawing action of the wheel surface to increase or decrease the feed rate of grinding rolls.

The single-ring wheel grinder was Heim's first attempt to design a machine for centerless grinding of roller-bearing rolls at high production rates. As with all new concepts, significant operational problems were experienced as explained by Heim in 1921:

I felt that there was a better way of producing small cylindrical work such as we used in our line of manufacture, and to that end I constantly had it in mind and pondered over the question, and some time early in 1911 I came to the conclusion that I had a new apparatus that would produce this type of rollers in a much faster manner. Still, not being sure about it, and having a very limited amount of capital, not wanting to get it wrong on it, I pondered over it for quite a long time, but finally, as our business increased, and the urgency of production also multiplied, it became very imperative that we have something that would enable us to fill our orders, and I finally decided that we would try out the idea which I had in mind, and to that end I purchased a ring wheel chuck, 14-inch ring wheel chuck, which would hold an abrasive wheel 14 inches in diameter and about four or five inches deep, I should say, and which had a hole through the center of possibly six, seven or eight inches in diameter, and I made drawings and produced a special apparatus which I could employ in connection with this wheel and chuck I mounted on a frame in our shop, and with this apparatus I ground rollers much more successfully—or much faster, I should have said— much faster than in the other lathe. It, however, was not altogether a success, as it was not dependable. Some of the time it would run beautifully, and some of the time it would be freakish. It was like a mule. It was undependable. We, however, ground rolls on the machine for several months, and it produced good rolls when it was working all right, and it was an improvement, decided improvement, over our other method of grinding, where we had to grind them singly or on a string. Well, we would set this machine up for a certain size roll, and sometimes it would be set up for a few moments and would be all right, and possibly run for half an hour or an hour, or maybe two hours, and all at once the rolls would start to cut through there like lightning, bound down the factory and through the windows, and we were afraid of it. Another time it would be going just as nice as could be, and the first thing these rolls would come right back, slide out of it, in your face if you were in the way, so we always cautioned to stand at one side, for fear it might start to kick. That destroyed our setting, destroyed our setting entirely every time that occurred, because it would injure the face of the wheel, and we would have to go all over it and reset the machine again to get it in shape.[2]

Heim tested his single-ring wheel grinder through May and June 1913 and, as a result of the problems encountered, determined that the machine lacked control over rotation of the work. Heim realized he needed a second wheel, a "regulating" wheel to control rotation of the roll while being ground by the grinding wheel as he later recalls in 1921:

*The only thing I had in mind was that the apparatus which I had provid-*ed (single-ring wheel machine) *was a failure in so far that while it would grind a roll beautifully when it was working all right, when it was not, it was very freakish, and that immediately placed before me the fact that we should provide some means which would positively determine a uniform rotation of the work which we were endeavoring to grind, and pondering over this, I conceived the idea of two wheels, one a regulating wheel for the blank being ground, and the other a grinding wheel, and that I could some way mount this blank or work between the two wheels and cause it to pass through properly.*[3]

Heim found that in order to achieve uniform rotation, he needed to slow the regulating wheel speed down to a point at which the wheel surface imposed sufficient braking forces and gripping action that allowed positive control of roll rotation. To test his regulating wheel concept, Heim modified a small universal grinder to add a second wheel and positioned a support *(carrier)* for the rolls between the wheels *(Fig. 6-3)*. For this machine, the carrier was placed to use the peripheries *(circumference)* of the wheels instead of the ring *(side)* faces. Heim described the two-wheel grinding machine as follows:

We employed in this case a small tool grinder, universal tool grinder, which was then in our tool room as it is today. We made a mandrel with flanges on and with centers in it, and threaded this mandrel at one end so that we could mount on this mandrel a small grinding wheel, I should say something like three inches in diameter and ½ inch face. We fastened this wheel on this mandrel and placed a dog (mechanical device for gripping) *on the mandrel at the end, and then this mandrel with the wheel was placed between the centers on the work table and rotated by the belt which drives the spindle in the work head. This machine also has a wheel head on which there was spindle mounted, and which was capable of employing a wheel about six inches in diameter with a half inch face. We mounted such a wheel on this spindle or possibly employed the one that was on there at the time we tried it out. We then bent a piece of steel, I should say about an inch and a quarter wide, ¼ inch thick, and made an L-shape out of it, slotting it at both ends, and one end was bolted to the work table by the aid of a screw in the T slot, which on the other end was fastened a small piece of steel on which we had provided two sets of guides or strips of steel, leaving a space between the two sets of guides. This was bolted on the other end of the L-shaped piece and placed in position so that this piece would rest between the two wheels already referred to. We elevated this piece to the proper height and placed rollers on this support, and ground rolls successfully on this machine.*

In our operation of this machine just described (two-wheel, peripheral grinder), *we ground work leaving the carriage parallel with the axis of the grinding wheel, and we fed work along the guide uphill and then we ground it also horizontally, but we had to push it through. Then we moved the table, that is, the work table on which the regulating wheel was mounted, and swing it off a few degrees, and then trimmed the wheel providing a high side on the regulating wheel. I am not sure whether we ran the grinding wheel straight under these conditions or not, but I do know that we also trimmed off by the aid of a diamond the grinding wheel, causing that to have a high side, thus providing two high sides for the wheel, and under these conditions, mounting the work rest properly, the rolls would pass through very nicely. Of course, this machine was a small, frail tool grinder, not capable of grinding all kinds of rolls due to its frailness, but small rolls could be ground successfully on it.*[4]

In this explanation, Heim confirms that the converted universal grinder became his first two-wheel centerless equipped with a regulating wheel. And, even though he did not say "automatic feed," it appears that was his meaning when he said the rolls passed through very nicely after tilting *(swung off a few degrees)* the table supporting

Fig. 6-3: Replica of Heim's First Two-Wheel Centerless Grinder Equipped with a Regulating Wheel. May 14, 1921.

the regulating wheel. The tilting of the regulating wheel not only provided vertical force to rotate the roll but also a lateral *(horizontal)* force to automatically move the roll through the grinding zone. Additional pictures of Heim's first centerless grinder using the wheel peripheries can be seen in Appendix 2 *(A2-8-1 and 2)*.

Heim operated his experimental two-wheel centerless grinding machine into July 1913, at which time he was convinced that the regulating wheel would meet his needs for positive and uniform rotation of the rolls and reliably grind rolls to required tolerances. After completing his tests, Heim had to decide which configuration *(sides or peripheries)* to use in his first commercial machine. Factors influencing his decision included the time and cost to develop a new machine, plus his primary need to grind rolls small in diameter and short in length.

Heim found that small-diameter rolls ground on his peripheral wheel machine would sometimes flap up and cross whereas with the ring wheel machine, the sides of the wheels offered protection and better contained small-diameter rolls. Another influence was that the Gardner Machine Company manufactured a double-ring wheel grinding machine for simultaneously grinding flat sides such as disks. Gardner patented this machine *(909,882)* in 1909 with illustrations shown in Figures 6-4 and 6-5. Heim decided that the ring wheel configuration was the better choice, as he believed modification of the Gardner machine into a double-ring wheel centerless grinder would be much quicker and cheaper than designing a peripheral machine from scratch.[5]

In early July 1913, Heim sent a request to the Gardner Machine Company for their complete catalog of grinding machines. On September 24, 1913, Heim placed an order for the Gardner disk grinder that arrived at the B&RB on October 24, 1913. Heim assigned employees William Weed and James Bennett to help modify the Gardner machine. Weed, approximately 38 years old, was an experienced machinist who had been recently hired by the B&RB on August 24 and would later become responsible for the entire roll-grinding department.[6] Bennett was hired into the Heim Machine Company in 1908 when he was 16 or 17 years old, promoted to superintendent sometime around 1920, and in 1934, elevated to vice president of the B&RB.[7]

The Gardner double disk grinding machine provided many of the features Heim needed for his ring wheel centerless grinder. The machine was made with a heavy base and had two independently rotating disks, with each disk mounted on a sliding support. These features minimized the time required for modifications, and after about ten days *(early November)*, Heim commenced operation and testing of his first full-size, two-wheel centerless grinder for mass production of roller-bearing rolls. This machine, the modified Gardner disk grinder, using one wheel for grinding and the other to regulate work rotation, became the basis from which Heim would develop improvements and perfect his centerless grinder designs.

Heim called the modified Gardner disk grinder a "Double-Ring Wheel Centerless Grinder" that contained the two essential features required for mass production of rolls suitable for use in roller bearings. The first is precision grinding hardened steel rolls of short length. The second is automatic feeding a succession of rolls through the grinding zone at relatively high speeds.

Heim's invention employed a three-sided grinding zone composed of a grinding wheel, a regulating wheel, and a centerless work support *(carrier)* placed between the two wheels. The carrier consisted of a flat surfaced metal bar designed with vertical guides attached on both sides before and after the grinding zone. The guides served to retain the work on the bar as it slid along the length of the carrier. This design, which eliminated the need for fixed supports, facilitated the feeding of a continuous stream of cylindrical work through the machine.

Heim's machine uses two wheels, each with specific functions. One wheel rotates downward on the work at high speed and performs grinding action. The other wheel rotates upward on the work at a relatively slow speed and regulates the speed of work rotation through braking and gripping action. With the carrier positioned between the two wheels, both wheels rotate cylindrical work in the same direction. By rotating down on the work at a high speed, the grinding wheel forces the work down on the carrier and against the regulating wheel. This three-sided system not only keeps the work firmly on the carrier, but due to frictional *(braking)* forces at the surfaces of the carrier and regulating wheel, prevents runaway work rotation and allows the regulating wheel to control the speed of rotation.

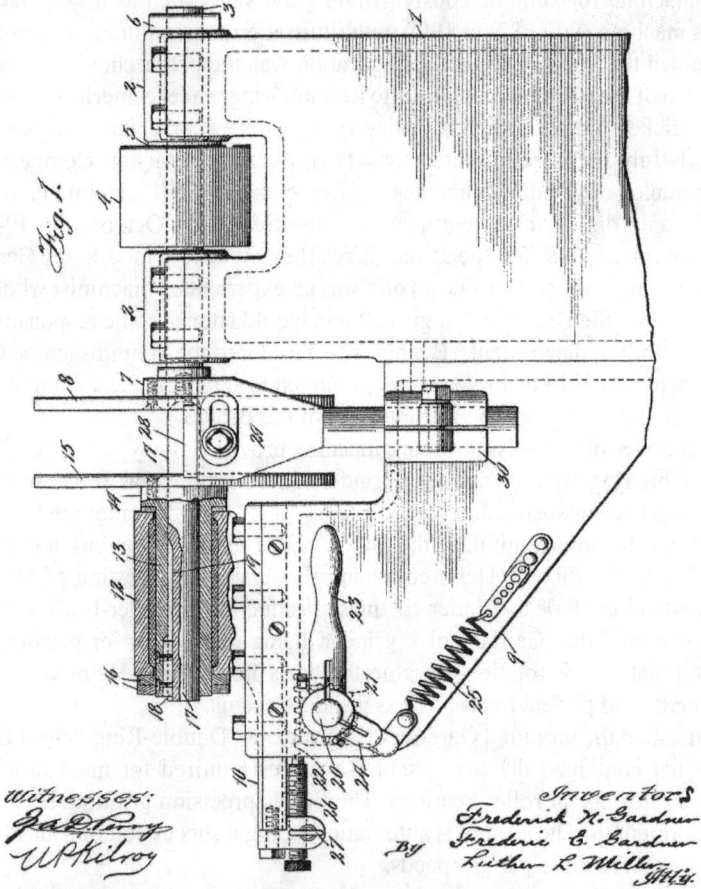

Fig. 6-4: Gardner Two-Wheel Disk Grinder—Top View. Patent 909,882, filed April 26, 1908.

Fig. 6-5: Gardner Two-Wheel Disk Grinder Advertisement in Machinery, *Aug. 1918.*

In his patents for centerless grinding, Heim does not provide the rotational speeds of the grinding and regulating wheels. Heim only mentions that the grinding wheel is rotating fast and down on the work and the regulating wheel rotates at a slower speed and up on the work. However, in an article published in *Machinery* in May 1923, Heim provided typical rotational speeds of 1,100 rpm and 28 rpm, respectively *(see Appendix 2, A2-5-3)*. According to product brochures *(A2-6)*, Heim installed 16-inch diameter grinding and regulating wheels. At the speeds stated in the article, that produced surface speeds at the point of contact with the work of 4,608 and 117 feet per minute. Typical grinding speeds at that time were in the range of 5,000 fpm.

The second feature, automatic feeding, involves dynamics other than rotation of two wheels in opposite directions and at different speeds. Whereas achieving precision grinding involves control of the vertical forces of work rotation, automatic feeding involves control of the lateral forces along the axis of the work. In other words, automatic feeding requires control of the horizontal forces employed to push or pull *(draw)* the work along the carrier through the grinding zone.

To achieve lateral *(forward or backward)* movement, there are two forces available. One force is gravity, which can be applied using the inclined carrier. The other is to use surface contact of the rotating wheels. By positioning the work relative to the wheel such that the rotating surface contacts at an angle other than vertical *(i.e., less than 90 degrees to the work axis)*, a horizontal force is generated, thus moving the work along the carrier. The combination of both vertical and horizontal forces creates a spiral action of work rotation and forward movement that is a fundamental characteristic of centerless cylindrical grinding. In Heim's first centerless grinder invention, the double-ring wheel machine, both gravity and wheel rotation are used to induce and control work feed rates along the carrier and are discussed in the section on the patent.

First Patents for the Centerless Grinder

On March 6, 1915, almost two years after conceiving the idea for two-wheel centerless grinding, Heim filed three applications for patents on inventions related to his bearing business. One was Heim's first invention for a roller thrust bearing designed to accommodate rotation of heavy axial loads. This bearing became patent 1,169,150 in January 1916 and was unique in that it employed rollers made from rings of different lengths with the purpose of minimizing wear and damage to the moving parts. As seen in Figure 6-6, the bearing rollers and socket plates *(races)* required precision machining to tight tolerances to ensure proper fit and rolling movement.

The other two applications describe a roll-grinding machine that Heim called a double-ring wheel centerless grinder. The first application, titled Feeding Device for Grinding Machines, that became patent 1,210,936, described the feed system installed on his centerless grinder to support and guide rolls through the grinding zone. The second application, titled Roll Grinding Machine, became patent 1,210,937 and described the machine and method for centerless cylindrical grinding. After filing the two applications, and perhaps at the request of the U.S. Patent and Trademark Office, the second application was divided into two applications, one for the machine *(1,210,937)* and the other *(1,281,366)* for the method of centerless grinding

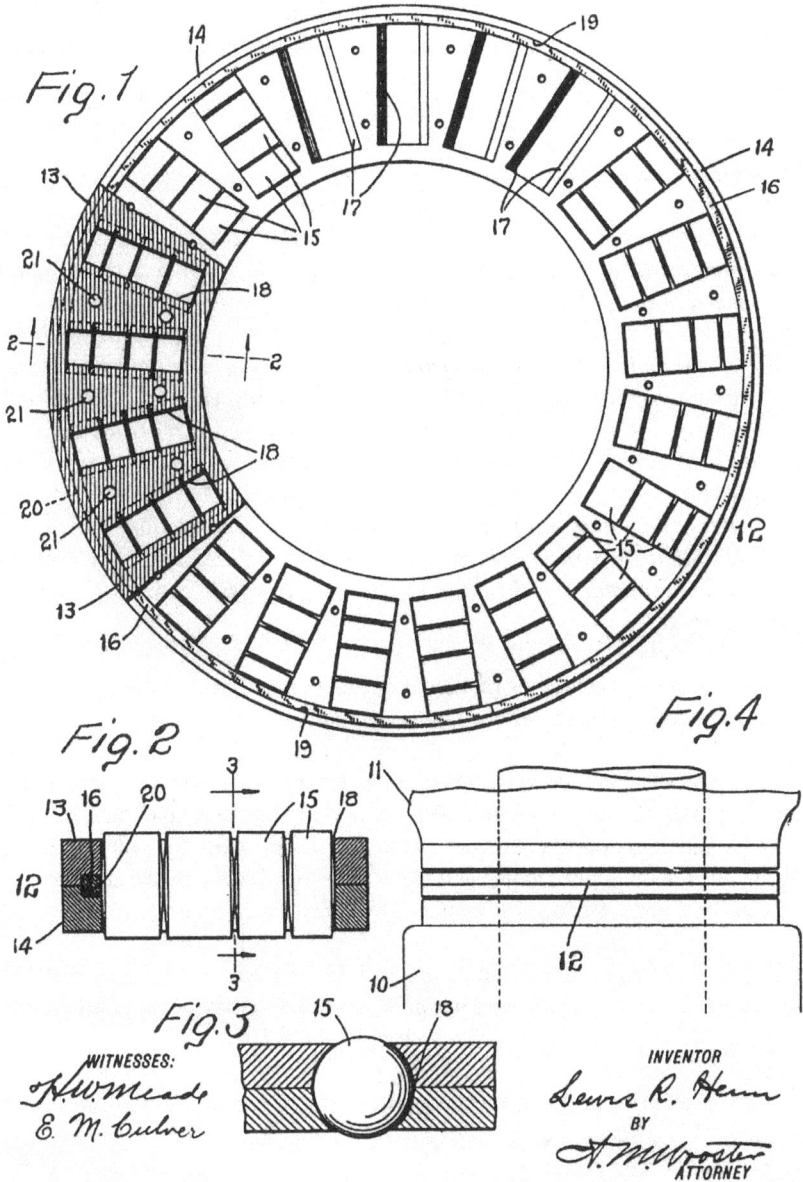

Fig. 1

Fig. 2

Fig. 3

Fig. 4

WITNESSES:
H.W. Mead
E. M. Culver

INVENTOR
Lewis R. Heim
BY
A. M. Brooster
ATTORNEY

Fig. 6-6: Roller Bearing for Heavy Thrust Loads. Patent 1,169,150, filed March 6, 1916.

that was filed the next year on July 3, 1916.

Heim built his first centerless grinder with a heavy, rugged frame and components *(Figures 6-7, 6-8, and 6-9)*. Heim learned, as did Charles Norton, that in order to grind metalwork to precise dimensions, vibration and movement of the machine and its parts must be kept to an absolute minimum and, thus, requires massive and rigid construction.

The unique and valuable feature of the Heim double-ring wheel centerless grinder is its ability to precision grind a continuous stream of hardened steel rolls for use in roller bearings. Taken from the machine patent *(1,210,937)*, Heim describes the functions of the wheels and the operation of his three-sided centerless grinding process:

As soon as a roll is picked up by the operative wheels the action is to draw the roll forward and pass it through between the operative faces. It will be noted that the operative face of the grinding wheel is traveling downward and the operative face of the regulating wheel is traveling upward. As the surface speed of the grinding wheel is much greater than that of the regulating wheel it follows that the rolls being operated upon will be held down upon the carrier, will be rotated and will also be drawn forward. The tendency of the regulating wheel to move the rolls backward with respect to the downward movement of the grinding wheel, or in other words, away from the carrier, is wholly overcome by the greater surface speed and greater effect upon the rolls of the grinding wheel. The Regulating Wheel however has this very important function, in that it insures constant and uniform rotation of the rolls (work) *so that I am enabled to produce with my novel machine, much more rapidly and economically than has heretofore been possible, rolls of the very highest grade.*

The patent clearly defines the rotational direction of both wheels relative to the work as well as the purpose and importance of the regulating wheel, *"to ensure constant and uniform rotation or the rolls"*. But what it does not define clearly are the mechanical principles involved to induce drawing action and continuous feeding of rolls. There is no mention of wheel speeds other than fast and slow and there is no explanation of why the wheels rotating in opposite directions induce and control roll feed rates. The section below, included in both the machine and method patents, is the only description for how drawing and automatic feeding work on the Heim centerless grinder:

The lateral adjustment of the standard [vertical bracket 43 in Figure 6-9] by means of the slide enables the operator to correctly position the carrier with relation to the operative wheels and to provide for different sizes of rolls and for changes in thickness of either the grinding or regulating wheel or both. It is of course necessary that the top of the carrier should be placed slightly above the horizontal line intersecting the axial line of the wheels, and it is obvious that by raising the carrier the amount of grinding action to which each roll will be subjected will be lessened, as the rolls will be drawn forward more rapidly. In connection with the vertical adjustment of the standard, the tilting adjustment of the carrier comes into use. As the conditions in grinding different rolls will vary it is impossible to arbitrarily determine any fixed heights for the standard or inclination of the carrier that will produce the best results under all the varying conditions of use. In other words, by a relative adjustment of the path of rolls on the carrier, and of the path of the operative surfaces of the regulating wheel, the rate of feed is varied as the thrust of the regulating wheel along the carrier is changed. In practice, I have found approximately the adjustment as to the height of standard and inclination of the carrier indicated in Figure 4 to produce perfectly satisfactory results.

As described in the above excerpt from the patent, the principles involved with movement of the rolls between the wheels were specifically written for the double-ring wheel centerless grinder equipped with an inclined carrier. It clearly states that the grinding wheel, rotating down on the work and faster than the regulating wheel, moves the rolls forward while the regulating wheel, rotating upward, holds the rolls back while imparting constant work rotation. It also states that the primary mechanism for controlling roll feed rates along the carrier is adjusting the vertical position and inclination of the carrier.

In a centerless grinder, the combination of rotation and lateral movement creates a spiral feeding action of the work. As discussed with the Gardner machine modification, spiral feeding requires that the rotating wheel surface contact the cylindrical work at an angle such that both vertical and horizontal forces are applied. In addition, inclining *(upward or downward)* the carrier to involve gravity influences lateral movement, adding another factor to the design of the feed system. It is not necessary to describe these factors in detail other than to say that the designer of a machine using the side *(ring)* surfaces of two rotating wheels has many options for positioning the carrier relative to the wheels in order to achieve and control spiral feeding of cylindrical work. But Heim stated in his patent that it was necessary to position the top of the downward inclined carrier slightly above the wheel centers *(Figure 6-9)*, which implied that positions other than above the wheel centers would not work, thus opening the door for others to develop alternative feed system designs to circumvent Heim's patents.

In the second type of centerless grinder, the two-wheel machine using the peripheral wheel surfaces, a different method is employed for inducing and controlling work feed. This method involves tilting the face of the regulating wheel relative to the plane of the carrier to achieve angled contact. To obtain automatic drawing action to pull the work forward through the grinding zone, the face of the regulating wheel must be tilted forward such that wheel surface is moving upward and forward when contacting the work. To achieve this, the shaft rotating the regulating wheel must be tilted downward relative to the direction of the work along the carrier. In Heim's original two-wheel prototype centerless grinder *(Figure 6-3)*, in order to induce forward feeding of the roll, he offset *(tilted)* the table supporting the regulating wheel and shaft a couple of degrees. By writing the patent specifically for Heim's double-ring wheel centerless grinder, and omitting arrangements using the wheel peripheries, the attorney, A. M. Wooster of Bridgeport, Connecticut, created opportunities for competitors to introduce their own inventions for two-wheel centerless grinding.

Heim's double-ring wheel centerless grinder equipped with a "regulating wheel" and an "automatic feed system" was used extensively at the Ball and Roller Bearing Company. His machine ground perfectly round cylindrical work to tolerances within a quarter of a thousandths *(0.00025)* inch. The feed rates on his machine, for a typical roll diameter of 5/16 inch, ranged from 6 to 7 linear feet of rolls per minute. The machine had the ability to produce about 100 rolls for every roll produced using center-type grinders. With his invention operational, Heim threw all other grinders aside and

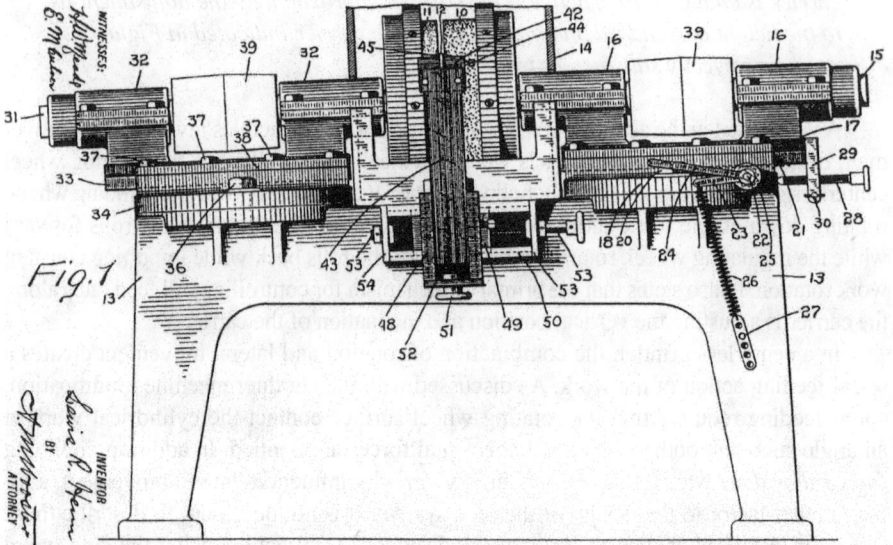

Fig. 6-7: Heim Original Double-Ring Wheel Centerless Grinder (Front View). Patent 1,210,937, filed March 6, 1915.

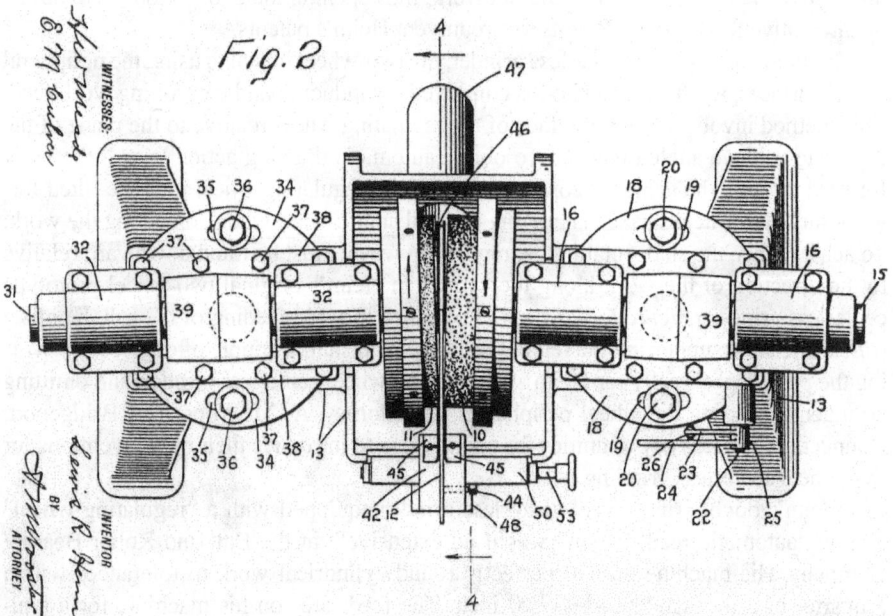

Fig. 6-8: Heim Original Double-Ring Wheel Centerless Grinder (Top View). Patent 1,210,937, filed March 6, 1915.

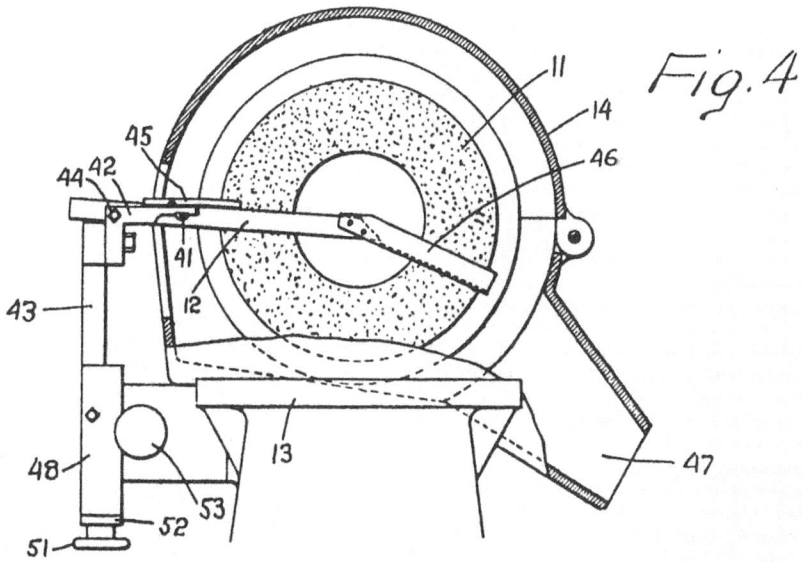

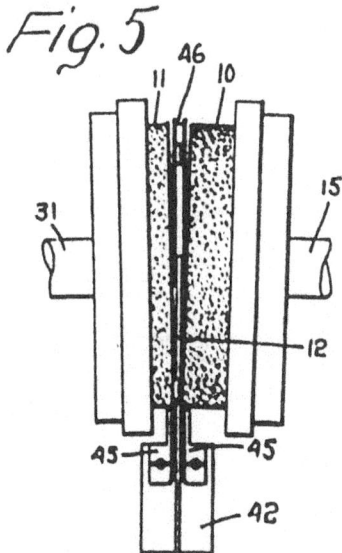

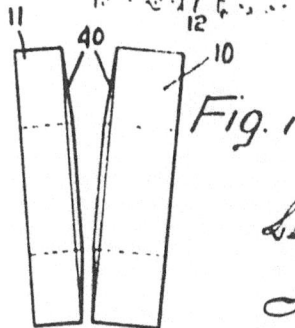

Fig. 6-9: Heim Original Double-Ring Wheel Centerless Grinder. Roll Feed System with
Carrier and Wheel Rotation Details. Patents 1,210,936 and 1,210,937, filed March 6, 1915.

used his new double-ring wheel centerless grinder to produce all rolls used in roller bearings sold by the Ball and Roller Bearing Company.[8] Heim was granted patents 1,210,936 *(Feeding Device for Grinding Machines)* and 1,210,937 *(Roll Grinding Machine)* on January 2, 1917, and the method patent 1,281,366 *(Method of Grinding Hardened Rolls)* was granted on October 15, 1918. Pictures of Heim's double-ring wheel centerless grinder used at the Ball and Roller Bearing Company can be seen in Appendix 2 *(A2-8-3 and 4)*.

On July 13, 1917, Heim filed patent applications for three more inventions. All three inventions were distinctly different from each other and surprisingly, granted on the same day of May 7, 1918. One invention that became patent 1,264,928 was a device for feeding rolls for end grinding. The second that became patent 1,264,929 was for a machine for grinding grooves in round plates. The third application that became patent 1,264,930 was Heim's first invention for a centerless cylindrical grinder using the peripheral *(circumference)* surfaces of the grinding and regulating wheels.

Grinder for Round Plates

In addition to ball and roller bearings, The B&RB manufactured ball and roller thrust bearings that were assembled with races fabricated from two flat plates that sandwich the balls or rollers. Ball thrust bearings require races with circular grooves in which the balls can roll. To make the races, Heim developed a machine for grinding grooves in round plates *(patent 1,264,929)*. The machine was unique in that it was designed to spin a round plate against a horizontally oscillating grinding wheel to grind a groove with the required circular profile. Figure 6-10 shows a side view of the grinder with the round plate *(#24)* mounted on the rotating and oscillating assembly in the upper left and shown in the retracted position against the grinding wheel *(#15)*. Taken from his patent *(1,264,929)*, Heim described the benefits of his invention:

> *The machine herein described has several advantages in operation such as the universal adjustment of the oscillating work support both laterally and longitudinally without interfering with the drive and also the means for varying the oscillation. In work requiring accuracy, such as for bearings, a delicate and accurate feed of the grinding wheel is desired, which is obtained herein by the rack and pinion adjusting means described, while the grinder can itself be adjusted to suit the requirements of the work.*

The development of this machine, his second grinder for bearings specifically designed to perform a special machining function, follows the pattern Heim set with hatting and laundry machines. Heim would see a need for a new type of machine and have the skills to design and build it.

Two-Wheel Peripheral Centerless Grinder

Heim built many double-ring wheel centerless grinders to manufacture the short, uniform length rolls used in roller bearings. Later, Heim found the need to grind other

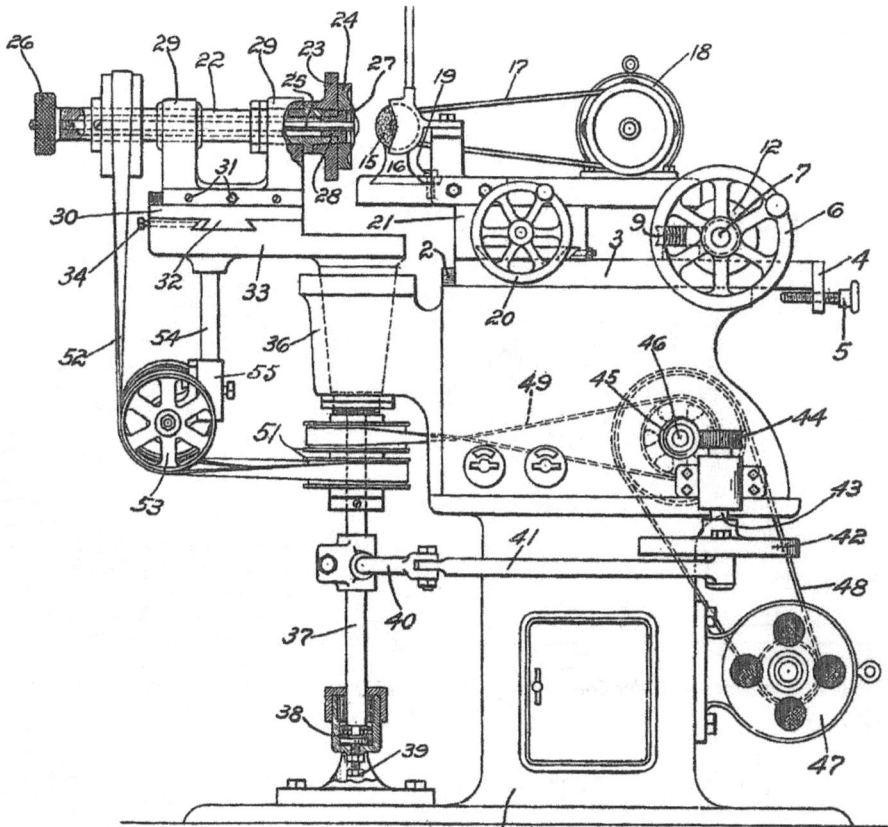

Fig. 6-10: Grinding Machine for Round Plates. Patent 1,264,929, filed July 13, 1917.

types of cylindrical work, such as tapered or conical rolls, plus headed items such as bolts that could not be ground on the double-ring wheel machine. It was this need to grind non-uniform cylindrical work that led Heim to develop a second centerless grinder using the wheel peripheries. On July 13, 1917, a little over four years after filing to patent the original double-ring wheel centerless grinder, Heim filed an application to patent his first invention.

The grinder had many similarities to the double-ring wheel design. The wheels were mounted on separate carriages that were both slideable and rotatable for positioning the wheel surface relative to the work carrier. The wheels were mounted on horizontal shafts and independently driven. The base of the machine, as done on the ring wheel grinder, was tilted toward the regulating wheel to help keep the work in contact with the wheel surface. To induce automatic feeding of the work, Heim used the same slightly downward-sloping carrier system used on the double-ring wheel machine.

The primary reason for using the wheel periphery was its flexibility in grinding different shapes. Combined with the rotating carriage, the machine had the ability to precision grind various cylindrical shapes as well as grind bolts up to the base of the

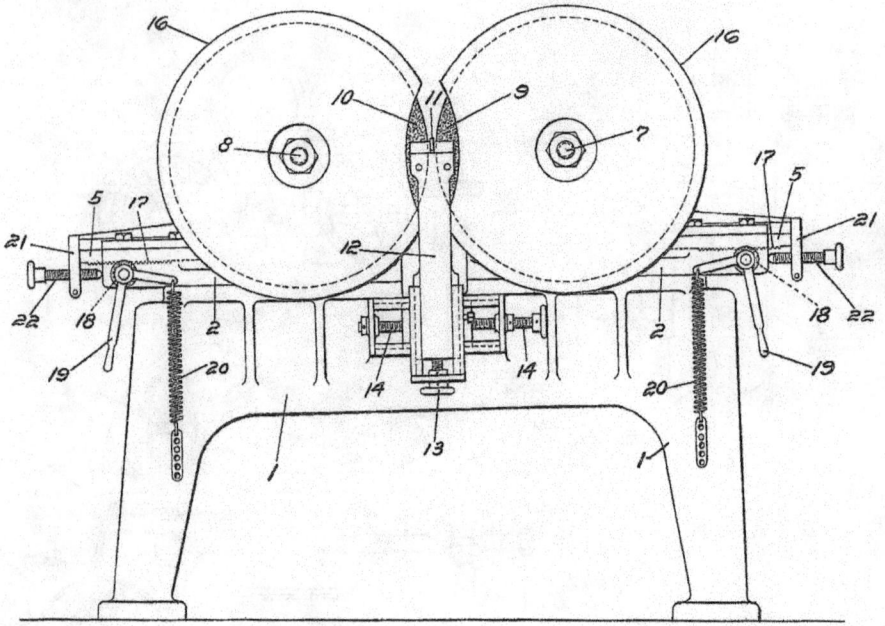

Fig. 6-11: Centerless Grinder Configured to Use Wheel Periphery. Patent 1,264,930, filed July 13, 1917.

head. Heim, in what later became patent 1,264,930, described its advantages as follows:

> *The employment of a peripheral grinding wheel, with or without a peripheral regulating wheel, has a number of practical advantages such for example as simplified feeding of the blanks, since they may be fed in either from the side or from above, better clearance, and it is much easier to maintain the wheels true. Moreover, headed articles such as bolts can be readily ground on this machine either on the shank or on the head, or both, and the shank can be ground close up to the head. Moreover, tapered or conical articles can be readily ground.*

Figures 6-11 and 6-12 illustrate Heim's design using the peripheral surfaces of the grinding and regulating wheels with Figure 6 showing the wheel carriages rotated and positioned for grinding bolts.

Heim developed his initial designs for centerless grinding from 1913 through 1917 that produced the double-ring wheel and peripheral wheel configurations as well as his initial concepts for supporting and feeding work through the grinder. Both machines were capable of precision grinding cylindrical work continuously at high production rates. Whereas the double-ring wheel machine was designed specifically to grind bearing rolls of uniform diameter, the peripheral machine could precision grind uniform and non-uniform *(tapered and conical)* diameter work plus headed or shoulder work such as bolts. During this period up through 1917 Heim filed five patent applications to protect his inventions. However, it was during the period from 1918 through 1925 that Heim perfected his centerless grinder around the peripheral wheel design. Photos of the Heim peripheral wheel centerless grinder used in advertising media can be seen in Appendix 4 *(A4-4-5 and 6)*.

Most of Heim's improvements were directed toward increasing grinding precision, expanding the types of work that could be ground, automating the grinding process, and simplifying operation of the machine. Improvements included devices for automatically feeding and removing work, carriers designed for non-uniform diameter work, and systems for truing the grinding and regulating wheels.

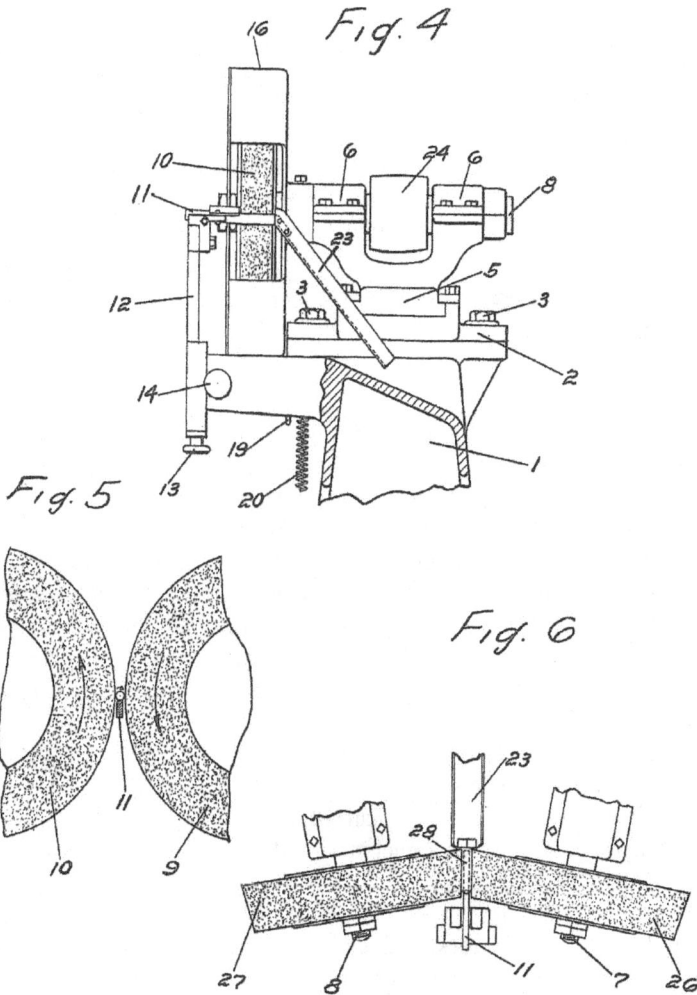

Fig. 6-12: Details of Centerless Grinder Configured to Grind Rolls and Headed Work Using Wheel Periphery. Patent 1,264,930, filed July 13, 1917.

Peripheral Wheel Centerless Grinder
Forces Affecting Rotation, Feeding & Grinding of Cylindrical Work

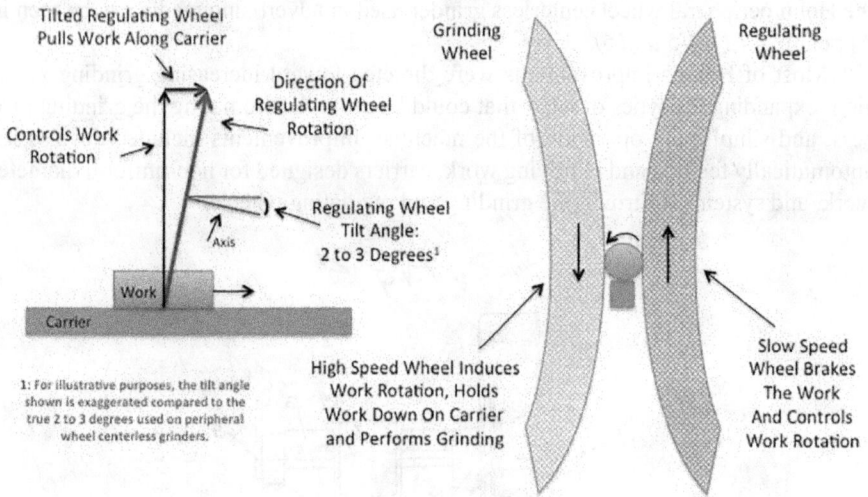

Tilted Regulating Wheel
Pulls Work Along Carrier

Grinding
Wheel

Regulating
Wheel

Direction Of
Regulating Wheel
Rotation

Controls Work
Rotation

Regulating Wheel
Tilt Angle:
2 to 3 Degrees[1]

Axis

Work

Carrier

1: For illustrative purposes, the tilt angle
shown is exaggerated compared to the
true 2 to 3 degrees used on peripheral
wheel centerless grinders.

High Speed Wheel Induces
Work Rotation, Holds
Work Down On Carrier
and Performs Grinding

Slow Speed
Wheel Brakes
The Work
And Controls
Work Rotation

Fig. 6-13: Illustration of Grinding and Work-Feeding Forces on Peripheral Wheel Centerless Grinder.

From 1920 through 1925, Heim filed 23 additional patent applications for improvements of his centerless grinder. While many patents were issued quickly, some of Heim's applications required seven to eight years for review by the U.S. Patent and Trademark Office. As such, it took until 1930 for the last of Heim's twenty-eight centerless grinder patents to be granted. *(See Patent List in Appendix 1, A1-3)*.

On February 18, 1921, Heim filed a second application to patent a centerless grinder using the peripheral wheel design. This machine contained three improvements that became standard on most future centerless grinding machines. The first change made was to the method for automatically feeding work between the grinding and regulating wheels. As found in his experimental peripheral wheel centerless grinder of 1913, tilting the face of the regulating wheel relative to the carrier induced horizontal forces to promote feeding of the work along the carrier. Whereas the regulating wheel axis and face of Heim's first peripheral wheel centerless grinder *(1,264,930)* was flat *(0°)* relative to the horizontal plane, thus necessitating the use of an inclined or positionable carrier, the improved machine was designed with a tilted regulating wheel paired with a flat and fixed carrier.

In the improved grinder *(1,579,933)*, Heim tilted the shaft driving the regulating wheel downward 2 to 3 degrees, thus changing the angle of contact on the work to draw and control lateral movement through the grinding zone while at the same time controlling work rotation. A diagram illustrating this effect is shown in Figure 6-13. Downward tilting of the regulating wheel shaft can be seen in the Heim grinder shown in Appendix 4 *(A4-4-7)*. By incorporating a tilted regulating wheel to induce and control automatic feeding, carrier systems designed to feed the work were no longer required. This change in the method for feeding work through the grinding zone expanded the capabilities of Heim's machine for grinding different types of cylindrical work. Instead of being designed for two functions *(support and feeding)*, carriers designed to support specialized work could now be attached to the grinder.

In addition to the regulating wheel change there were two other improvements that greatly enhanced grinding accuracy. The first improvement was to lower the centerline axis of the grinding wheel relative to the work and regulating wheel. The second improvement was to modify the carrier surface supporting the work from horizontal *(side to side)* to inclined or slanted downward toward the regulating wheel. Heim discovered that these modifications produced a significant improvement in grinding accuracy by allowing the work to float up and down between the grinding and regulating wheels when exposed to surface projections or other surface irregularities. Heim described his improvement as an *"automatic corrective action resulting in extreme accuracy in grinding cylindrical objects."*

Since the work is rarely supplied perfectly round, the ability of the work to float reduced pressure variations against the grinding wheel, and allowed the machine to grind out irregularities without creating flats or other damage to the surface of the work. An illustration of the slanted work support *(#71)* is shown in Figure 6-14 under the item being ground *(#17)*. The offset *(lowered)* grinding wheel is also shown in this figure and can also be seen on page 3 of Bulletin 110 *(see Appendix 2, A2-6-1)*, and on the Heim Centerless grinder No. 209 retained by the Smithsonian Institution *(A2-8-10)*.

In 1959, Robert Woodbury published a book titled *History of the Grinding Machine (MIT Press; Cambridge, Massachusetts)*. In the chapter on high-speed production

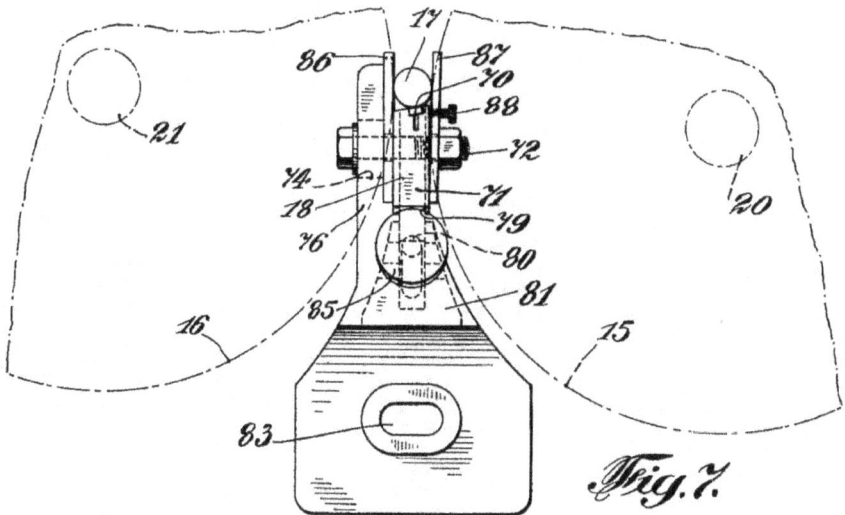

Fig. 6-14: Slanted Work Support and Offset (Lowered) *Grinding Wheel on the Heim Centerless Grinder. Patent 1,579,933, filed Feb. 18, 1921, Grinding Wheel: #15, Regulating Wheel: #16, Roll: #17, Slanted Work Support* (Blade): *#71.*

grinding, Woodbury acknowledges the tremendous appeal of centerless grinding for its high production rates at low cost. On page 153, Woodbury cites Heim for two critical inventions related to centerless grinding in which he states, "The critical invention for centerless grinding was the regulating wheel and work blade by L. R. Heim in 1915."

The term "work blade" refers to the slanted carrier described above. In this section, Woodbury discusses the importance of these features in relation to controlling work rotation and the position of the work in order to obtain high grinding accuracies. According to Heim, these improvements allowed his centerless grinder to achieve grinding accuracies within one ten-thousandth *(0.0001)* inches *(0.0025 mm)*.[9]

In addition to the two machines, Heim developed attachments that included roll feed devices, carriers for different types of work, and truing systems to shape and maintain the grinding and regulating wheels. Examples of these attachments are shown below with the illustrations taken from the patents.

1.) Roll Feed System

Heim's centerless grinder ground work at high speeds and in many cases faster than the operator could feed work to the machine. To improve production rates and alleviate the tedious task of feeding work to the grinder, Heim developed a roll feed system designed for the double-ring wheel centerless grinder. The system consisted of two devices. One was of a roll-feeding mechanism that was attached to the front of the grinder and successively placed unground rolls on the carrier. The other was a roll-feeding machine that supplied a constant stream of rolls in proper orientation to the roll-feeding mechanism. Both devices are shown in Figures 6-15 and 6-16.

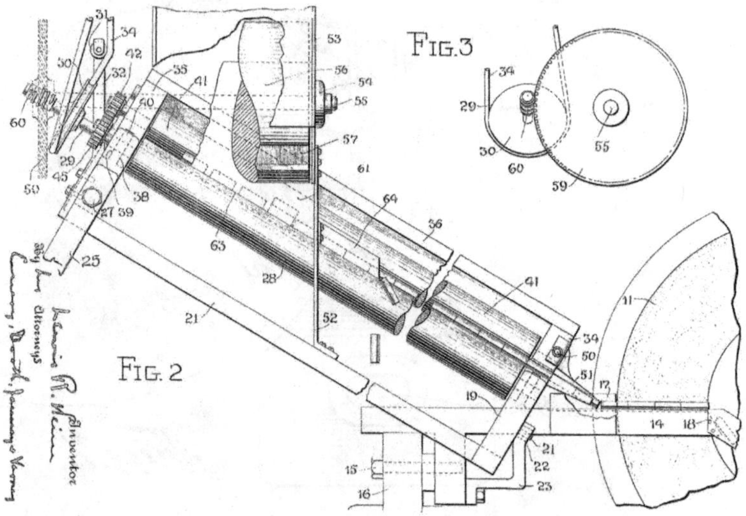

Fig. 6-15: Heim Roll Feeding Mechanism for Centerless Grinders. Patent 1,278,463, filed Oct. 18, 1917.

Fɪɢ.4

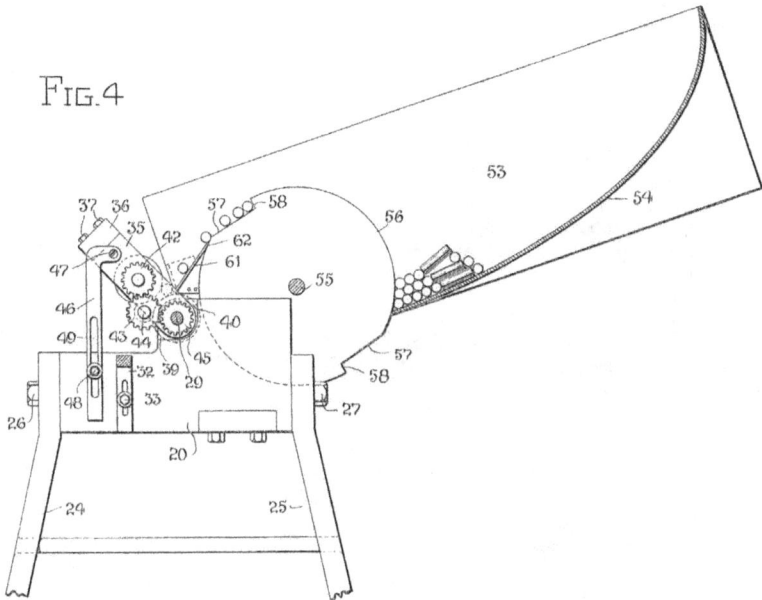

Fig. 6-16: Heim Roll Feeding Machine for Centerless Grinders. Patent 1,278,463, filed Oct. 18, 1917.

2.) Carrier Designs

Heim developed many types of carrier systems to support various types of cylindrical work. In some cases, Heim developed variants of his peripheral machine to accommodate specific carriers. By 1924, Heim was selling peripheral wheel centerless grinders designed with the two wheels mounted on separate supports that were independently slideable relative to the work. For this machine, Heim developed two principal carriers that were rigidly bolted to the front of the machine. The first carrier was designed for "through grinding," in which successive pieces of work rapidly slide along the carrier and through the grinding zone. The second carrier was designed for "spot grinding," in which the work is placed on the carrier, moved into the grinding zone for grinding, and then removed or otherwise ejected, all without being drawn through the grinding zone by the regulating wheel. With these carriers and the two wheels *(grinding and regulating)* mounted on slides, Heim's machine could rapidly and accurately grind a wide range of work diameters and shapes. Both carriers are shown in Figure 6-17 taken from Heim Centerless Grinder Circular HG1 *(see Appendix 2, A2-6-4)*. Additional photos of these and other carriers can be seen in Appendix 2 *(A2-8-7 and 8)*.

In through grinding, the work consists of uniform diameter cylinders such as the rolls used in standard roller bearings. In spot grinding, the work consists of non-uniform diameter work such as tapered rolls or irregular work such as bolts, poppet valves, or similar work that was generally referred to as "headed" or "shoul-

Sizing Work. In the Heim machine adjustments for sizing the work are easily and quickly accomplished.

Since both the regulating and grinding wheel slides are movable, the work fixture need not be disturbed once it has been properly located on the wheel housing. By means of the hand wheels the regulating and grinding wheels are brought into proper relation with the work fixture and each other.

Once the correct position of the regulating wheel has been established, the size of work is maintained by adjustment of the grinding wheel only, to compensate for wear and truing.

No. 3 Standard Work Fixture for *Through* Grinding

Work Fixtures. All work fixtures bolt solidly to the bridge of the wheel housing and the exchange of one fixture for another is quickly accomplished.

Fixtures for *through* grinding are equipped with rests arranged to receive renewable hardened wearing strips, and the guide plates are easily adjusted to accommodate the diameter of work being ground.

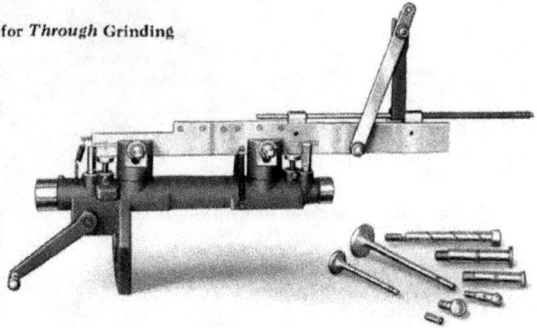

No. 22 Standard Elevating Fixture for *Spot* Grinding

Fig. 6-17: Standard Carriers (Work Fixtures) for "Through Grinding" and "Spot Grinding": 1921.

der work." For spot grinding, Heim designed a carrier for use on the peripheral centerless grinder without the need to change the tilt of the regulating wheel. The carrier was designed with a lever-actuated gear and cam system to raise and lower the work between the two wheels for grinding. When in the raised position the operator would place a new blank on the carrier, actuate the lever to lower it between the two wheels for grinding, then actuate the lever again to raise the carrier for removal of the ground work. To limit or prevent non-uniform or headed work from being drawn by the regulating wheel, Heim designed the carrier with integral stops in the rear section.

Heim developed three variants of the gear and cam carrier that were disclosed in patents 1,585,982, 1,585,983, and 1,585,984. The carrier from patent 1,585,982 is shown in Figure 6-17 as the No. 22 Standard Elevating Fixture and was designed with an integral stop and lever for ejecting ground work through the front of the machine. All variants of the gear and cam carrier significantly increased production rates of the centerless grinder when compared to other grinders at that time. To convey this performance to prospective customers, Heim published actual production

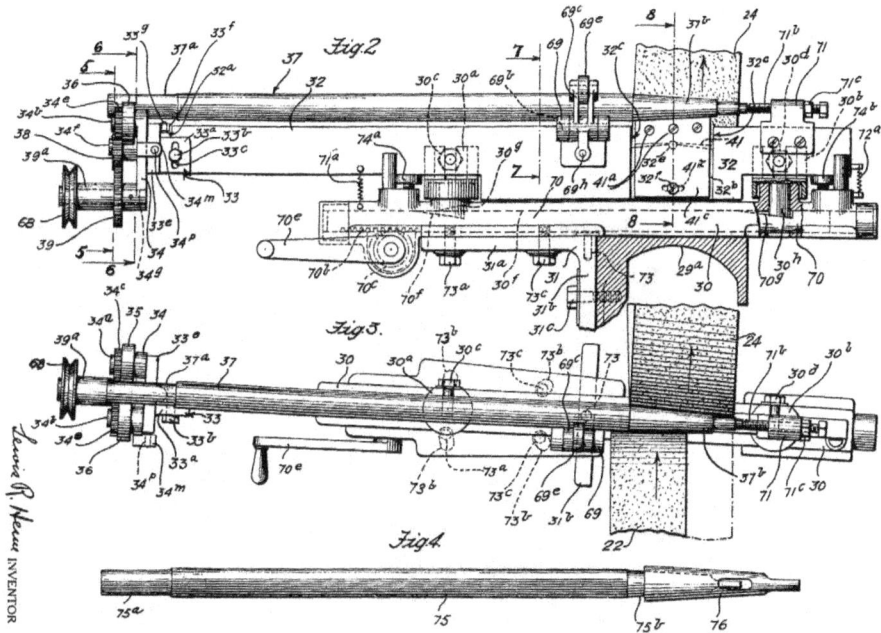

Fig. 6-18: Carrier (Work Fixture) *for Long Tapered Rods Patent 1,590,190, filed Oct. 23, 1923.*

rates in some advertisements. One such advertisement published in *American Machinist* in June 1925, cited the production rates for straight and tapered drill bits and can be seen in Appendix 2 *(A2-7-17)*.

As previously mentioned, Heim developed many types of carriers to support different types of work. An example of one such carrier and its centerless grinder was disclosed in patent 1,590,190 in which Heim developed a carrier to support long rods with tapered tips. The carrier, shown in Figure 6-18, used a manually actuated gear and cam system for feeding into the grinding zone but was modified for the length and shape of the work. The carrier required an extension to support and allow rotation of the rod during grinding and also to hold the rod at an angle relative to the grinding wheel necessary for grinding the tapered tip. The invention also contained another new feature in which the grinding wheel had the ability to oscillate back and forth across the work. This feature was introduced to reduce uneven wear of the grinding wheel, in order to reduce the frequency of wheel truing.

3.) Carriers and Methods for Grinding Rolls with Convex, Concave, and Ring-Like Shapes

Not all roller bearings are designed to use uniform diameter or tapered rolls. Some bearings use rolls with non-uniform convex or concave profiles while other bearings use rolls that look like rings where the diameter is greater than the length. Rings can be uniform or non-uniform in diameter, with many shapes difficult to pass through a centerless grinder since they are unstable on the periphery. To address this need, Heim developed several devices, carriers, and methods to grind these types of rolls in his centerless grinder. Illustrations of these are shown in Figures 6-19 through 6-22.

4.) Truing Devices

By the 1920s, all grinding machines used grinding wheels made from stone-like materials consisted on an artificial abrasive bonded by a cement-type material. In

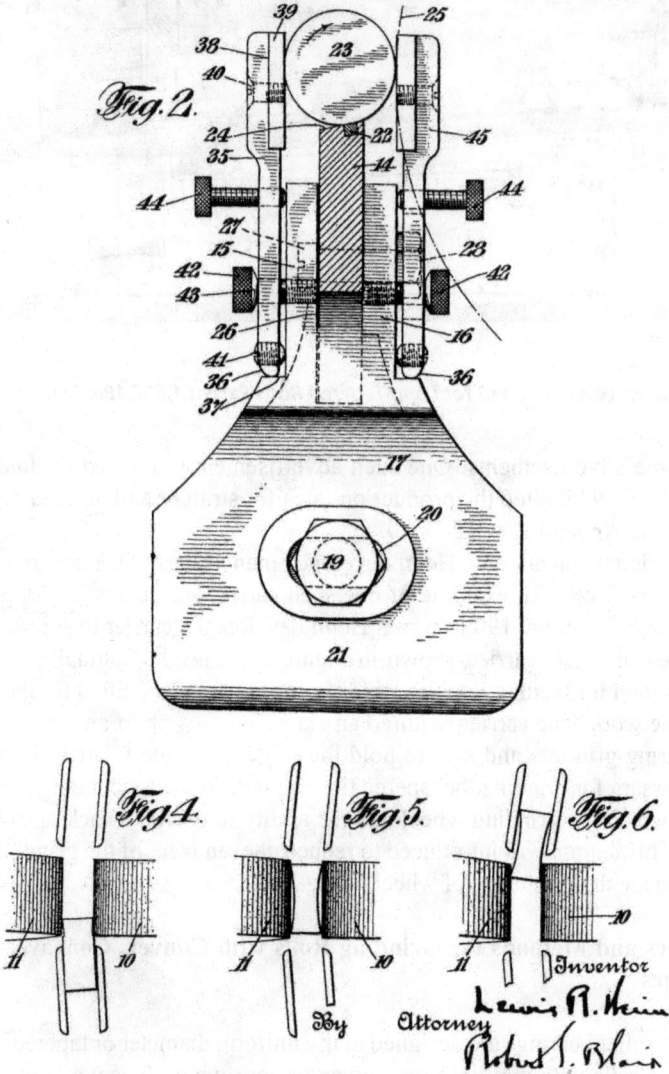

Fig. 6-19: Carrier and Method for Grinding Rolls with Convex and Concave Profiles. Patent 1,741,236, filed April 26, 1922.

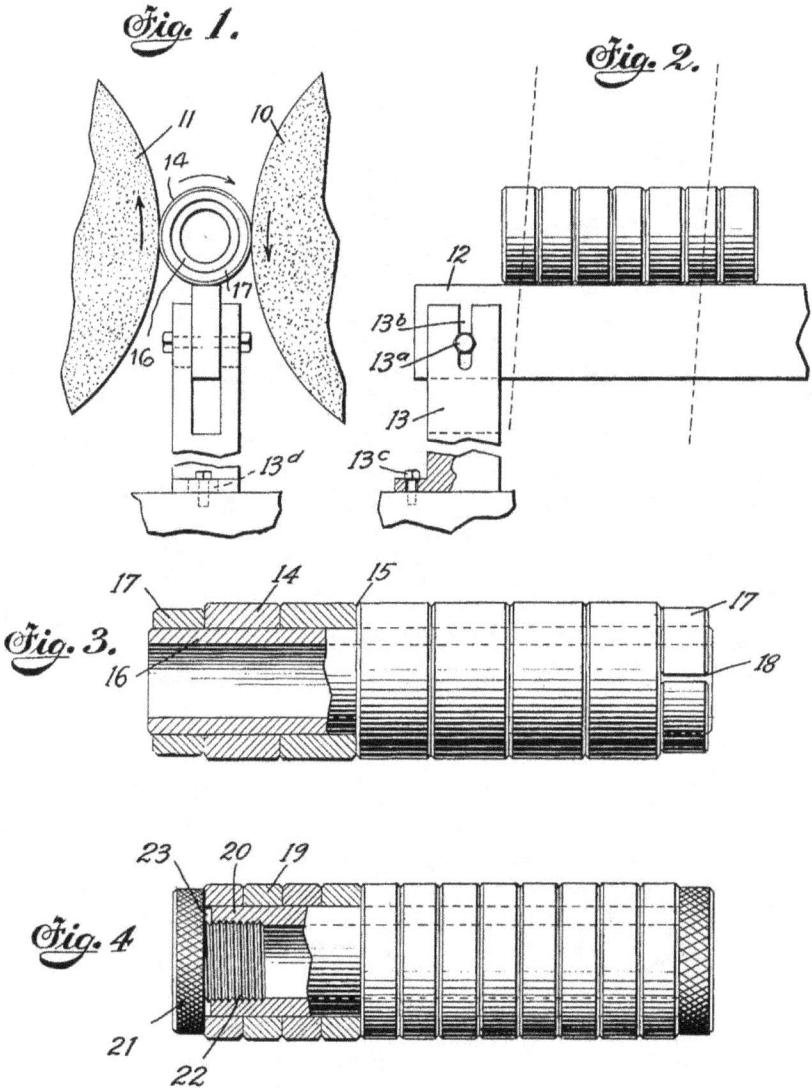

Fig. 6-20: Device and Carrier for Grinding Rings of Uniform Diameter. Patent 1,647,129, filed Oct. 17, 1921.

Heim's centerless grinders, both wheels *(grinding and regulating)* were made from these materials and required periodic "truing" to maintain their surface profile for the work being ground. Truing devices use a very hard material such as a synthetic diamond to shave off material on the wheel. Heim developed truing devices and methods that he installed on each wheel of his machines, with some having cam and follower systems to accurately shape the face of the wheels. Figures 6-23 and 6-24 illustrate Heim's truing systems for peripheral wheel grinding machines.

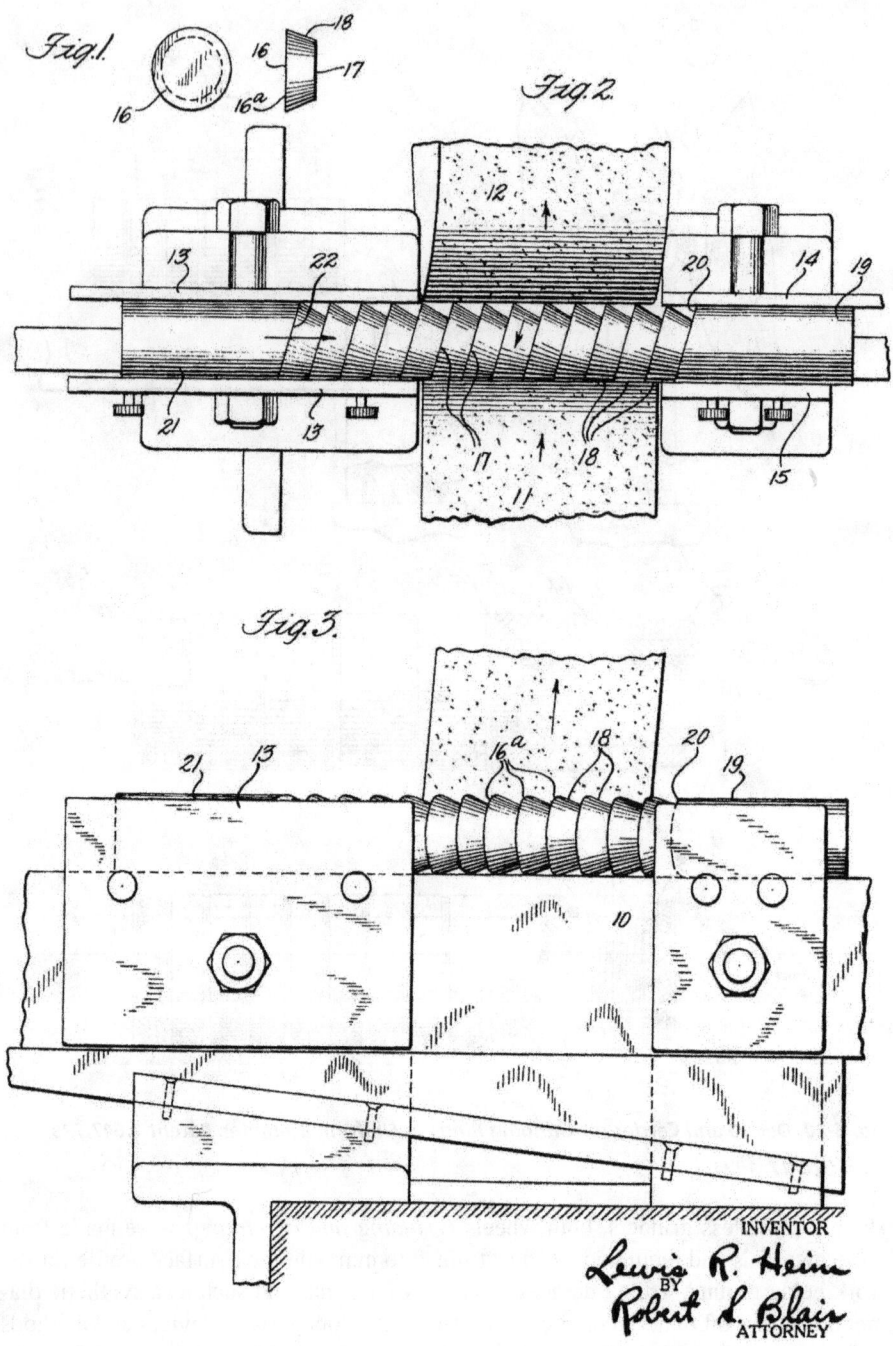

Fig. 6-21: Carrier, Device, and Method for Grinding Tapered Rings. Patent 1,647,131, filed
Jan. 29, 1925.

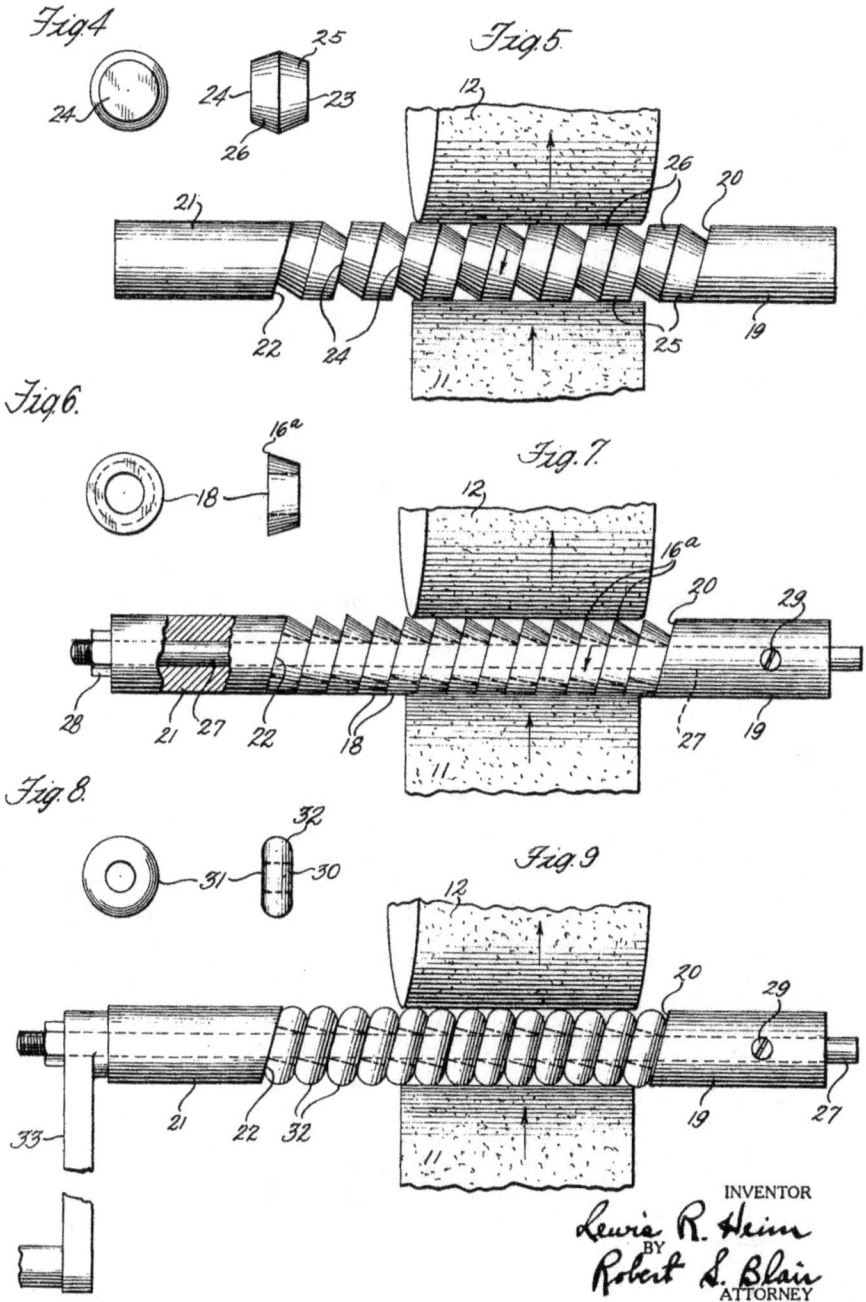

Fig. 6-22: Devices and Methods for Grinding Various Types of Non-Uniform Rings. Patent 1,647,131, filed Jan. 29, 1925.

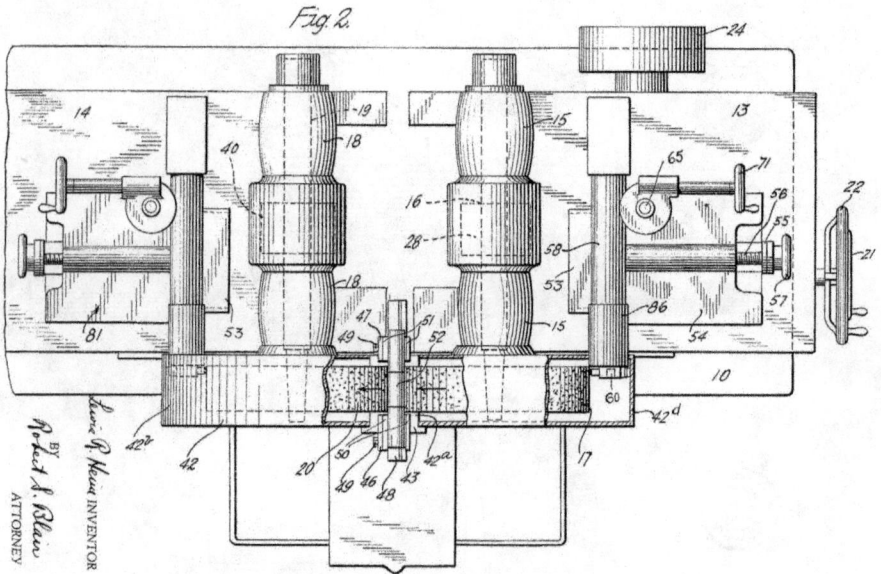

Fig. 6-23: Truing Systems for Peripheral Wheel Centerless Grinder. Patent 1,772,544, filed Feb. 16, 1923.

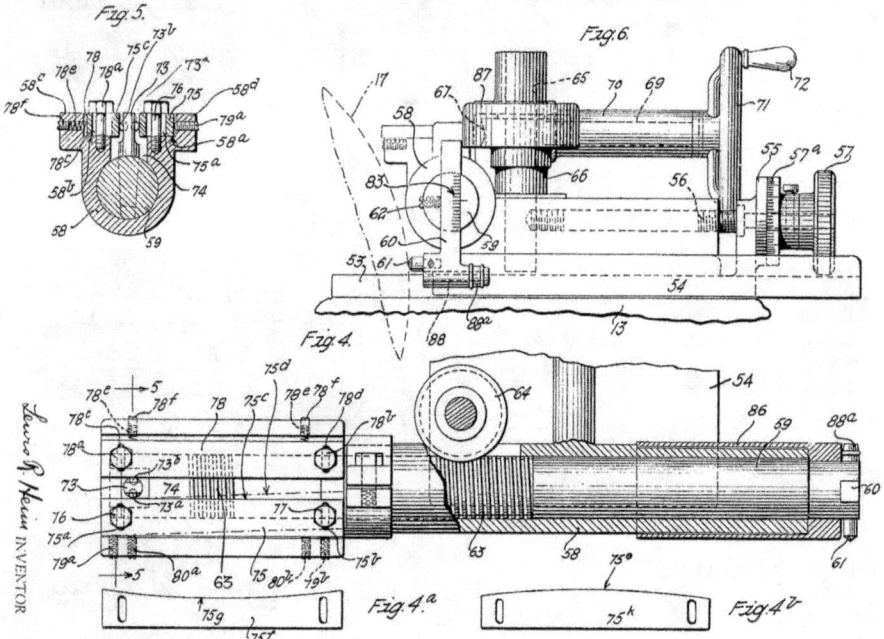

Fig. 6-24: Details of Heim Cam and Follower Truing System for Peripheral Wheel Centerless Grinders. Patent 1,772,544 , filed Feb. 16, 1923.

The Automated Centerless Grinder

To appreciate Heim's accomplishments in developing the centerless cylindrical grinder, a review of the last two patents filed in October 1923 illustrate his inventive genius. Patents 1,733,088 and 1,733,089 describe Heim's most versatile and automated centerless grinders that integrated into one machine many of his best improvements. Prior to the development of these machines, Heim's centerless grinder was designed for through-feed grinding using one type of carrier and spot-feed grinding using another carrier that contained integral stops and manually operated feed and ejection systems.

In order to improve automation capabilities, Heim developed two machines equipped with a system of helical gears that were "attached" to the rear of the machine and performed two functions. One was to reciprocate *(oscillate)* the grinding wheel back and forth across the work with the purpose of reducing uneven wear of the grinding wheel. The other function was to eliminate manual operation of the carrier when grinding non-uniform diameter and headed work. The machine in patent 1,733,088 was equipped with a carrier that was automatically raised and lowered to allow loading, grinding, and ejection of the work. This machine could grind headed work at high production rates but not tapered work. Figure 6-25 shows the helical gear system attached to the back of this grinder. A similar picture showing the entire machine and attachment can be seen in Appendix 2 *(A2-8-7)*. The carrier system, with its cam-operated actuator for automatically grinding headed work, is shown in an advertisement *(A2-7-16)* published in *American Machinist* in May 1925.

The machine in 1,733,089 was modified from its predecessor *(1,733,088)* with the purpose of grinding not only headed work but also tapered, concave, or convex work. Heim developed a fixed carrier that would pivot to align tapered work with the wheels

Fig. 6-25: Heim Helical Gear System for Automated Grinding of Shoulder Work per Patent 1,733,088.

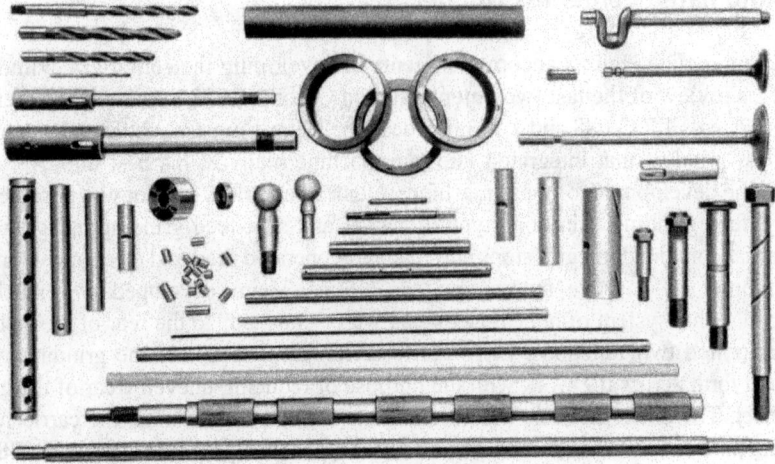

Fig. 6-26: Samples of Work Ground on a Heim Centerless Grinder. 1922.

and another carrier that allowed grinding of concave or convex work. To automate loading, grinding, and unloading of the work, Heim modified the carriage supporting the regulating wheel to allow movement away and toward the carrier. Driven by a modified set of helical gears, the regulating wheel carriage could be actuated for timed withdrawal from the carrier. Using this system, work could be loaded onto the carrier from the top and then automatically unloaded when the regulating wheel retracted, allowing work to fall off the carrier into a collection hopper.

Both machines were equipped with Heim's truing system for shaping and maintaining the respective surfaces of the grinding and regulating wheels. The truing system allowed the operator to regulate the amount of material removed from the wheels, which reduced the time required to reposition each wheel relative to the work. To accommodate grinding work of varying diameter, the truing system was equipped with the cam and follower system shown in Figure 6-24 that allowed the operator to shape the wheels for parallel, tapered, concave, and convex surface profiles. For a complete description of the machine of patent 1,733,089, see the article in Appendix 2 *(A2-5-4)* titled "Heim Improved Centerless Grinding Machine" published in *American Machinist* in December 1924.

Heim's original double-ring wheel centerless grinder using the side surfaces succeeded in grinding constant diameter rolls accurately at high volume. However, it was the two-wheel peripheral design that offered the greatest flexibility in the types of cylindrical work ground and eventually became the standard design for future centerless grinders. Its arrangement allowed for greater flexibility in the design of attachments such as feed systems, work supports *(carriers)*, and automation devices while achieving precision grinding at high production volumes. With the peripheral wheel arrangement, centerless grinding of cylindrical work consisting of straight, tapered, convex, and concave rolls, cylindrical work with multiple diameters or headed sections *(bolts)*, and work consisting of short-length rolls that look like rings, was possible with a single machine. Samples of cylindrical work ground in Heim's machine are shown in Figure 6-26.

HEIM CENTERLESS CYLINDRICAL GRINDER

Taking the Industry by Storm

The Heim Centerless Cylindrical Grinder has captured the country. Heim performances did it. Take your pencil and convert these facts into terms of dollars and cents. Then you'll begin to realize the need of Heim equipment in your plant.

The Heim positively increases production 50 to 500 per cent. It turns "lost time" into output. On short run work as well as on quantity production it shows the same results. As few as 25 pieces may be profitably handled on the Heim.

And here's another eye-opener. Heim Grinding Wheels are trued in 5 seconds, by the watch. You do not have to remove the work or disturb the guide plates. The set-up stays put. The Heim way of truing wheels helps you to do eight hours actual grinding in eight hours' time. Write for the Heim story.

THE BALL AND ROLLER BEARING COMPANY
DANBURY, CONN., U. S. A.

MACHINERY'S GRINDING SECTION FOR AUGUST—135

Fig. 6-27: Heim Advertisement for Centerless Grinders in **Machinery** *August 1922.*

Expansion of the Ball and Roller Bearing Company

Heim developed his centerless grinder in the Heim Machine Company that shared the same building as the Ball and Roller Bearing Company. Whereas Heim was a partial owner of the Ball and Roller Bearing Company, he owned 100 percent of the Heim Machine Company. In the latter months of 1915, Heim sold his entire laundry machinery business to John E. Fidler of New York City, which is documented by the absence of the Heim Machine Company in the 1916 Danbury Business Directory.[10] Heim sold only the equipment and intellectual property related to manufacturing laundry machines while retaining the balance of assets, including the property, buildings, and some machines.

In March 1916, the investors of the Ball and Roller Bearing Company voted to increase capital stock to $1 million and used the funds to expand production capacity. Heim transferred the remaining assets of the Heim Machine Company plus the patent for his centerless grinder to the Ball and Roller Bearing Company in exchange for capital stock. Heim's investment was valued at $290,000 with a split of $190,000 for the Heim Machine assets and $100,000 for the centerless grinder patent.[11] Adding in the $10,000 value of the original investment brought the combined investment for Heim to $300,000 and increased his ownership of the B&RB from 10 percent to 30 percent. Heim's partners, Roth and Barrett, retained 70 percent of the B&RB with an additional investment of $610,000.

The original owner of the two buildings at 20–22 Maple Avenue in Danbury was the machinist Joseph Nutt. In 1886 he built a two-story house and two-story machine shop, both of which were of wood-frame construction. During the period after 1909 up to 1919, the Nutt buildings housing the Heim Machine Company and the Ball and Roller Bearing Company were expanded five times. First, there was a third story added to the Nutt machine shop. After that there were three two-story wood-frame additions built on the north side of the property. Built on the west side and behind the house was a brick building designed with furnaces for hardening steel used to make bearings. All these additions were made to accommodate the growing production of both bearings and the Heim centerless grinder. However, the factory at 20–22 Maple Avenue eventually could not accommodate increasing demand for both bearings and grinders. This led Heim and his partners, around 1920, to expand production facilities by building a brick factory across the street from the B&RB at 17–21 Crosby Street.

Impact on Industry

Prior to centerless grinding there was no relatively inexpensive way of grinding cylindrical work, particularly the rollers used in bearings. Heim's centerless grinder provided significant savings over production with centered-type grinders. In many manufacturing applications, Heim's machine did the work of 6 to 20 men and reduced labor cost of finished work by 80 to 90 percent.[12]

From 1913 through mid-1918 Heim used his invention exclusively at the Ball and Roller Bearing Company for making bearing rolls. In mid-1918 Heim initiated commercial sales and sold centerless grinding machines to several companies in Massachusetts. At first, most customers were skeptical that his centerless grinder could grind cylindrical work in production quantities accurately and inexpensively. But within a short period Heim proved his machines could automatically grind successively fed work to precision tolerances and word began to spread about its capabilities.

Heim's centerless grinder had an immediate and widespread application in the automobile industry for grinding small cylindrical engine parts such as pistons, piston pins, valve tappets, poppet valves, shifter rods, cam rollers, as well as roller bearing casings and bearing rolls. The machine had a tremendous appeal for its high production rates while significantly reducing manufacturing costs. Buyers included automobile companies like Oldsmobile, Reo, Nash Motors, and Ford; industrial manufacturers such as International Harvester; and bearing manufacturers such as New Departure and the Torrington Company.[13] By 1926, Nash Motors had standardized solely on Heim grinders, buying 13 machines for two of its plants.[14]

Starting around 1920, Heim advertised his peripheral wheel centerless grinder and popularized its many features. One such advertisement, shown in Figure 6-27, is titled "Taking the Industry by Storm," which is what two-wheel centerless grinders were doing at that time. Centerless grinding, with its ability to precision grind cylindrical metal parts at high production rates, was a disrupting technology that attracted much attention, not only from buyers of grinding machines but from machine tool manufacturers that sold competing machines.

Chapter 7

Centerless Grinder Competitors

Crucible Grinding and Polishing Machine

As discussed in the first chapter, steel was a critical component of the Industrial Revolution and, along with its alloys, was used widely in the manufacture of machine tools and mechanical bearings. By 1870, the three primary methods for mass-producing steel were the Bessemer, open-hearth furnace, and crucible processes. The Bessemer and open-hearth processes were the most common types used and produced low-cost but imperfectly refined steel. Whereas steel produced using these processes was suitable for tonnage applications such as rails, buildings, and bridges, only the crucible process was able to produce highly refined steel required for making cutting tools and specialty high-strength machine parts.[1]

The crucible process involved melting pieces of low-quality steel in a sealed container called a crucible. The crucible was historically made from clay minerals such as kaolinite *(an aluminum silicate)* but was later made from ceramic materials such as silicon carbide or other non-metallic materials capable of withstanding very high temperatures without deforming. Heated from the outside in a very hot fire, the crucible can withstand temperatures over 3000° F *(1649° C)*, which allows production of cast iron, steel, and wrought *(pure)* iron with respective melting temperatures around 2100°, 2700°, and 2800° F.

Steel making using the crucible process was pioneered in England in the late 1700s with one of the first being the Naylor and Sanderson Steel Mill in Sheffield that made cutlery and other steel products. Naylor and Sanderson later became Sanderson Brothers and Company, which specialized in high-quality specialty tool and alloy steels. In 1876, due to tariffs imposed by the United States, the Sanderson brothers bought an existing iron-making plant in Syracuse, New York, and founded the Sanderson Brothers Steel Works *(Figure 7-1)*. The brothers modified the plant for crucible steel production and later, in 1883, it was the first steel mill in North America to use gas-fired furnaces to produce tool steel.[a][2]

Sanderson was not the only manufacturer of crucible steels. Hussey, Wells and Company of Pittsburgh, Pennsylvania, in 1860 began the first American manufacture of crucible steel. About the same time, Park, Brother and Company, also located in Pittsburgh, commenced production of crucible steel. Up through the late 1890s, other companies such as Sanderson joined in production of crucible steel to meet rising demand. However, as separate operations making small batches of steel, they could not keep up with demand. In 1900, in what some called "the great consolidation," Sanderson and twelve other companies, mostly from western Pennsylvania, formed the Crucible Steel Company of America.[3]

SANDERSON BROS. STEEL WORKS, SYRACUSE, N. Y.

Fig. 7-1: Image of Sanderson Brothers Works from the Crucible Steel Company Products Catalog: 1913.

After the consolidation, Crucible Steel continued to develop advanced steels and improved manufacturing processes. In 1906, it was the first company to install electric arc furnaces for melting steel, and later, the first to develop advanced high-strength alloy steels by adding elements such as chromium, nickel, tungsten, and vanadium. In the early 1900s, the Sanderson plant developed vanadium high-speed steel[b] and free-machining stainless steels.[c] [4, 5] These steel alloys were used in machine tools for cutting and shaping metal and for applications requiring high strength and corrosion resistance such as plain and rolling bearings:

a. Tool Steel: Refers to a variety of carbon and alloy steels that are very hard, abrasion-resistant steel with the ability to hold a cutting edge at elevated temperatures. These properties make the steel suitable for making into tools, especially blades and bits used to cut and shape metal.

b. High-Speed Steel: This steel is a subset of tool steels, commonly used in tool bits and cutting tools. It is often used in power-saw blades and drill bits. It is superior to the older, high-carbon steel tools used extensively through the 1940s in that it can withstand higher temperatures without losing its temper (hardness).

c. Free-Machining Steel: Is steel that forms small chips when machined. This increases the machinability of the material by breaking the chips into small pieces, thus avoiding entanglement in the machinery. This enables automatic equipment to run without human interaction.

In their 1913 product catalog, the company described the products made at the Syracuse foundry as follows:

> *The products of the Sanderson Brothers Works of the Crucible Steel Company of America consist of High Grade Steels made by the crucible process, including High Speed and Self-Hardening Steel, High Speed Polished Drill Rods, Carbon Steel Polished Rods and Wire, Permanent Magnet Steel, a number of brands of Fine Tool Steels, Die Blocks, Sheet Steels and many Specialties. The excellent quality and careful finish of the products are specially noteworthy.*[6]

As mentioned, one of the products made at the Sanderson Works was high-speed polished drill rods with diameters ranging from 0.0135" to 0.5" and stocked in 3-foot lengths. The rods were made true and accurate to size and were recommended for high-speed taps, reamers, twist drills, straight drills, punches, and other similar purposes.

As a result of manufacturing under molten conditions, Crucible Steel needed a process to remove the unusable outer skin that contained oxidized compounds. Sometime around 1910 to 1911, two employees, Edmund French and George Stephenson, developed a grinding and sizing machine to economically remove the skin and irregular surface to achieve required rod diameters and surface finish. The French and Stephenson *(F&S)* machine used two independently rotating and beveled 8-inch grinding wheels: one was a high-speed wheel rotating downward on the work and the other was a slower-speed wheel rotating upward on the work with both wheels rotating the work in the same direction. The higher-speed wheel rotates in the range of 1,500 rpm *(approximately 3,140 fpm)*, which is slow for typical grinding machines. A guide rest *(carrier)*, supported at both ends, extends in a horizontal plane just below the wheel centers and guides the rods *(called "bars" in the patent)* between the grinding wheels. Each wheel is mounted on independent shafts, with the slower wheel shaft positioned slightly in front of the faster wheel. With the offset or staggered position of the wheels, the rods fed into the machine first made contact with the upward-rotating wheel followed by contact with the downward-rotating wheel.

The F&S machine *(Figure 7-2)* looks remarkably similar to the Heim double-ring wheel grinder. Even with all its similarities, the F&S machine was a completely different machine that was not suitable for precision grinding the rolls used in roller bearings. Two workmen are required to operate the machine, one at the front end feeding the rods and the other at the rear receiving the ground bar. In operation, a bar is inserted on the guide rest *(carrier)* at the front *(open)* side of the grinding wheels until the leading end of the work engages the surface of the first wheel rotating upward. Upon contact, the workman feeding the rods must hold the work down on the carrier against the upward motion of the first wheel. As the work moves forward, it engages the second wheel, rotating at a higher speed and downward on the work to keep it pressed against the carrier and, in combination with the first wheel, automatically pulls the work through the grinding zone to be received by the second workman.

The novel aspect of the French and Stephenson machine was not just the ability to grind and size round bars, but to induce an automatic spiral feeding *(drawing)* action

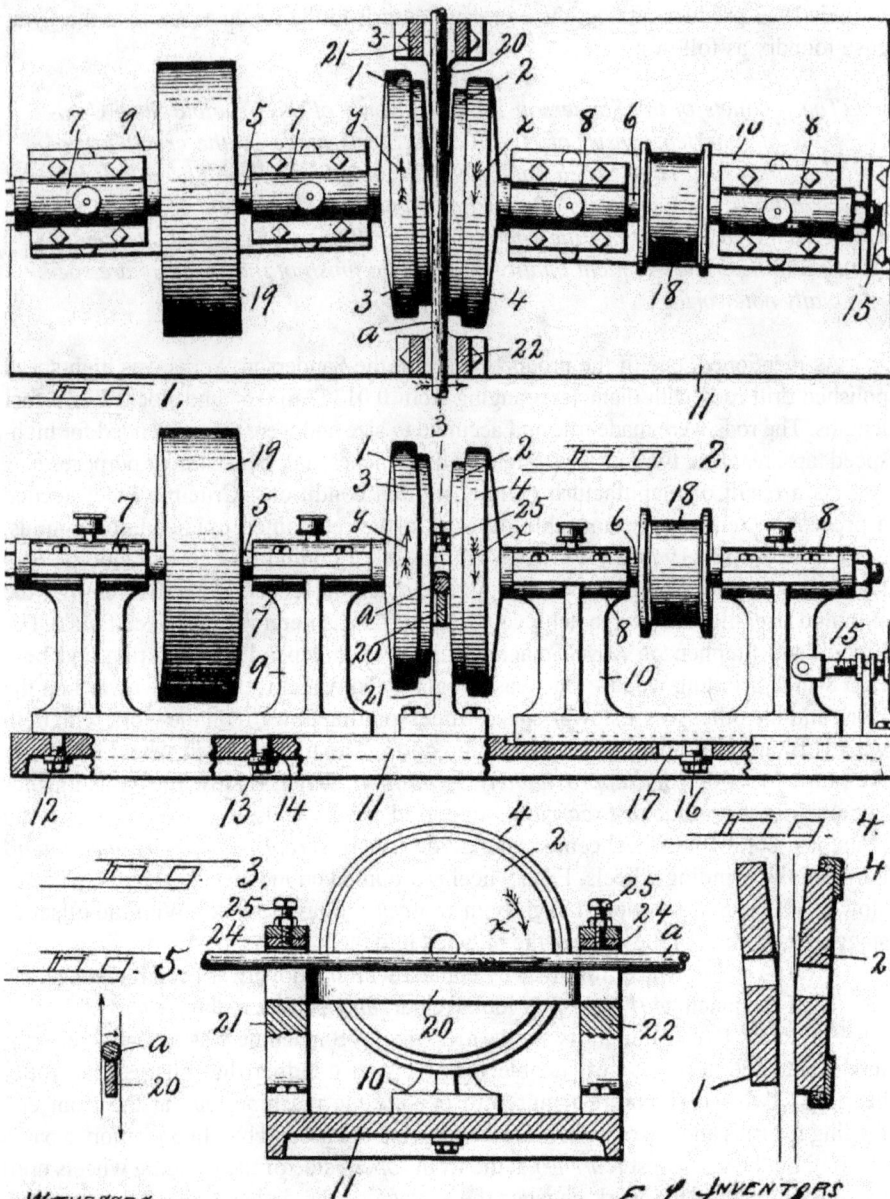

Fig. 7-2: French and Stephenson Round Bar Grinding Machine. Patent 1,111,254, filed May 16, 1911.

of the work between the grinding wheels. According to French and Stephenson, this drawing action was induced as a result of the opposed wheels rotating in opposite directions *(up and down)* relative to the work and at different speeds with the carrier positioned below the wheel centers. On September 22, 1914, three years after filing, the U.S. Patent and Trademark Office granted French and Stephenson patent 1,111,254 for their invention.

French and Stephenson received their patent six months before Heim submitted his application for the double-ring wheel centerless grinder. Sometime just after granting, Heim learned of the F & S patent and after reviewing it, determined that it was not capable of manufacturing rolls for bearings. But since it did incorporate the feature for automatic drawing and feeding, the same process used in his centerless grinder, Heim obtained a license from French and Stephenson to use their patented features. The license was obtained sometime in late 1914 or early 1915.[7]

Heim never saw the French and Stephenson round bar grinding machine in operation until April 1921, just prior to the start of his first patent infringement lawsuit. Heim and John Henry Roth visited the Syracuse plant of the Crucible Steel Company to witness operation of the French and Stephenson grinding machine for round bars. Roth, who was treasurer, had obtained extensive experience operating grinding machines when he was in charge of the B&RB grinding room in 1913 and assisted in evaluating the F & S machine against the Heim centerless grinder.

Heim's primary reason for witnessing operation was to determine the capabilities of the F & S machine and confirm that it could not grind or be modified to precision grind rolls for use in roller bearings. Edmund French conducted the demonstration and confirmed to Heim and Roth that the purpose of his machine was solely to remove the thin skin of decarbonized steel formed when the bars were made at very high temperatures.

Heim also discovered that the F & S machine was a very poor grinder. During the demonstration, French ground unhardened steel bars that were ¼-inch in diameter with a smooth surface. Heim examined the ground rods and found the surface finish rougher and diameter accuracy far less than before grinding. French admitted that grinding to produce smooth and straight bars was not an objective of his machine. When asked that a ¼-inch diameter bar be reduced by 5/1000 of an inch, the operators had to pass the bar through the F & S machine more than 20 times to grind away the required material. The reason for the poor grinding ability of the F & S machine was the lack of control over the rotation of the bar. With both wheels contacting and rotating the bar in the same direction, the only force resisting free rotation was the carrier on which the bar was supported. After the demonstration Heim offered to furnish French a machine that would remove the same amount of material in two passes *(one roughing and one finishing)*, producing a smooth and polished bar.

When examining the carrier system for its ability to feed short rolls, Heim found that the offset *(staggered)* wheel arrangement would not facilitate grinding short rolls and that rolls contacting the first wheel would flip up, turn somersaults, and fly off the carrier. Heim concluded that the F & S machine could not be modified or converted into a suitable machine for grinding bearing rolls and subsequently used the findings to defend his patents.[8]

The Need for Centerless Grinders

During the early 1900s up to the end of World War I *(1918)*, the automobile industry transitioned from high-wheeled motorized buggies *(Oldsmobile Curved Dash and Ford Model A)* to the standard touring car design with a front-mounted internal combustion engine, rear-wheel drive, and a mid-mounted manual transmission *(Ford Model T)*. This period experienced rapid changes in manufacturing techniques, driven by the need to produce many vehicles and lower production costs. Early adopters of new manufacturing processes included Ransom Olds and Henry Ford. Ransom Olds developed "stage" manufacturing for mass-producing the Oldsmobile Curved Dash two-seat runabout on a stationary assembly line. In 1908, Henry Ford introduced the Model T, a low-priced, affordable car that could be purchased by average folks. In 1913, Henry Ford introduced the modern "moving" assembly line on which he assembled the Model T. The moving assembly line allowed mass production at faster rates and much lower cost than stationary manufacture. The Model T sold for $850 in 1909, dropped below $550 in 1915, and below $300 in the early 1920s. The price drops between 1908 and the early 1920s were due to many factors such as improved assembly line efficiencies and advancements in machine tools to make automobile components at lower cost.

The automobile industry was not the only industry to benefit from new manufacturing technologies. Other industries such as the burgeoning airplane industry, even though far behind the automobile industry in using mass-production processes, were also experiencing rapid growth. By their nature, airplanes required higher levels of reliability in systems using moving parts than did automobiles. Components such as piston engines and other mechanical systems required modern grinding processes to manufacture precision-made interchangeable parts.

In 1917, as the United States was entering World War I, the military lacked suitable engines for its aircraft. The military wanted reliable engines with high power-to-weight ratios that could be mass-produced and delivered quickly. Since the automobile industry had the most advanced manufacturing processes at the time, the military issued contracts to many automobile companies for airplane components. The Hall-Scott Motor Company and Packard Motor Car Company, which had experience building high-horsepower aircraft engines, quickly developed a design for what became the Liberty engine. The Liberty *(L-12)* was a 27-liter water-cooled V-12 overhead cam engine that developed 400 horsepower. It was known as a powerful *(at that time)* and reliable engine and installed in the British-built single-engine bomber, the DH.9A. Ford, Cadillac, Marmon Motor Car, and others that had mass-production expertise, built more than 22,000 engines for the war effort.[9]

Another technology that got a boost in World War I was the development of armored vehicles, such as tanks. With the advent of trench warfare and both sides dug in, a stalemate developed, causing both the Allies and Germany to develop tanks that could breach enemy defenses. As the United States entered the war in 1917, it had no tanks of its own, so the army built a copy of the French Renault FT, a 7.5-ton, two-man tank that was dubbed the M1917. Later, in mid-1918, Ford began work on a 3-ton, two-man tank that became the M1918. The tank was small, equipped with one .30-caliber machine gun and powered with two Ford Model T engines. As with automobiles and planes, tanks required bearings, especially those designed for heavy loads. The Ball and Roller Bearing Company, with bearings designed for heavy radial and thrust loads, was a prime contractor to the army for all US-made tanks.[10]

As World War I ended in November 1918, the automobile industry, which was the largest buyer of machine tools, had the greatest impact on the development of new machines and manufacturing processes. Machine tools had to mill, saw, drill, and grind increasingly higher-strength hardened alloy steels that were more difficult to cut and shape. At the urging of automobile companies, machine tool manufacturers, especially manufacturers of grinding machines, had to develop new technologies to achieve greater dimensional tolerances and at the same time increase production rates to lower the overall cost of manufacture. Small cylindrical components made up a large portion of the moving parts of automobiles, trucks, airplanes, and machines. Up to the 1920s, grinding of cylindrical parts was still using slow, labor-intensive and high-cost grinding processes. Centerless grinding, with its ability to precision grind cylindrical work at high production rates and much lower cost, became a valuable and must-have machine tool.

Single-Wheel Centerless Grinder

At the beginning of the twentieth century, antifriction *(rolling)* bearings were well under way to replacing journal bearings in machines of all kinds. Roller bearings were preferred over ball bearings in applications requiring heavy radial loads as well as applications involving axial loads. However, the machines and techniques required for precision grinding of rolls at low cost were still unavailable. As the artificial grinding wheel containing man-made abrasives was just becoming widely available, grinding machines for high production rates had not yet been invented, requiring use of slow and costly centered grinding. Before 1918, there were no two-wheel centerless grinders commercially sold. The only class of machine considered centerless was that using a single-wheel configuration.

In 1901, Samuel S. Eveland invented the very first centerless grinding machine that became patents no. 747,541 and 747,542 in 1903 and belonged to a class of machines called single-wheel centerless grinders. The grinder was composed of a metal block called a holder *(#6 in Figure 7-3)* containing a triangular groove *(#7)* along its length that held cylindrical work up against the face of a fast-rotating emery grinding wheel. Contact with the grinding wheel caused the work to rotate within the holder while the triangular groove securely retained the work to prevent "flipping off" as described by Heim when grinding work in his "open groove" work holder. Friction generated between the rotating work and the holder produced enough braking action to allow grinding of the work circumference. In order to grind cylindrical work on Eveland's machine, the operator manually pushed work through the grinding zone by inserting successive pieces of work into one end of the holder.

Eveland invented his machine solely for use at the Standard Roller Bearing Company that he founded in 1898 to make various types of roller bearings. Eveland's machine was never sold commercially to any other company. The following comments made by

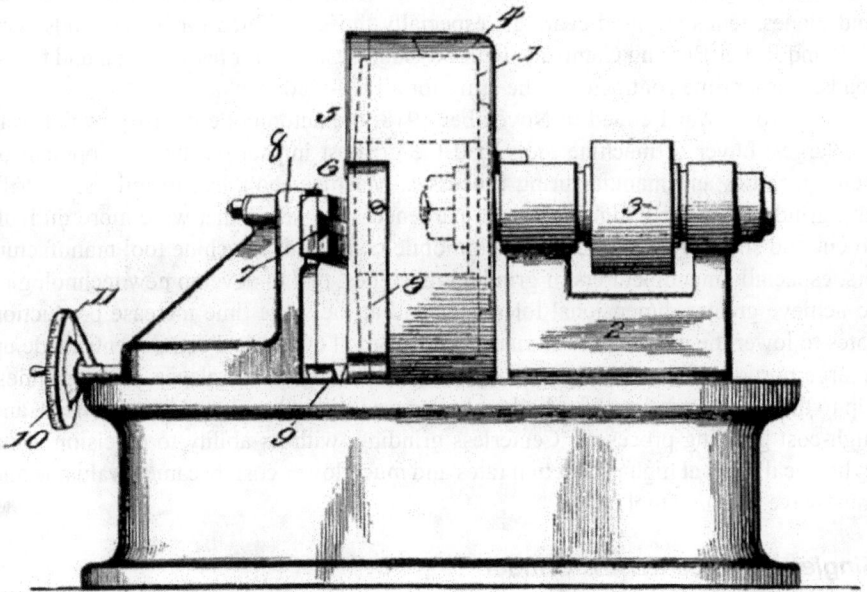

Fig. 7-3: Eveland Single-Wheel Centerless Grinder. Patents 747,541 and 747,542, filed Dec. 26, 1901.

Eveland in 1921 summarize the origins of his single-wheel centerless grinder:

> *I founded the Standard Roller Bearing Company about 1898, and was thoroughly conversant with grinding machines and methods and grinding machinery makers, throughout the United States, as well as in England, France and Germany, which countries I visited for the purpose of securing any machines suitable for the purpose of making all or any part of roller bearings. I never heard of any centerless grinder or machine at all similar to the one shown in my patents, either here or abroad, mine being the pioneer or original in the field and being developed exclusively from my original design and for our specific purpose. Before I had developed this machine, and during its development, I was in active negotiations with the Emery Wheel Manufacturers and with the grinding machine manufacturers, in this country, in an effort to secure from them some form of a machine that would do better, quicker or equally as accurate work, but none of them, at any time, offered machines other than their standard types of center grinding, and from my experience and intimate acquaintance with the machinery business, I am now and always have been satisfied that nothing of a similar character, or in other words, a centerless grinder, had been made prior to the making of my own machines.*

The machines covered by the patents were designed and built by me when I was Vice-President and General Manager of the Standard Roller Bearing Company of Philadelphia, for the purpose of securing a machine for grinding rolls of varying lengths for use in radial journal roller bearings and similar radial rolls used in roller thrust bearings, and for the further purpose of making such rolls very accurate as to size and shape and to grind them cheaply as compared with the cost of grinding similar rolls upon the only other grinding machines then obtainable from grinding machinery makers or others, all of which machines required the rolls to run on centers and which machines ground one roll at a time, thereby making the grinding of rolls by the then known methods extremely costly as compared to the cost of doing similar and equally accurate work upon my machines covered by my patents above referred to, as my machines did not require the rolls to be centered and they turned out many thousand rolls per day on each machine of small sizes and many hundred on larger sizes, which gave an output much greater than was possible on any ordinary grinding machine then procurable in the market for a similar purpose.[11]

In 1912, Robert Grant invented a second type of single-wheel centerless grinder that incorporated a feature for automatic feed of the rolls that replaced manual feed required in the Eveland machine. The Grant machine *(Figure 7-4)* used a flat plate holder embedded with a channel to hold the work, similar to that used by Eveland in his grinding machine. The holder mounted on a sliding support for positioning against the peripheral surface of a grinding wheel. The roll, as with the Eveland machine, rotated within and was braked by the holder to promote grinding along the circumference of the roll.

The novel aspect of the Grant invention was the ability to promote automatic feed of the work through the grinding zone. This was accomplished by rotating the holder a few degrees from horizontal. Rotating the holder off the horizontal face of the grinding wheel caused the grinding wheel to impart two forces on the roll. One was a downward force that rotated the roll within the holder and performed grinding. The second was a lateral motion that caused the roll to move along the length of the holder. Grant filed his patent application *(1,106,803)* in September 1912 and subsequently assigned it to the Grant Automatic Machine Company of Detroit, Michigan.

The Grant single-wheel centerless grinder was similar to the single-wheel grinder developed by Heim in 1913. Whereas Heim used the ring *(side)* surface of the grinding wheel, Grant used the periphery surface. To feed the work through the grinding zone, Heim used gravity and Grant used angled contact of the work against the grinding wheel. As with the Heim single-wheel machine, the Grant grinder was prone to erratic behavior and failure that made the machine undependable. A former salesman, Arthur Kirkland, who worked for the Detroit Machine Tool Company that sold a similar single-wheel centerless grinder said that between April 1, 1919, and January 31,1920, the company sold fifty to sixty machines at a price of $1,100 each. When asked in 1921 if there was general satisfaction among his customers who bought the machine, Kirkland made the following comments:

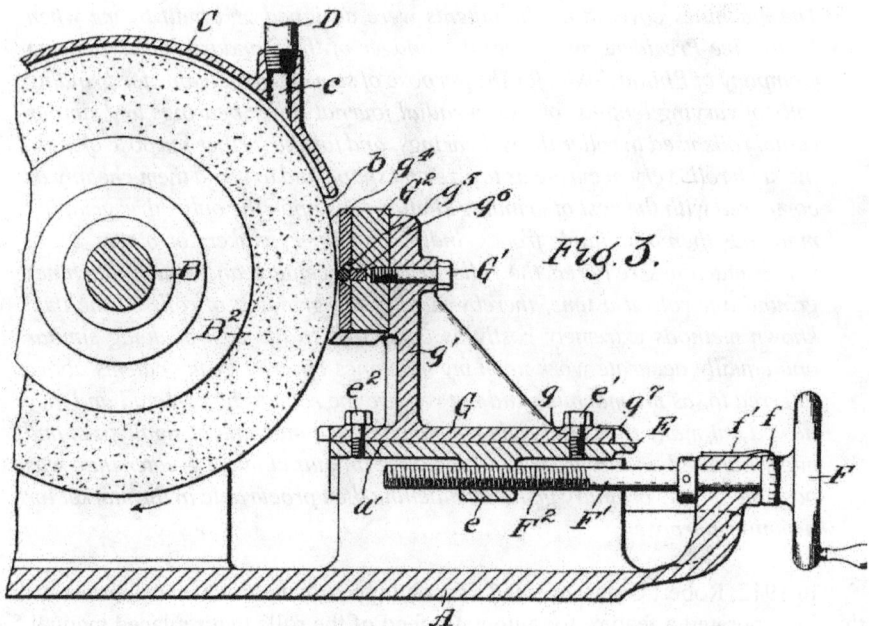

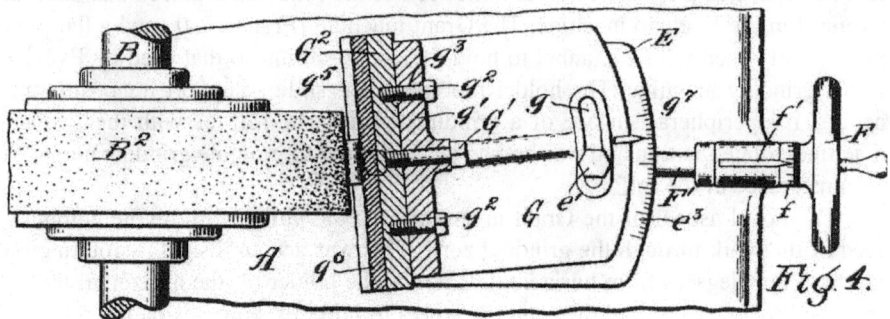

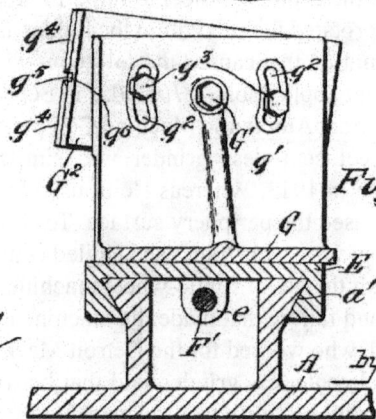

Fig. 7-4: Grant Single-Wheel Centerless Grinder. Patent 1,106,803, filed Sept. 21, 1912.

No, I should say there was not. The machine was freakish. At times it would do the work, and then when you wanted the work the worst way, then this freakish action would take place and it would grind with flats, taper or the work would be irregular.

When asked about the success in selling the machine, Kirkland related:

At the time we sold that machine, there were practically no other centerless grinders on the market. There was a field for such a machine and it answered the purpose until the two-wheel grinder came out. After that time, why the Detroit machine became obsolete, and the sales diminished.[12]

Introduction of Two-Wheel Centerless Grinders

As World War I progressed, the United States government procured tremendous amounts of machined metal parts, with many companies dedicating most of their production to fulfilling government orders. The Ball and Roller Bearing Company was one of the few companies that made rolling bearings for heavy loads that were in high demand by the military. Ball thrust, roller thrust, and journal roller bearings were used in army tanks and navy pumps and periscopes. Between 1917 and 1919 nearly 100 percent of B&RB production went to orders servicing the government.[13]

In 1917, U.S. bearing manufacturers formed a group to aid bearing manufacturing for World War I. The group was called the Ball and Roller Bearing Manufacturers Association, which later became the American Bearing Manufacturers Association. The purpose of the group was to standardize the industry by developing codes and regulations used to design, manufacture, and order mechanical bearings. Heim represented the B&RB at the association meetings and was required to commit time in addition to his work responsibilities.

Sometime during 1917, Heim began to suffer from exhaustion due to the many obligations and long working hours. The extent of his condition was such that it at times prevented Heim from going to work and persisted for more than two years into 1919.[14] The illness also prevented Heim from working on improvements to his centerless grinder. After filing four patent applications in 1917 there were no new applications for four years until December 1920. Thereafter, Heim spent a great deal of time improving his grinder, leading to 23 additional inventions.

Also during 1918, Heim's brother Alfred sold his business, the Danbury Troy Laundry, and went to work at the Ball and Roller Bearing Company. The business was sold to one of the company directors, James Durkin, and Durkin's partner, John McKenney. The reason for the sale is unknown but may have been related to Lewis's illness in that the sale freed up Alfred to help manage the B&RB. In 1921, Alfred was promoted to vice president of the B&RB and so recognized in the 1922 and 1923 Danbury City directories.

In 1918, the last year of World War I, the B&RB operated seven double-ring wheel centerless grinders, with each machine capable of producing between 5,000 and 10,000 roller-bearing rolls per day. To aid in timely manufacture, the B&RB ran

Fig. 7-5: Battery of Seven Heim Double-Ring Wheel Centerless Grinders at the Ball and Roller Bearing Company in Danbury, 1921.

the machines frequently and kept a very large stock of various sizes to expedite orders. Figure 7-5, a photograph from 1921, shows the battery of belt-driven double-ring wheel centerless grinders installed in the Maple Street factory of the B&RB.

As the war wound down, American industry began to shift away from war production and back to servicing domestic markets. The year 1918 was also when Heim began selling his double-ring wheel centerless grinders. His first machine was sold on July 24, 1918, to the Duckworth Chain Manufacturing Company of Springfield, Massachusetts. The next sale was eleven months later on June 28, 1919, for two machines to the R. B. Phillips Manufacturing Company of Worcester, Massachusetts. Four additional machines were sold for a total of seven ring wheel grinders.

By 1921, sales began to accelerate as the automobile industry discovered the mass-production and cost-saving features of two-wheel centerless grinders. That year, sales of centerless grinders by the B&RB reached $48,000 *(around 12 machines)*. In 1922, sales nearly tripled to $138,000 for 34 grinders and in 1923, sales reached $248,000 for around 62 machines. Between 1918 and 1923, the B&RB realized an average profit of $28,000 per year.[15] However, even though sales of centerless grinders by the B&RB seemed to be good, they should have been far higher. Starting around 1917, other companies began making and selling two-wheel centerless grinders and took a large percentage of the market away from Heim.

At first, Heim advertised his centerless grinder for making cylindrical parts of uniform diameter, with many directed at the automobile industry and listing engine parts that could be ground. Figure 7-6 is one of the early advertisements placed by Heim and references B&RB Bulletin 110 *(Appendix 2, A2-6-1)* that describes the machine and grinding capabilities. The bulletin lists the six patents in effect at the time *(shown below)* protecting Heim's invention from competing machines.

The Heim Centerless Cylindrical Grinder
The machine is manufactured under the following U.S. Letters Patents:

No.: 1,111,254	No.: 1,164,930
No.: 1,210,936	No.: 1,278,463
No.: 1,210,937	No.: 1,281,366

Other patents applied for.

Notice that the first patent on the list is No. 1,111,254 issued to French and Stephenson. Knowing that the French and Stephenson machine looked remarkably similar to Heim's machine, Heim included the French and Stephenson patent to prove he had the right to use its patented features. Heim also included the same list of patents on the nameplates of his centerless grinders. *(See Appendix 2, A2-8-9 and 10.)*

Between 1917 and 1926, seven U.S. companies developed various designs for two-wheel centerless grinders using the primary principles of the Heim invention. The eighth filing shown below was by a foreign company that first filed in Europe. The first patents granted to each company for their respective grinding machines are listed below in order of the filing dates:

Patent No.	Filing Date	Inventor	Company
1. 1,264,129	April 28, 1917	Milton Reeves	Reeves Pulley Co.
2. 1,483,748	July 30, 1920	Francis C. Sanford	F. C. Sanford Mfg. Co.
3. 1,733,086	April 23, 1921	Edward Sobolewski	Detroit Machine Tool Co.
4. 1,599,584	June 14, 1921	Walter J. Peets	Singer Mfg. Co.
5. 1,639,958	June 28, 1922	Charles H. Norton	The Norton Co.
6. 1,575,558	Jan. 18, 1924	Sol Einstein *et. al.*	Cincinnati Milling Co.
7. 1,863,832	Feb. 12, 1926	William Chapman	General Motors Co.
8. 1,777,607	May 6, 1929	Carl G. Ekholm	LMV, Sweden

Each inventor developed a centerless grinder with unique features to make it appear novel and different from the Heim machine. However, upon examination, all applications incorporated most or all of the four primary features of operation described in Heim's original patents *(Nos. 1,210,937 and 1,281,366)* as follows:

1. Centerless Support: A support system for cylindrical articles *(work)* to be ground, generally consisting of a horizontal or inclined metal bar that supports the work without centers and allows the work to slide in between and past two rotating wheels for the purpose of being ground to precision dimensions in high volume. In Heim patents, the term "carrier" is used to

Heim Centerless
Cylindrical Grinder

Fig. 7-6: Heim Centerless Grinder Advertisement Published in Machinery, October 1921.

112

reference the centerless support system.

2. Grinding Wheel: A grinding wheel made from an abrasive material rotating at high speed and downward on the work to perform a cutting action on cylindrical work moving across its surface.

3. Regulating Wheel: A regulating wheel made from an abrasive material located on the opposite side of the cylindrical work from the grinding wheel, rotating at a much slower speed and upward on the work to control rotation of the work against the grinding wheel.

4. Natural Drawing *(Feeding)* Action: A drawing *(spiral feeding)* action resulting from angular contact by the regulating wheel on the work that controls both work rotation and lateral movement across the surface of the grinding wheel.

Even though all the competing inventions listed above contained the essential patented features developed by Heim, they were granted letters patent by the U.S. Patent and Trademark Office (USPTO). One would believe that the primary function of the USPTO is to deny patents to such applications; but, in reality, the USPTO defaults to the inventors and the courts to resolve disputes. A short description of each machine and its features follows, to illustrate how the competing inventors tried but could not circumvent Heim's novel invention.

Reeves Pulley Company

The first company to introduce a two-wheel centerless grinder using Heim's patented method was the Reeves Pulley Company of Columbus, Indiana, which was founded in 1888 by Marshall T. Reeves and his two younger brothers, Milton O. and Girnie L. Reeves. The company manufactured gas engines, steam traction engines, pulley-driven devices, and many other products. Of the three brothers, Milton was the most mechanically inclined and invented many types of machines, with his most famous being the variable speed transmission.[16] The Reeves transmission *(Figure 7-7)* used pulleys incorporated with a system to vary the pulley diameter, allowing quick and smooth changes in speeds and was an early version of the modern-day continuously variable drive. Reeves's variable speed transmission was a huge success and was used in automobiles, motorcycles, machine tools, and many other devices. An advertisement of the Reeves transmission published in 1923 can be seen in Appendix 2 *(A2-7-13)*.

Reeves was also one of the first in the U.S. to manufacture automobiles and other motorized vehicles. In the late 1890s, Reeves built one of the first automobiles in the U.S. that was called a "motocycle" or motorized four-wheel bicycle, also known as a "horseless carriage." The motocycle used the belt and pulley variable speed transmission that Reeves believed gave him a competitive edge against Henry Ford's Quadricycle, which had only one speed. In 1911, Reeves founded the Reeves Sexto-Octo Company that produced six and eight-wheeled luxury automobiles. The purpose of the

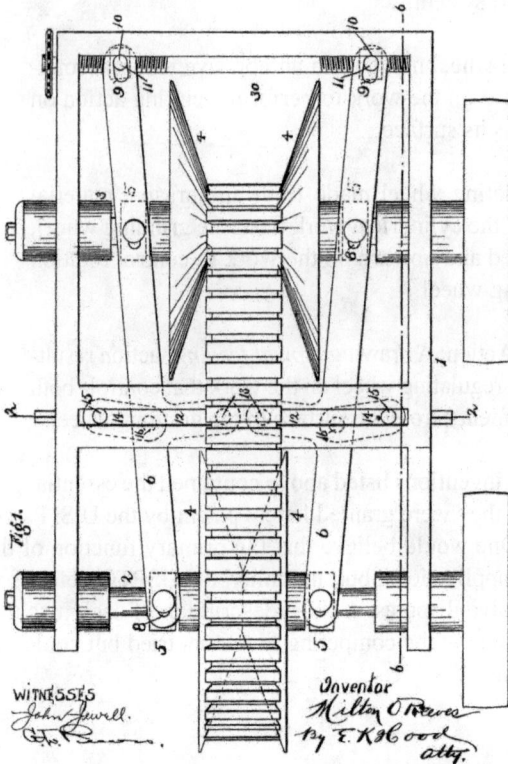

Fig. 7-7: Reeves Variable Speed Transmission. Patent 630,407, filed Feb. 25, 1899.

extra wheels was to smooth out the ride when traversing rutted roads and to extend the life of the tires. However, with the added wheels and required extended length, the Reeves designs were expensive and never caught on with luxury car buyers. And it did not help to have the eight-wheel model named one of the ugliest cars ever produced. With its high price and odd looks, the cars sold poorly, forcing Reeves to abandon automobile production.

With his keen interest in machines, Milton Reeves also designed machine tools and on April 28, 1917, filed a patent application for a two-wheel centerless "Grinding Machine" that "accurately grinds cylindrical objects such as rollers for roller bearings at small expense."

Reeves called his version of the regulating wheel a "Propeller Wheel," as it mostly controlled the rotation and axial feeding of the article being ground. Reeves stated that the faster-moving wheel performed the grinding action but could also be used to affect axial feed by tilting its shaft either upward *(retard axial feed)* or downward *(accelerate axial feed)*. Reeves's application became patent 1,264,129 on April 23, 1918, and contained illustrations showing his invention in three configurations. As shown in Figures 7-8 and 7-9, these were:

   a. Perpendicular *(Figs. 3, 4, and 5)*
   b. Peripheral *(Fig. 6)*
   c. Offset Ring Wheel *(Fig. 7)*

Even though Reeves patented more than one configuration, none matched the Heim double-ring wheel arrangement consisting of opposed obliquely positioned beveled wheels.

A second aspect of the Reeves patent application was that one of the three configurations, the two-wheel peripheral arrangement, was identical to Heim's second centerless grinder that became patent 1,264,930. Since Reeves filed two and one-half months before Heim, Reeves was actually the first to file for a centerless grinder using the peripheral configuration. Reeves eventually standardized on the "perpendicular" configuration using the ring *(side)* surface of the propeller wheel and peripheral surface of the grinding wheel and filed five additional applications that were later granted letters patent. Reeves started selling his machine sometime before June 1919 and sold through the Gardner Machine Company of Beloit, Wisconsin. The Reeves machine could grind cylindrical work up to 3 inches in diameter and 5 inches in length, and in 1922 the company introduced an improved machine with a drop-in work rest *(carrier)* that allowed the machine to grind shoulder *(headed)* and tapered work. Advertisements for the Reeves centerless grinder published in both *American Machinist* and *Machinery* magazines can be seen in Appendix 2 *(A2-7-9 and 13)*.

It is interesting to look at the timing for filing by Reeves of his first centerless grinder application. The filing was almost four months after Heim was granted a patent for his first centerless grinder *(double-ring wheel)* and two and one-half months before Heim's filing of an application for his second *(peripheral)* centerless grinder on July 13, 1917.

Since four months is too short a time period to develop a complex machine and then write a patent application, Reeves no doubt learned of Heim's machine well before the granting of Heim's patents. How Reeves found out is not known but a likely source may have been a salesman eager to sell bearings touting precision dimensions of the rolls manufactured by the B&RB on a new type of grinding machine. Another possibility is through Heim's advertising of its *"unusual facilities for the production of small hardened and ground cylindrical rolls . . . ."* Reeves was not only a machinist and inventor, he was also a businessman and always looking for something new to sell. So it is highly probable that when Reeves learned of Heim's machine he was able to quickly design and patent his concept.

F. C. Sanford Manufacturing Company

The second company to build a centerless grinder was the F. C. Sanford Manufacturing Company of 2060 Fairfield Avenue, Bridgeport, Connecticut. Founded in 1906 by Francis *(Frank)* C. Sanford when he was around 34 years old, Sanford Manufacturing made metalworking machinery and stamped metal goods.[17] On July 30, 1920, Sanford filed for what became patent 1,483,748 for a centerless grinder that was:

> ... *designed for precision work and, more especially, for grinding to true cylindrical form and to predetermined gage, bodies such as rollers for antifriction bearings and for like uses, requiring great accuracy of form and dimensions, and uniformity of product.*

This application was the first of three inventions for centerless grinders that were later granted patents with the latter two being 1,611,135 and 1,611,136. As shown in Figure 7-10, Sanford designed his machine to use the periphery surface of both wheels and called his regulating wheel a "Governing and Feeding Wheel." To promote and control

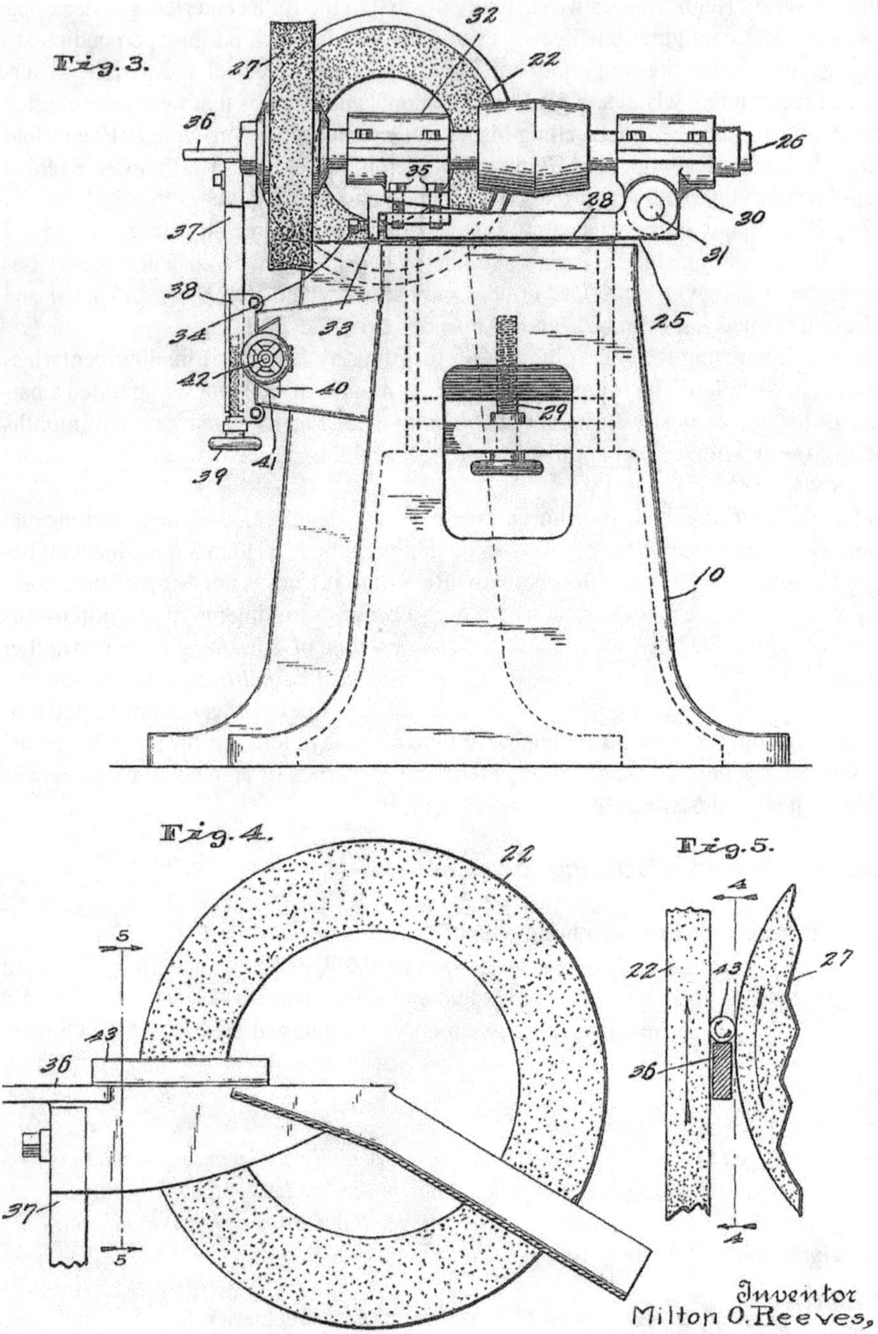

Fig. 7-8: Reeves Centerless Grinder Using a Combined Radial and Peripheral Wheel Configuration. Patent 1,264,129, filed April 28, 1917.

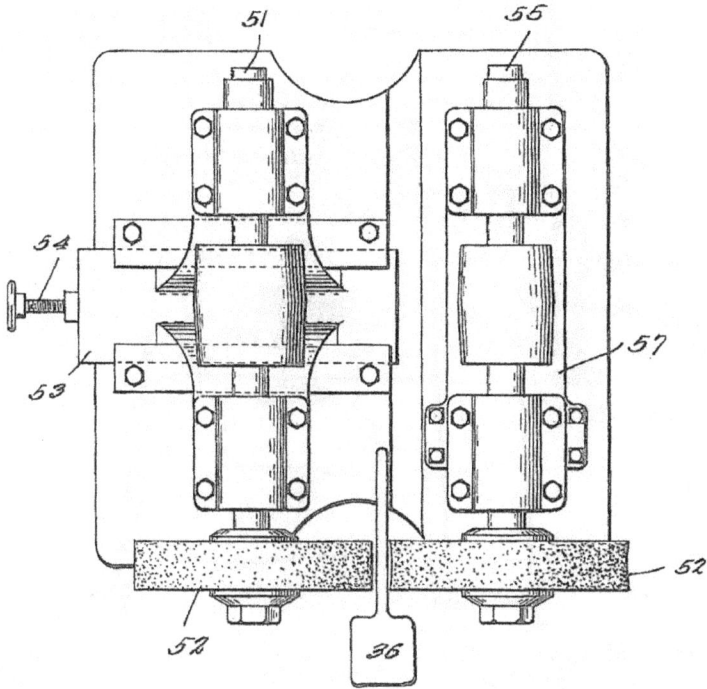

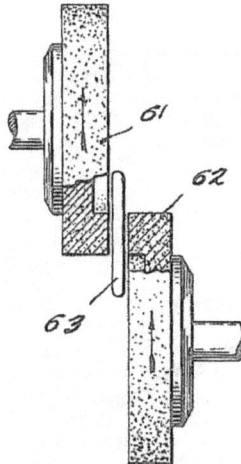

Fig. 7-9: Reeves Centerless Grinder Using Peripheral and Offset Radial Configurations.
Patent 1,264,129, filed April 28, 1917.

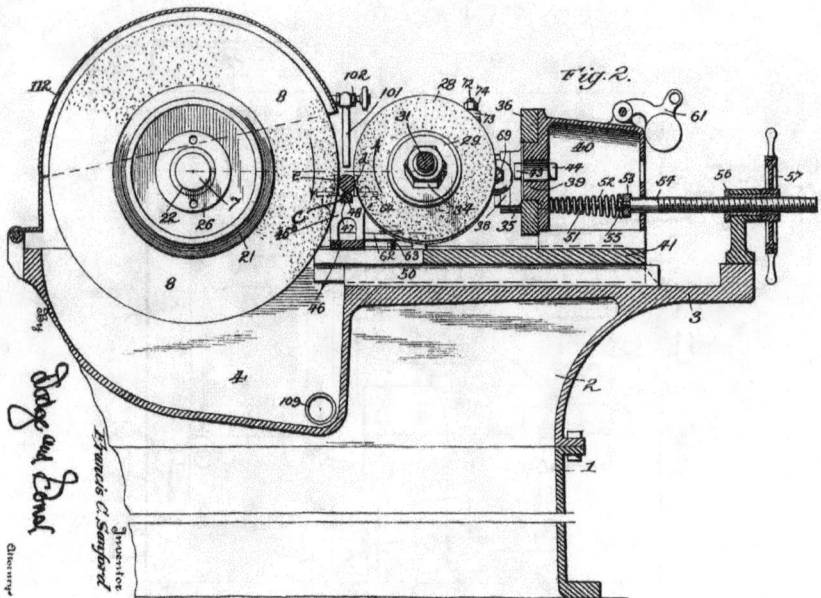

Fig. 7-10: F. C. Sanford Centerless Grinder. Patent 1,483,748, filed March 1, 1920.

the feed rate of the work past the grinding wheel, Sanford tilted the face of the governing and feeding wheel such that the wheel surface rotated both upward and forward. Other differences from Heim involved placement of the carrier below the wheel centerlines, the use of a smaller "governing" wheel, and the ability to vary the tilt angle of the governing wheel surface to control the rate of feed through the grinding zone. Whereas Heim used an adjustable carrier system *(position and tilt angle)*, Sanford used a flat and fixed carrier system that was not able to tilt. To accommodate various types and diameters of work, Sanford supplied multiple carriers that could be easily changed out.

Detroit Machine Tool Company

The third company to build and sell centerless grinders was the Detroit Machine Tool Company located at 6523–45 St. Antoine Street in Detroit, Michigan. Detroit first sold single-wheel centerless grinders based on the Grant design *(1,106,803)* and then developed its own version that was filed for a patent *(1,364,006)* on March 1, 1920. After the introduction of the two-wheel centerless grinders and the resultant decline in sales of their single-wheel machine, Detroit developed its own version of a two-wheel grinder.

On April 23, 1921, Edward Sobolewski submitted a patent application for a two-wheel centerless grinder in which the wheels were positioned vertically with the grinding wheel placed above the regulating wheel. Detroit first called the lower wheel a "work-rotating" wheel and later in the application as a "regulating" wheel that controls the rotation and "effects an axial feeding movement." In the Detroit arrangement, the regulating wheel also supports the work vertically and uses "work-holding members" located on both sides of the work to retain the work in position between the wheels. Figure 7-11 from the patent *(1,733,086)* illustrates the Detroit centerless grinder showing the grinding and regulating wheels positioned one above the other. Even though the machine looks different, the primary features of the Detroit machine are identical to the Heim centerless grinder.

Detroit began selling its two-wheel centerless grinder around the time it filed its first patent application. In an *American Machinist* article published in June 1921, Detroit described its machine as being capable of grinding diameters up to 3 inches at lengths up to 16 inches. Grinding cuts for its machine ranged from 0.0001 inches up to 0.015 inches per pass.[18] In 1923, Detroit introduced the Model 4B and by 1925, the 4C that provided improvements in construction, variable-speed motor drives, and work feed rates up to 89 feet per minute.[19]

Singer Manufacturing Company

The fourth centerless grinder sold to industry was invented by Walter J. Peets and patented by the Singer Manufacturing Company of Elizabeth, New Jersey. The Singer machine *(Figure 7-12)* was similar to Sanford in that it used the two-wheel peripheral configuration and placed its work carrier below the wheel centerlines. However, the Singer machine was the only design that did not use the natural drawing action of the wheels. Instead, Singer designed its machine with two sets of grinding and regulating wheels to increase production rates. In order to feed and remove the work, Singer employed a retractable governing *(regulating)* wheel. Work to be ground is dropped onto the carrier from above. When grinding is completed, the governing wheel retracts from the carrier, allowing the work to fall off into a collection bin.

The Norton Company

The fifth centerless grinder sold was patented by Charles Norton of the Norton Company in Worcester, Massachusetts. Norton filed his first application for patent on June 28, 1922, and filed four additional applications within the next seven months. As seen in Figure 7-13, Norton used a peripheral wheel configuration with the regulating wheel axis located well above both the work carrier and the grinding wheel axis. Norton developed his machine with a carrier called a "work supporting shoe" that used a concave surface with a curvature equal to the required radius of the finished work. According to Norton in patent 1,639,958, a specific feature of his concave carrier was that *". . . irregularities of outline are removed first and the grinding proceeds until the work just fits the curve of the shoe."* Further, Norton also claimed that:

> *I have also found that in a machine of this type it is of some advantage that the work be supported by such a steady rest below a line joining the centers of the two abrasive wheels, and that the wheels be so mounted as to permit them to be brought into contact for truing one another.*

Even though Norton used a concave carrier, the principles for centerless grinding

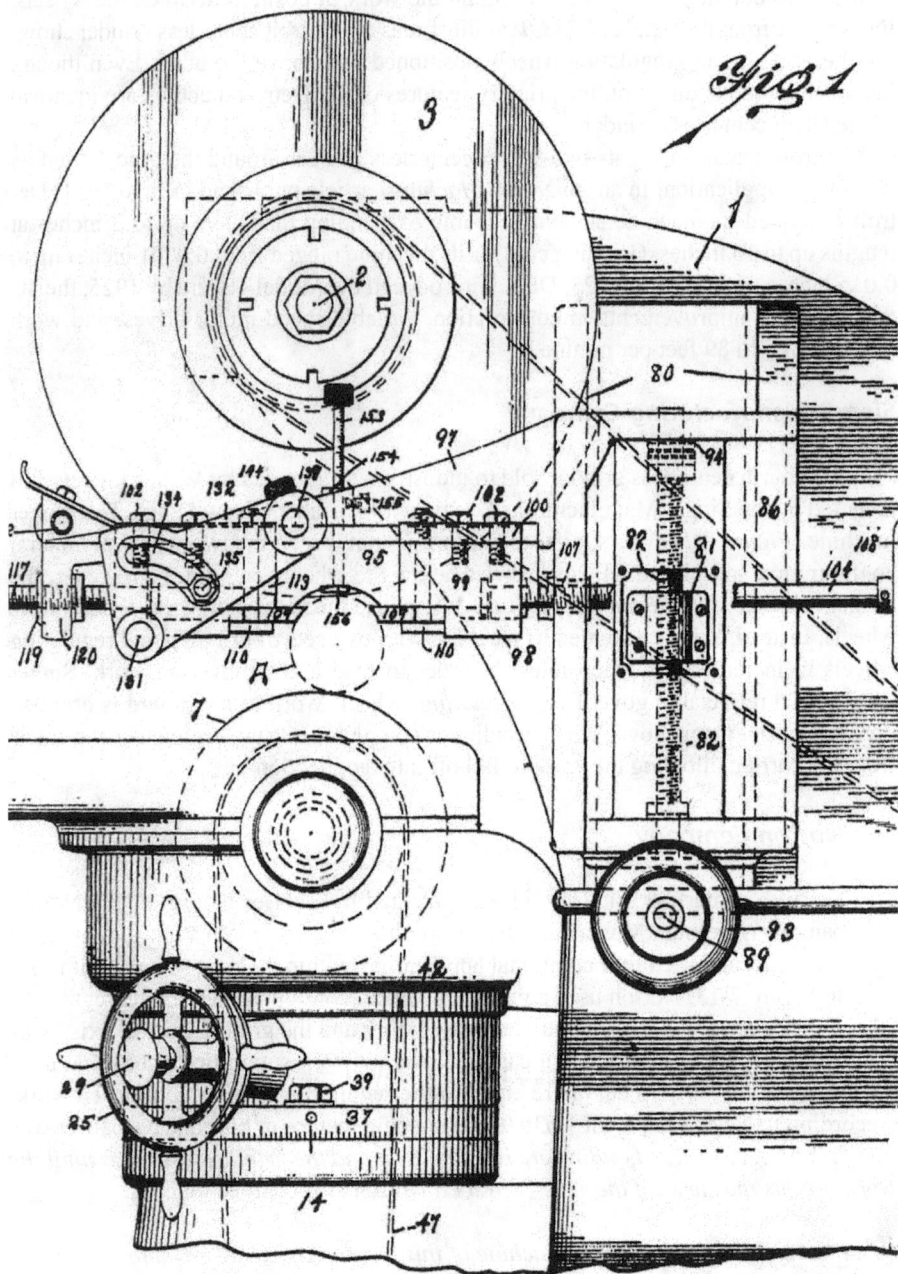

Fig. 7-11: Detroit Machine Tool Centerless Grinder. Patent 1,733,086, filed April 23, 1921.

were the same as those developed by Heim. The work is fed axially and drawn through the grinding zone using the same natural drawing action of the wheels. Norton, probably in an attempt to distinguish his machine from Heim, tilted the grinding wheel instead of the regulating wheel to promote drawing of the work.

The Cincinnati Milling Machine Company

The sixth manufacturer of early centerless grinders was the Cincinnati Milling Machine Company of Cincinnati, Ohio. Cincinnati Milling was owned by the Geier family, Frederick A. Geier and his son, Frederick V. Geier. Frederick A Geier was not a machinist or an engineer. His specialty was sales, marketing, and management. Geier got his start in machine tools when he bought an interest in the Cincinnati Screw and Tap Company in 1887. His partners were German immigrants Fred Holz and George Mueller, both engineers and machinists. In 1878, Holz developed an improved milling machine, with several new innovations resulting in high demand by other companies. Geier bought into Cincinnati Screw and Tap and later changed its name to the Cincinnati Milling Machine Company.

After the end of World War I, there was another economic recession followed by the Depression of 1920–1921. Geier's son, Frederick V., convinced his father to diversify into the manufacture of grinding machines. They saw the potential of grinding as a key manufacturing process for the automobile industry. In 1921, they bought the Cincinnati Grinder Company, which made center-type grinding machines. Apparently Geier knew Henry Ford personally, probably through his company's business relation-

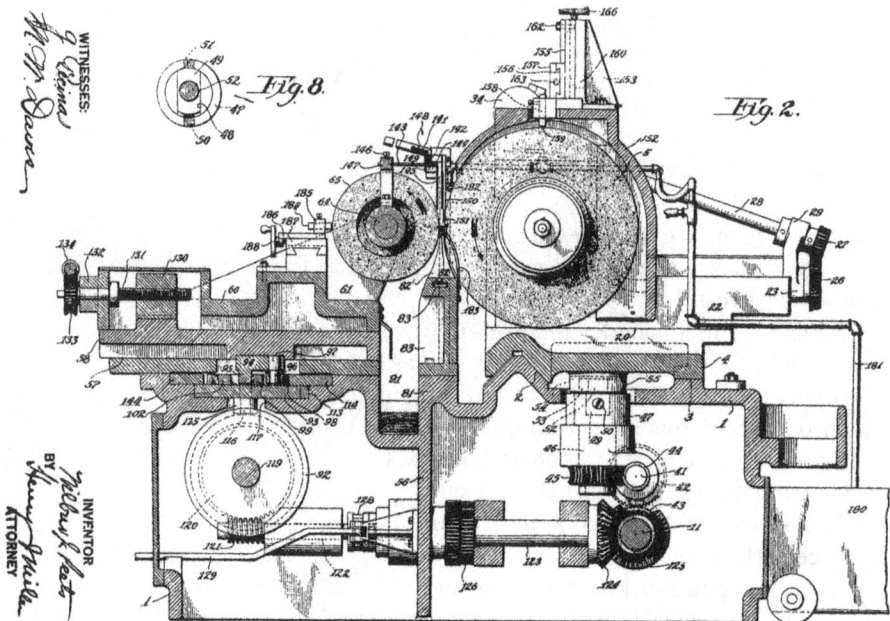

Fig. 7-12: Singer Manufacturing Co. Centerless Grinder. Patent 1,599,584, filed June 14, 1921.

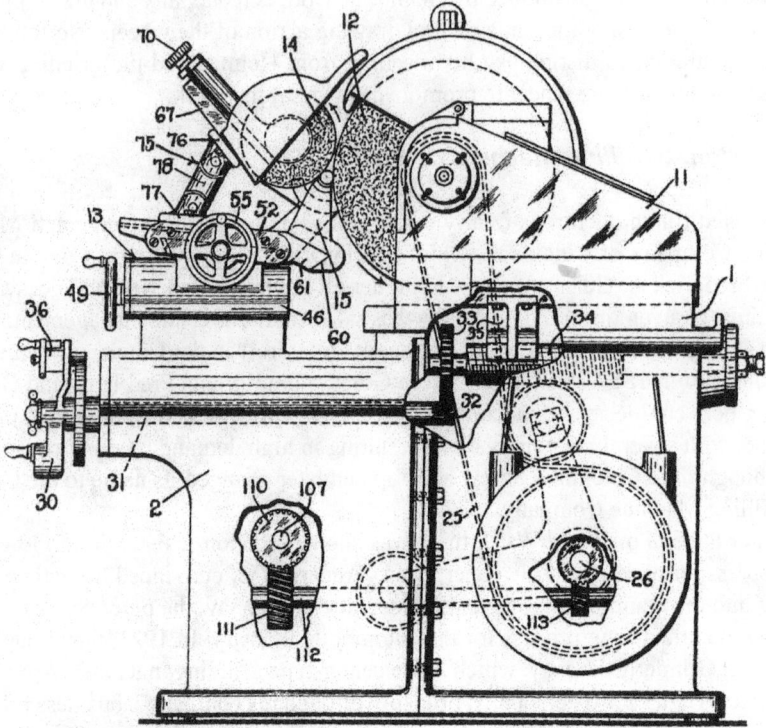

Witnesses
Harold W. Eston
Leah A. Sessions

Inventor
Charles H. Norton
By Clayton R. Jenks
Attorney

Fig. 7-13: Norton Company Centerless Grinder. Patent 1,639,958, filed June 28, 1922.

ship of selling milling machines to the Ford Motor Company, and Ford let Geier know that he was more interested in centerless grinding due to its ability to mass-produce parts at lower cost.[20]

Cincinnati first manufactured single-wheel centerless grinders, with many of them installed in automobile plants. By late 1921 Cincinnati was manufacturing the two-wheel machines.[21] Since Cincinnati Milling did not have its own design, Geier obtained a license from Frank Sanford that allowed Cincinnati Milling to immediately begin production. With the Sanford technology, Cincinnati Milling was able to deliver its first ten centerless grinders to Ford Motor by February 17, 1922.[22]

In the July 1922 issue of *American Machinist*, Cincinnati published an article on its centerless-type automatic grinding machine. The machine was able to grind straight work up to 3 inches in diameter and 15 inches in length. The Cincinnati machine had all the characteristics of the Sanford centerless grinder such as the smaller regulating wheel, multiple work rests, and placement below the wheel centers. Cin-

122

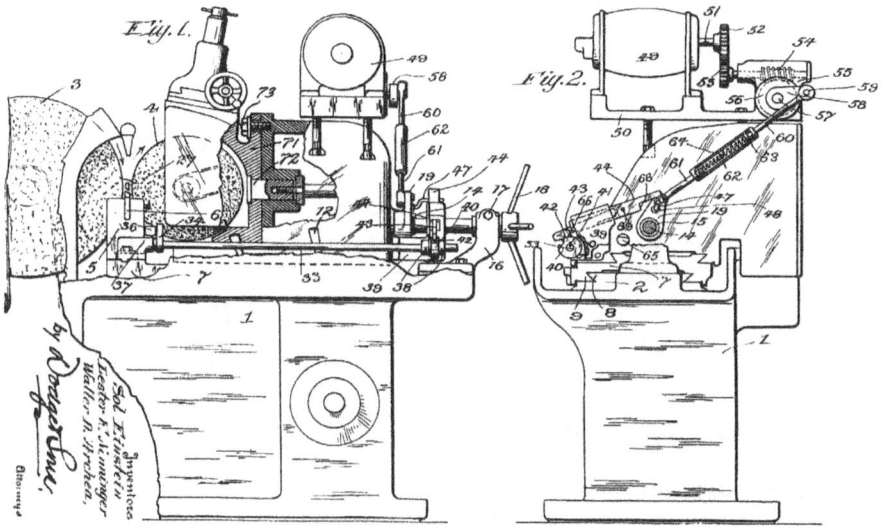

Fig. 7-14: Cincinnati Milling Machine Centerless Grinder. Patent 1,575,558, filed Jan. 18, 1924.

cinnati, as did Sanford, called its regulating wheel a "governing or control-wheel" but never mentioned that its machines were made under the Sanford license.

On January 18, 1924, Cincinnati Milling filed an application for a centerless grinding machine of its own design *(Figure 7-14)* developed by employees Sol Einstein, Lester Nenninger, and Walter Archea. According to the patent *(1,575,558)*, Cincinnati's:

> . . . *invention pertains to centerless grinding machines for grinding articles of circular cross section, particularly such as are provided with a head, a shoulder or shoulders, or both. The machine may also do taper grinding, and the mechanism to which the invention is more specifically directed, that is to say, mechanism for automatically ejecting the work when ground to the desired extent, is capable of use in the grinding of the several classes of objects indicated.*

The primary invention of Cincinnati's machine was the mechanism for automatically ejecting headed or shoulder work after grinding. The automatic ejecting device was brought into action only after the work had been ground and after the governing or control-wheel had been retracted a sufficient distance. This allowed the machine to mass-produce not only uniform and tapered cylindrical work but also cylindrical work with a head or shoulder. Cincinnati introduced its centerless grinder many years after Heim and the other manufacturers, and with its new design attempted to distinguish itself from the competition.

General Motors

The seventh centerless grinder was patented by William H. Chapman of the General

Motors Corporation in Detroit, Michigan, with the first patent application filed on February 12, 1926. There is nothing remarkable about the Chapman design as it embodies the key inventive features of Heim with a fast and downward-rotating grinding wheel and slower, upward-rotating regulating wheel. Chapman called the regulating wheel a "control wheel" and designed it to have a width greater than the grinding wheel *(Figure 7-15)*.

The patent claims that flats are prevented when the work first contacts the regulating wheel and is rotated before contacting the grinding wheel. Chapman designed the carrier with inclined plates that he calls "jaws" that properly guide the work toward and away from the wheels. The machine was designed to allow the regulating wheel to tilt in either direction so that work could be fed from either side. Also included were truing devices, with the one surfacing the regulating wheel designed to be parallel with the carrier, regardless of the tilt angle of the regulating wheel. The patent was issued six years after filing on June 21, 1932.

Lidkoping Mekaniska Verkstad AB, Sweden

The eighth manufacturer and the first foreign company to manufacture a centerless grinder was Lidkoping Mekaniska Verkstad *(LMV)* of Lidkoping, Sweden. LMV, established in 1874, was an old manufacturer of machine tools. When Sven Wingquist founded SKF *(Svenska Kullagerfabriken AB)* in 1907 to manufacture ball bearings, LMV was drawn into the manufacture of grinding machines. LMV employed an engineer by the name of Carl Gustav Ekholm who was sent to the

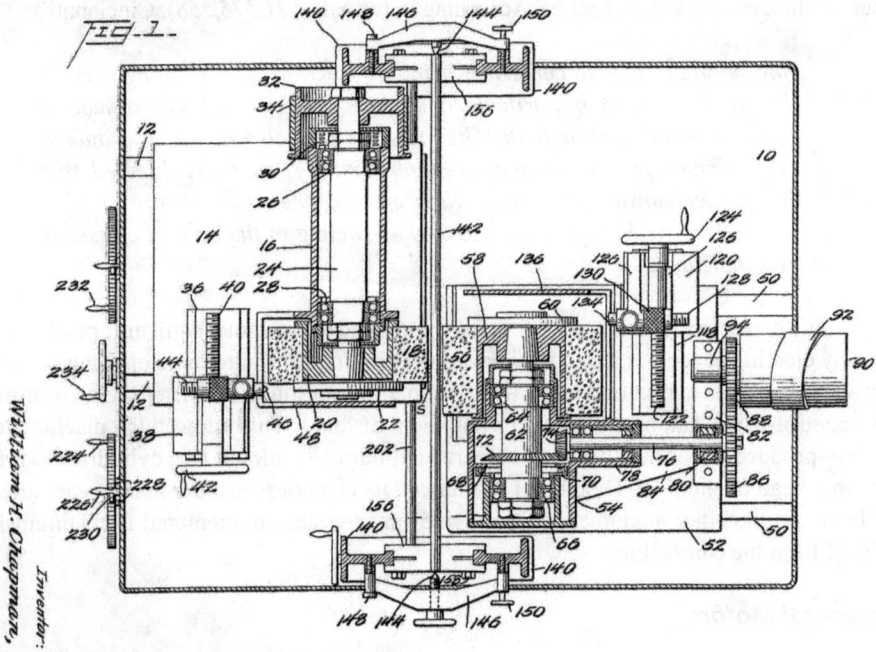

Fig. 7-15: General Motors Centerless Grinder. Patent 1,863,832, filed Feb. 12, 1926.

124

Strom Bearing Company in Chicago in 1921 to assist in the start up and operation of ball-making machines supplied by LMV. While on assignment, Ekholm was able to extend his stay for six months to study machine tools made by U.S. manufacturers, at which time he learned of the two-wheel centerless grinder and how it was gaining a foothold in the automobile industry. Ekholm made arrangements and visited both the Cincinnati Milling Machine Company and the Detroit Machine Tool Company to see their centerless grinders. Whereas Cincinnati was secretive about the design and operation of its machine, Detroit gave Ekholm a demonstration of the machine, probably with the hope of selling to LMV. Ekholm later visited Ford and Nash Motors and witnessed operation of Cincinnati centerless grinders. Following up his factory visits, Ekholm obtained catalogs for centerless grinders built by Heim and Sanford.

Ekholm returned to Sweden, and, using the acquired information, designed, built, and tested centerless grinders for LMV. By the end of 1922, LMV sold its first Ekholm-designed centerless grinder to Schmid-Roost, a maker of bearings located in Switzerland. At first, Ekholm's grinder performed poorly but after several design changes, his machine met the needs of manufacturers. By 1925, LMV was selling centerless grinders throughout Europe.[23] However, with all the machines sold, it was not until May of 1929 that Ekholm filed his first patent application for a "Machine for Grinding Bodies of Revolution" *(1,777,607)*. As seen in Figure 7-16, the Ekholm machine looked very much like other two-wheel centerless grinders.

The primary feature on which the patent was granted was the way the regulating wheel functioned. In the Ekholm machine, the regulating wheel is not driven at a slow speed; instead, it is allowed to freely rotate but purposely slowed or braked to control the rotation of the work. In the absence of a driven regulating wheel, frictional contact of the rotating grinding wheel causes the work to rotate, which in turn contacts and rotates the regulating wheel. Without braking action, the work and hence the regulating wheel would accelerate to speeds at which little to no grinding would occur. In operation, there is little difference between using a driven or braked regulating wheel, since both control rotation of the work against the grinding wheel and achieve similar grinding action as the original Heim two-wheel centerless grinder.

LMV was acquired by SKF in 1929 and incorporated into Karolin Machine Tools *(KMT)* of Germany in 2000. The name was later changed to KMT Precision Grinding AB and in 2009 founded a division in Boston, Massachusetts. KMT manufactures a line of through-feed and in-feed centerless grinders in addition to many other CNC grinding machines.

Impact of Competition

As the 1920s progressed, industrial companies increasingly recognized the value of Heim's two-wheel centerless grinding process as a must-have technology for the low-cost, mass production of small and precision-made cylindrical metal parts. Automobile and related companies started purchasing centerless grinders and trying them out. And some companies, such as the big automobile companies, bought centerless machines from more than one manufacturer or in the case of General Motors,

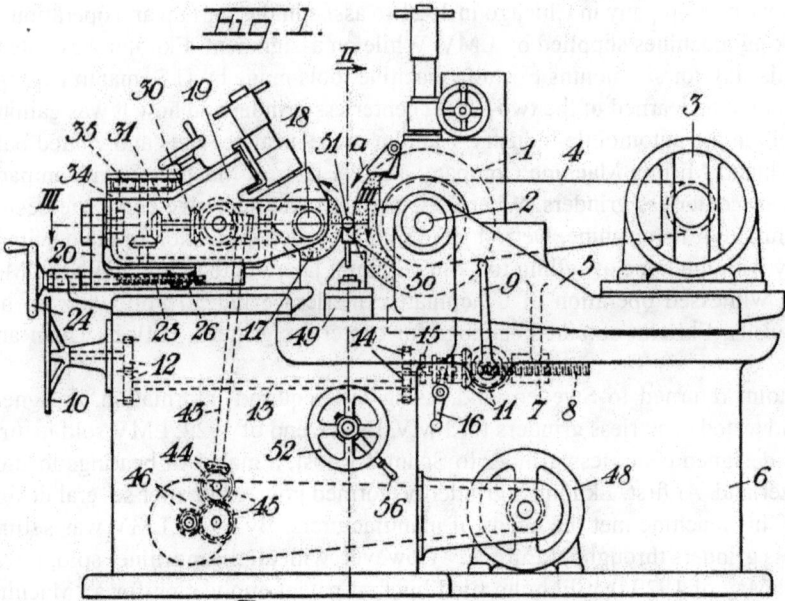

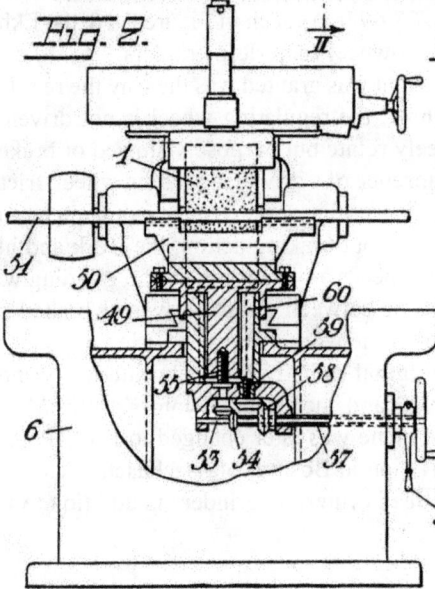

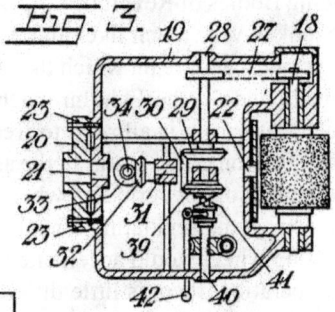

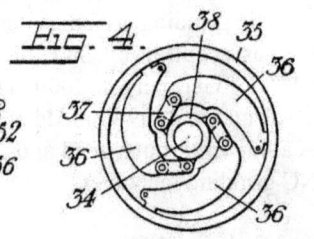

Fig. 7-16: Lidkoping Mekaniska Verkstad (LMV) Centerless Grinder. Patent 1,777,607, filed May 6, 1929.

built machines of its own design. As the number of manufacturers of two-wheel centerless grinders increased, competitive pressure significantly reduced sales by the smaller and lesser-known Ball and Roller Bearing Company.

Chapter 8

Ball and Roller Bearing Versus F. C. Sanford

Production Grinding

The period between 1918 up through the beginning of the Depression is known as the Vintage Era of the automobile that saw not only fundamental changes in automobile design but also rapid improvements in manufacturing technology. The internal combustion engine expanded in size and power with the introduction of multi-valve cylinders, overhead camshafts, and larger V-style 8, 12, and 16-cylinder engines. This era saw the development of many of the mechanical components seen in automobiles today such as hydraulic brakes, automatic transmissions, and torque converters. Competition among automobile companies became fierce, with the number of U.S. companies decreasing from more than 175 to around 70 by 1925. Many smaller companies could not stay competitive against larger companies and falling prices.

By the early 1920s, the automobile industry was the largest customer of the machine tool industry. Grinding operations on the parts to make a single car numbered in the hundreds. In order to produce interchangeable parts, production grinding was employed to meet the demand of manufacturing precision parts in large quantities. In 1922, Fred B. Jacobs of the University of Michigan College of Engineering, published a book titled *Production Grinding* in which he describes the grinding machines and processes used by established manufacturers. In the preface, Jacobs makes the following statement on the importance of grinding:

> *Production grinding is widely practiced to meet the demand for cutting and finishing repetition parts in large multiples. In the automobile, motor truck, agriculture and similar industries, production grinding has increased output and has effected economies never before attained by other machining methods. Large production and precision are not generally associated as being practically obtainable, but the cutting of metals by grinding makes this possible.*[1]

In *Production Grinding*, Jacobs provides detailed descriptions of the grinding machines and processes used by manufacturers of automobiles, farm equipment, rolling bearings, paper, and other metal products. Two-wheel centerless grinders were a recent innovation, and in 1921 were only beginning to be deployed in large manufacturing operations. The following are some examples of how centerless grinders were used at that time cited in Jacobs's book.

Fig. 8-1: Detroit Machine Tool Single-Wheel Centerless Grinder Operating at the Packard Motor Car Company in Detroit, Michigan: ca. 1921.

The Packard Motor Car Company of Detroit used single-wheel centerless grinders made by the Detroit Machine Tool Company *(Fig 8-1)* to make small cylindrical parts, rear-axle torque arms, and gear shafts. In 1921, the Ford Motor Company employed single-wheel centerless grinders from the Cincinnati Milling Machine Company and two-wheel centerless grinders from F. C. Sanford to make rolls used in roller bearings. Ford stated that the purpose of centerless grinding was to eliminate the time consumed in setting up the work and where a large number of similar cylindrical parts were ground, centerless grinding was both productive and economical.

Another company using centerless grinders in 1921 was the Cleveland Tractor Company, which developed many of its own production processes to ensure the exact duplication of all parts. During WWI, the company developed a highly efficient endless-track tractor equipped with internal combustion engines for both farm and industrial use. Cleveland acknowledged that grinding played an important part because of the accuracy obtained and the subsequent reduction in the cost of production compared with other machining methods.

Cleveland used Sanford centerless grinders for finishing straight cylindrical work, including piston pins made of case-hardened steel. A description of its centerless grinder mentions that the "feed wheel" *(regulating wheel)* is set at a slight angle that causes it to *"pull the work along with the operator required only to feed the work to the machine."* It goes on to say that when making piston pins, five passes were made to reduce the pin to the desired size. At six seconds per pass, it took 30 seconds to finish grind each pin and was far quicker than previous grinding methods.

The Sanford Centerless Grinder

On March 2, 1918, William Weed, who managed the roll-grinding department at the Ball and Roller Bearing Company, resigned and returned to his hometown of Bridgeport, Connecticut. Two days later *(March 4)*, Weed was hired as a foreman at the F. C. Sanford Manufacturing Company in Bridgeport, Connecticut.

Weed and Frank Sanford were longtime friends, having known each other for more than twenty years, since they were teenagers. Sanford also knew Heim, as they had cottages next to each other on the beach in Fairfield, a town just west of Bridgeport. Being neighbors close in age *(40s)* and in similar businesses, they probably had many discussions on their businesses, machines, and manufacturing.[2] It was highly likely Sanford knew quite a bit about the Ball and Roller Bearing Company, and its products and machining processes, including Heim's centerless grinder.

Sometime in 1918, around the same time he began selling centerless grinders, Heim was informed that the F. C. Sanford Manufacturing Company was building a centerless grinder using his patented concepts.[3] Since Heim knew that Sanford had just hired William Weed, he arranged a meeting with Sanford to discuss the issue. As a witness to the meeting, Heim brought his cousin, Charles Waters, who also worked at the B&RB. The intent of the meeting was to advise Sanford that he owned patents for centerless grinding and that he would protect his interests against infringement. Sanford denied that he was building or planning to build a machine using Heim's invention and alluded to possibly using the Detroit single-wheel centerless grinder. Sanford's account of the meeting was somewhat different and more colorful. When asked about the meeting, Sanford responded with the following:

> *Well, I remember his call, yes, very well. He called at my office one day with a gentleman, I do not just recall what his name was; he came into my office and he said, "I understand, Sanford, that you are putting a centerless machine on the market." I said, "No, I am not." He said, "I hear you intend to." I said, "Well, I am not sure; I cannot answer that question." He said, "Well, there is one thing I want you to understand—" this is the language he used—"there is one thing I want you to understand, and for your own information, we have been good neighbors and good friends; I do not want to have any trouble with you. I am the sole inventor of the centerless method and I hold controlling in all of the centerless patents of any use. The method is mine absolutely. $10,000.00 or $40,000.00 does not go far in a lawsuit, and I will spend that or more." I said, "Mr. Heim, if I decide to put a centerless machine on the market it will be after the United States Government gives me*

the right to. If they give me that right, that patent right to do that, I feel that I have then just as good a right to put a machine on the market as anyone else." I think during the conversation he asked me if I had Mr. Weed working for me, and I said, "Yes," and he said "Be very careful." I think that ended the conversation.[4]

There is an interesting aspect of Sanford's statement on patents. Sanford seemed to believe that obtaining a patent was an automatic right to ownership of an invention and would eventually learn that was not always true.

One of the purposes of a patent, other than to disclose and protect the inventive features, is to describe the invention in such a way as to reduce the ability of others to obtain similar patents. In other words, a good patent attorney will claim a broad range of features in such a way that prevents others from infringing the patent. Some patent attorneys refer to this intent as patenting the "forest" instead of a "few trees." However, if a patent is written too broadly and lacks specificity, it can open the door to similar inventions. In the case of the Heim machine patent *(1,210,937)*, Sanford, Reeves, and others either tried to exploit the broadness of the patent when developing their centerless grinders or believed that a patent issued by the government gave them protection. Heim initially patented a double-ring wheel configuration using the radial surfaces of wheels rotating in opposite directions and equipped with an inclined carrier positioned above the wheel center. In an attempt to circumvent Heim's patents, competing machines were developed and patented using different configurations of the wheels and carrier. Whereas Reeves patented three configurations using combinations of ring and peripheral arrangements, the others patented various peripheral wheel arrangements.

Shortly after Weed started at Sanford Manufacturing, Sanford began work designing his first centerless grinder and on April 12, 1918, produced a drawing *(blueprint)* containing the dimensions and details of his first machine, designated the "Model A." The "A" machine *(Figure 8-2)*, first produced in May 1918, was somewhat light in construction for a precision grinder and incorporated many of the features found in the Heim grinder, including a high-speed grinding wheel, a low-speed regulating wheel, an adjustable carrier, and automatic feeding of the work. Sanford called the regulating wheel a "governing or feed" wheel whose function was to control rotation and feed the work through the grinding zone.

During the summer of 1918, Sanford received an order to make 40,000 bullet ejectors for the U.S. military. Bullet ejectors are small round solid cylinders similar to bearing rolls. The order had a quick delivery time and Sanford accepted the order with the intention of manufacturing them on his "Automatics," the name he gave his Model A centerless grinders built during the summer. Apparently, the Automatics were not capable of achieving required grinding tolerances, forcing Sanford to grind the bullet ejectors on centered machines. Since grinding on centers required much more time, Sanford realized he could not meet his contracted delivery date. With the deadline looming, Sanford, on September 16, placed an order with the Ball and Roller Bearing Company to grind 800 bullet ejectors. After receiving the first order from the B&RB, Sanford placed two additional orders for thousands more. Sanford would later claim that many of the ejectors ground by the B&RB were defective, forcing Sanford

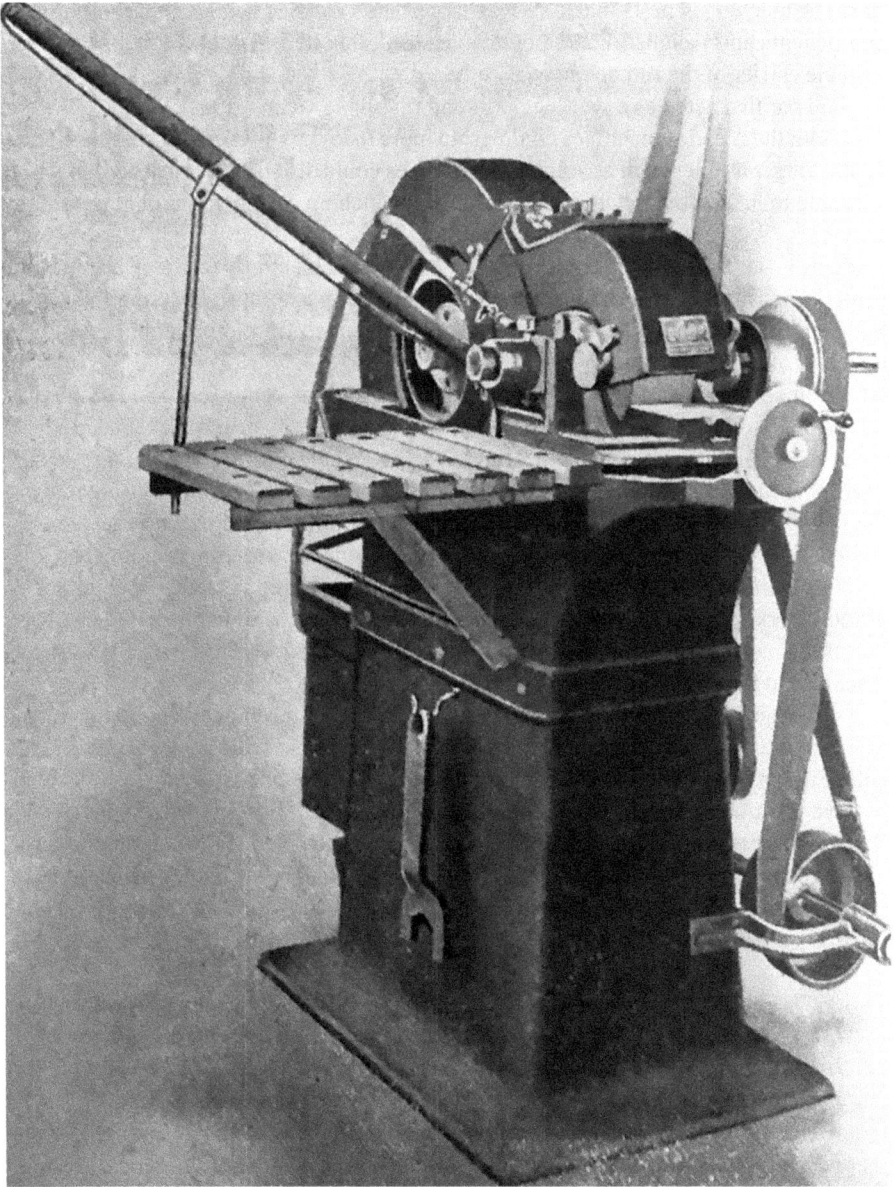

8-2: Model A Sanford Centerless from the Product Bulletin.

to complete his order using his centered grinders. But it was the inability of his A machine that forced Sanford to purchase ejectors from Heim and return to the design board to build a grinder capable of better precision.

The second centerless grinder built by Sanford was designed in late 1918 and called "Model B." In this machine, Sanford made changes that corrected inadequacies of the A machine. The B machine was larger and designed with heavier, more rigid components and weighed 2,200 pounds versus 950 pounds for the Model A. The adjustable carrier of the A machine was replaced with a fixed and rigid carrier fitted with a round bar that acted as a work support and revolving wear strip.

Both the A and B machines could grind more than the rolls used in bearings. Larger and longer articles such as solid bars and pipes could also be ground. The B machine was able to handle diameters ranging from 1/10 inch up to 6 inches and lengths up to 20 inches. With extensions to the carrier, Sanford claimed that lengths up to 16 feet could be ground. The production rate of grinding was essentially the same as the Heim centerless grinder, with rates in the range of 5 feet per minute. Information on the Sanford Model A and B centerless grinders was sourced from product bulletins, with a picture of the B machine shown in Figures 8-3 and 8-4.

From February 1919 to early 1920, Sanford sold his centerless grinders directly to customers without the use of a sales agent or middleman. Initially, the Model A sold for $1,650, and once available, the Model B model sold for $2,500. Sanford sold his first B model to the Precision Machine Company of New York City sometime in the spring of 1919. The machine contained springs on the support slides that were installed to prevent damage from oversized work passing between the grinding and governing *(regulating)* wheels. The springs were weak, resulting in chattering and poorly ground work. After removal of the springs plus a few other adjustments, virtually all the problems disappeared, allowing the Precision Machine grinder and subsequent machines to perform well.[5]

On page 219 of the 1919 Bridgeport Directory, Sanford advertised his company as a manufacturer of special machinery and roller bearings. With the B machine available, Sanford had two machines selling at different price points to increase his competitive position. Sanford also undercut Heim on a sale of bearing rolls to the Hendee Manufacturing Company of Springfield, Massachusetts. Hendee Manufacturing was the original name of the Indian Motorcycle Company before the name change in 1928. Heim, in 1921, recalled the impact of the Sanford machine on sales of rolls by the B&RB:

> We have always been manufacturing roller bearings to a greater extent than ball bearings, and we not only furnish a much larger percentage of roller bearings, I should judge about 75 percent, but in addition to that we furnish our customers with cylindrical rollers for various purposes, and when Mr. Sanford's machine was placed on the market, and I understand he had placed quite a good many, it naturally diverted some of that business from us. I have been informed that one customer—for instance, I would like to refer to, who had furnished us considerable business, and who the Sanford Company took away from us, and that is the Hendee Manufacturing Company, through a cut price proposition. Of course, there are one or two other roller bearing companies that have started up since, and who have adopted this machine, and who have batteries of those machines in their plant, and I guess they have got some of our work all right. We have not got very much just now.[6]

Sanford

PRECISION CENTERLESS CYLINDRICAL GRINDER
MODEL B

MANUFACTURED BY
THE F. C. SANFORD MANUFACTURING COMPANY
BRIDGEPORT, CONN.

RUSSELL, HOLBROOK & HENDERSON, INC.
30 CHURCH STREET   NEW YORK
SOLE SELLING AGENTS

Fig. 8-3: Model B Sanford Centerless Grinder (Rear View) from the Product Bulletin.

In January 1920, Sanford hired the services of Russell, Holbrook and Henderson *(RHH)* located in New York City as sole selling agents of his centerless grinder. Russell, Holbrook and Henderson in turn, employed Arthur Kirkland to be the exclusive salesman for the Sanford centerless grinder. Kirkland formerly worked at Detroit Machine Tool selling its "single-wheel" centerless grinder that was similar to the machine patented by Robert Grant *(1,106,803)*. During his employment *(April 1, 1919 through*

January 31, 1920), Detroit sold 50 to 60 single-wheel centerless grinders priced around $1,000 each. However, customers in general were not satisfied with the performance and dependability of the Detroit grinder and only one customer placed a repeat order.[7] Even with its reliability problems, there was a market for the single-wheel centerless grinder since it was still more economical than centered grinding. After the two-wheel centerless grinders became available, sales withered and the Detroit single-wheel machine eventually became obsolete.

While working at Detroit Machine Tool, Kirkland called on the Hyatt Roller Bearing Company in Orange, New Jersey. At Hyatt, Kirkland witnessed operation of the Sanford centerless grinder and became convinced that the two-wheel machine was superior to the Detroit single-wheel machine. Shortly thereafter on February 1, 1920, Kirkland joined Russell, Holbrook and Henderson *(RHH)* to sell the Sanford

Fig. 8-4: Model B Sanford Centerless Grinder (Front View) from the Product Bulletin.

(Letterhead of)

A. M. WOOSTER

BRIDGEPORT, CONN. February 24, 1920.

F. C. Sanford Mfg. Co., Inc.,
  2060 Fairfield Avenue,
  Bridgeport, Conn.

Gentlemen:

This is to call your attention to patents to

French and
Stephenson, No. 1,111,254, Sept. 22, 1914
Heim   1,210,936, Jan. 2, 1917
Heim   1,210,937, Jan. 2, 1917
Heim   1,264,930, May   7, 1918
Heim   1,281,366, Oct. 15, 1918

all of which are controlled by The Ball and Roller
Bearing Company of Danbury, Connecticut. I am
informed that you are placing on the market a
centerless grinding machine which is apparently
an infringement of claims in each of these patents.
I am requested by Mr. L. R. Heim, President of The
Ball and Roller Bearing Company, to formally no-
tify you of this infringement and request that you
desist at once from any act in infringement of these
patents or either of them, whether the infringing
acts consist in making, selling, advertising, or using
infringing machines, and report to me the number
of infringing machines you have made, have sold,
to whom they have been sold, and how many machines you have in process of construction.

This matter should receive your prompt attention. Unless the matter is satisfactorily adjusted,
suit for infringement will be brought against you.

Very truly yours,

A. M. WOOSTER

Fig. 8-5: Wooster Letter: B&RB Notice of Infringement to Sanford: Feb. 24, 1920.

machine and was based in Detroit, Michigan. During his eleven months at RHH,
Kirkland sold 74 Sanford centerless grinders split between 24 A-type and 50 B-type.
A typical advertisement for Sanford centerless grinders listing RHH as the selling
agents can be seen in Appendix 2 *(A2-7-1)*.

By February 1920 Heim had enough evidence that Sanford was not only infringing on his centerless grinder patents but was taking away a significant amount of business from the B&RB. For on February 24, the Ball and Roller Bearing Company,
through Bridgeport lawyer A. M. Wooster, sent a Notice of Infringement to Sanford
Manufacturing, citing five centerless grinder patents under which Heim protected

his inventions. Wooster, who was the patent lawyer that wrote the original centerless grinder patents for Heim, demanded that Sanford cease manufacture and sales of centerless grinders and turn over a list of companies using his machines *(Figure 8-5)*.

Sanford turned the Wooster letter over to his legal counsel, F. W. Smith of Bridgeport and William Dodge of Washington, DC, who conducted an investigation and assessment of the Heim patents against the Sanford centerless grinder. Upon completion of their investigation, Sanford's patent attorneys notified the B&RB that they believed there was not only no infringement, but the Heim patents were invalid due to prior art. At the time of their response to Heim, Sanford's attorneys determined that the Sanford machine could be viewed as infringing the "structure" of the peripheral centerless grinders developed by Reeves and Heim. Given that Reeves's patent application *(1,264,129)* was filed earlier than Heim's application *(1,264,930)*, Sanford decided to take a license with Reeves, which was subsequently obtained in July 1920.

Outside of the notice sent to Sanford, Heim included warnings about infringement in his advertising. These warnings were aimed not only at the infringing manufacturers but also at buyers of two-wheel centerless grinding machines. A typical warning used by Heim taken from Bulletin 110 *(see Appendix 2, A2-6-1)*, published in late 1920, is shown below:

The machine is manufactured under the following U.S. Letters Patents:

No. 1,111,254	*No. 1,264,930*
No. 1,210,936	*No. 1,278,463*
No. 1,210,937	*No. 1,281,366*

There are several manufacturers, who as we understand, have started to place machines on the market embodying many of the patented features of this machine. We give warning that we intend to protect our rights to the limit, whether the infringement is by reasons of manufacture, or use of such machines.

After receiving Heim's infringement letter, Sanford continued to sell grinders using his newly hired sales representative, Russell, Holbrook and Henderson. In April 1920, Kirkland received a request from a prospective customer for the patent number of the Sanford machine. Most customers knew that purchasing machines that infringed another patent potentially exposed them to litigation and financial penalties. As a result, some companies did their own investigation to ensure that the manufacturer rightfully owned the patents to new technologies.

Kirkland notified Sanford, who provided a patent number for submittal to the customer. Upon receiving the number, the customer traveled to Washington, D.C. to confirm patent ownership at the U.S. Patent and Trademark Office. The customer found that the patent number provided had no connection to a grinder but described some type of lock nut device. When notified of the customer's finding, Sanford responded that the number provided was not the patent number but the serial number assigned to the patent application. Apparently, the answer satisfied the customer as the issue was dropped. Sanford gambled on his response to the customer, but had the customer pursued the matter further, he would have found that no application had been filed. Sanford at that time was still writing his application that was submitted three months later on July 30.

Sanford's response to Heim's Notice of Infringement confirmed that Sanford had no intention of stopping sales of centerless grinders. With the warning ignored, Heim backed up his promise to sue Sanford and hired the patent law firm of Emery, Varney, Blair & Hoguet of New York City. Heim's new attorney, Robert Blair, filed the original Complaint on July 21, 1920, and cited infringement of Heim patent 1,210,936 for the roll feed system and patent 1,264,930 for the peripheral centerless grinding machine *(see Appendix 3, A3-1)*.

The third patent, No. 1,210,937, which was the machine *(device)* patent for the double-ring wheel centerless grinder, was purposely omitted. Apparently, Blair found errors in 1,210,937, which was originally written by A. M. Wooster of Bridgeport, and needed time to rewrite it and file for a Reissue patent. Reissue of a patent is generally made to correct errors such as a defective specification or claims that are too narrow or too broad. In his submittal for a reissue, Blair reduced the total number of claims from 25 to 21 and revised the wording of the claims to clarify the inventive features of the patent. The application was filed on September 8 and granted on January 25, 1921, as Reissue Number 15,035. Blair then filed a Supplemental Bill of Complaint on March 3, 1921, that added Reissue Patent 15,035 to the list of patents infringed.

Undaunted by the pending litigation, Sanford took a defiant position against Heim's accusation of patent infringement. From Sanford's point of view, he had a machine that was different from Heim's, which was selling very well and making sizeable profits. Sanford felt he had every right to make and sell his centerless grinders and in early 1921, with confidence on his side, conducted an extensive advertising campaign to promote his machines. Many of Sanford's advertisements were in two popular trade magazines, *American Machinist* and *Machinery*. In these advertisements, which can be seen in Appendix 4, Sanford cited the features, accuracy, and production capabilities of his machines and in one advertisement, included a list of customers and their favorable comments *(A4-3-4)*. In the midst of the Sanford advertising campaign and two months after filing of the supplemental complaint, on May 16, 1921, the patent infringement trial between the Ball and Roller Bearing Company and the F. C. Sanford Manufacturing Company began in the District Court of Connecticut in South Norwalk with Judge Edwin S. Thomas presiding.

Trial One: 1921

Both Heim and Sanford hired legal counsel outside Connecticut that presumably were some of the best firms in the business of patent law. With Heim's firm from New York City, Sanford hired William W. Dodge of Washington, D.C. as well as retaining his Bridgeport lawyer, F. W. Smith.

Blair prepared the Complaint identifying the defendant, the patents infringed, harm caused, and compensation demanded consisting of triple the value of damages sustained.[8] Blair also included allegations that infringement by Sanford induced others to infringe *(i.e., Reeves Pulley, Detroit Machine Tool, and Cincinnati Milling)* and that

Sanford conspired with others to infringe Heim's patents.

One of the first tasks conducted by Blair was to identify other companies supporting Sanford's defense. To obtain this information, Blair included questions regarding third-party financial and legal support in the plaintiff's interrogatories to the defendant. Sanford's answer identified Reeves Pulley Company as the only third party involved and relegated only to providing "Patent Counsel" should Sanford request it with no direct financial support. Given that Reeves had skin in the game with the Sanford license and his own business selling centerless grinders, plus much more experience with patents, it is possible that Milton Reeves arranged for Sanford to be defended by the Washington, D.C. law firm of William Dodge.

Dodge and Smith prepared the Answer to the Complaint for the Defendant, citing the following three defenses:

1. Non-infringement
2. Non-patentability as to the Reissue Patent 15,035
3. Anticipation by prior patents *(prior art)*

With the first defense, Non-infringement, the defendant cited differences in the machine design and operation and even though it performed similarly to Heim's, the machine achieved its performance using different and unique features. Differences in wheel rotation, carrier placement, and inclination were the primary proofs cited as Non- infringement.

For the second defense, Non-patentability, the defendant claimed that the Reissue patent was invalid due to introduction of new material, broadening of the claims, laches *(unreasonable delay in the changes)*, and intervening rights *(causing another party, Sanford, to infringe due to Heim's patent changes)*.

For the third defense, Anticipation, the defendant claimed that the principles used by Heim in his centerless grinder were old in the art and practiced by many before him. The defendant in its Answer, listed more than 50 U.S. patents and 10 foreign patents issued prior to Heim's for similar inventions. In addition, the defendant cited grinding machines used commercially prior to Heim that allegedly incorporated Heim's patented features.

Plaintiff's Case:

In presenting the plaintiff's case for infringement, Blair and assistant Haynes had to address all three defenses. The plaintiff's challenge was to prove that despite the differences, the Sanford machine incorporated the Heim patented features. For the second defense, the plaintiff had to show that the Reissue patent did not violate any of the requirements associated with revisions, such as new material or broadening the claims. And in the third defense, prior art, the defendant inundated the court with patents, articles, demonstrations of earlier machines, and testimony, putting the plaintiff on the defensive to prove that none of the prior art embodied Heim's centerless grinder invention.

Blair set the stage for his case with testimony of an expert witness by the name of Dr. Harold Pender. Pender had a long list of accomplishments in academia and business that qualified him to testify on centerless grinding. Pender received his BA and PhD at Johns Hopkins University, worked at Westinghouse Electric and other companies, was made a professor of theoretical and applied electricity at Massachusetts Institute of Technology, and at the time of the trial was professor-in-charge of the electrical engineering department of the University of Pennsylvania. Outside of his work in electrical engineering, Pender had an extensive background in physics and mechanics, and he wrote the chapter on the "Principles of Mechanics" for *The Handbook for Electrical Engineers*. Pender's experience with patents included applications recently filed for inventions on mechanical drives.

Blair called on Dr. Pender to testify on the following aspects of the plaintiff's case:

1. Explain the importance of precision grinding in the manufacture of rolls used in roller bearings.
2. Describe the operation of the Heim Centerless Grinder and its inventive features.
3. Show that the Sanford Centerless Grinder, despite its design differences, incorporated Heim's inventive features and thus infringed the Heim patents.
4. Demonstrate and prove to the court that machines cited as prior art were distinctly different, did not incorporate Heim's unique combination of inventive features, and could not mass-produce precision-ground rolls equivalent to Heim's.

Pender opened up his testimony describing the basis for and factors influencing Heim's invention for centerless grinding. Pender described the various types of roller bearings, why it was necessary to make bearing rolls out of hardened steel, and why it was imperative to grind rolls to a high degree of precision. Pender stated that rolls out of round or tapered by as much as a half-thousandths inch *(0.0005)* created friction, leading to early bearing failure.

After outlining the requirements for grinding bearing rolls, Pender reviewed the machines and methods used to grind rolls prior to Heim's invention. Pender described how rolls were ground on "centered" grinders such as a lathe or with chucks and how the process, even though satisfactory in precision, was slow, tedious, and costly. After describing centered grinding, Pender made some brief comments on the Detroit single-wheel centerless grinder in which he stated that the single-wheel machine was not capable of achieving high levels of precision, as did centered grinding or the Heim machine.

After completing his discussion of centered and single-wheel machines, Pender turned to the patent in suit with a statement that the object of Heim's centerless invention was to grind hardened rolls with speed, accuracy, and reliability. Pender then launched into a detailed description of the machine components and operation using the illustrations contained in the "machine" patent, Reissue 15,035, that was called Patent Z at the trial.

Pender used Figures 4, 5, 6, and 7 from the patent *(see Chapter 6, Figure 6-9)* to explain how the rolls moved along a carrier in between two wheels and how the fast,

downward-rotating wheel performed grinding and the slower, upward-rotating regulating wheel controlled both rotation of the work against the grinding wheel and the
rate of feed along the carrier. Pender explained the purpose of the carrier, its position
and inclination relative to the wheels, and how it was adapted to support many rolls for
simultaneous grinding. Pender's explanations covered the entire machine and its operation, but with emphasis on the principles described in the first two claims, as follows:

> *Claim 1: A machine of the character described comprising a grinding wheel
> adapted to be rotated at a relatively high speed, a regulating wheel adapted
> to be rotated in the opposite direction at a relatively low speed, a carrier for
> rolls between said wheels adapted to support a plurality of rolls for simulta-
> neous action thereon, means adapted to adjust relatively the paths of rolls on
> said carrier and of the operative surface of said regulating wheel and thereby
> adjust the rate of feed of rolls along said carrier with a given rate of drive of
> the machine, means adapted to guide the rolls up to the operating surfaces
> of the wheels, and means adapted to receive the rolls as they pass out of the
> grip of said surfaces.*

> *Claim 2: A machine of the character described comprising a grinding wheel
> adapted to be rotated at a relatively high speed, a regulating wheel adapted
> to be rotated in the opposite direction at a relatively low speed, a carrier for
> rolls between said wheels adapted to support a plurality of rolls for simul-
> taneous action thereon, means permitting relative angular adjustment in an
> upright plane of said carrier and the path of travel of the active portion of
> said regulating wheel whereby the rate of feed of rolls along said carrier is
> varied with a given rate of drive of the machine, means adapted to guide the
> rolls up to the operating surfaces of the wheels, and means adapted to receive
> the rolls as they pass out of the grip of said surfaces.*

After his initial presentation Pender asked Judge Thomas if he understood the
principles behind two-wheel centerless grinding, to which Judge Thomas responded:

> *I take it the action of one wheel going faster in one direction draws it through
> and the other one draws it back; that the one drawing it through goes faster
> than the other, and it naturally draws it through.*

Thomas apparently focused only on wheel speeds and did not understand the relation of the carrier position to the feeding action. The judge's failure to grasp the
operational concepts forced Pender to dedicate considerable time and effort to ensure
the judge fully understood the principles of centerless grinding as well as the function
of the major components of the Heim and Sanford machines.

The explanations continued into the next day *(Tuesday, May 17, 1921)* with live
demonstrations of the Heim and Sanford machines at the Norwalk Iron and Foundry
Company. By the end of the trial, with all the testimony, exhibits, demonstrations, and
arguments, as seen in the two opinions *(decisions)*, Judge Thomas thoroughly under-

stood how the machines were built and operated.

As previously discussed, if a patent if written too narrowly or too broadly and/ or lacks specificity, it can open the door to others to patent similar inventions. Heim, in his original patent *(1,210,937)*, described only the double-ring wheel configuration using the radial surface *(sides)* of the wheels with an inclined carrier positioned between the two wheels and above the wheel centers. The original patent appeared to be somewhat narrow since it was only written around a double-ring wheel arrangement in which the wheels rotated in opposite directions. There was no mention of alternative configurations such as the peripheral wheel design or component arrangements, such as locating the carrier below the wheel centers. On the broad side, the patent never described the rotational speeds of the wheels, using only terms such as rotating relatively fast and relatively slow.

For the Heim patent describing the machine and process *(1,210,937)*, Sanford, Reeves, and others attempted to exploit holes in the patent specification when developing and patenting their respective centerless grinders. Milton Reeves patented three configurations using combinations of ring wheel and peripheral wheel surfaces and a flat carrier. Sanford developed and later patented a machine using wheel peripheries mated to a flat carrier fixed in place and mounted below the wheel axes. Detroit Machine Tool patented a machine in which the wheels were mounted in the vertical position without the use of an independent carrier.

Heim's specification around the ring wheel arrangement with an inclined carrier became the focal point of Sanford's first claim of Non-infringement. Sanford claimed that his machines *(Models A and B)* were unique in that the wheels rotated in the "same direction" and that the machine did not require an inclined carrier to make the rolls advance between the wheels. People mechanically inclined would readily notice, wheel rotation and carrier position aside, that it was the relationship of surface contact between the wheels and work and the resulting action that made the invention. However, the judge, based on his comments, clearly had difficulty understanding the concepts of centerless grinding and was easily confused by the differences in the two grinding machines. As will be shown, Dodge succeeded in sowing a great deal of confusion around the direction of rotation of the wheels and the differences between Heim's ring wheel machine and the peripheral wheel machine of Sanford. This confusion can be seen in the opinions of Judge Thomas.

After Pender's initial testimony, Blair obtained testimony from Heim and other employees of the Ball and Roller Bearing Company that supported the case for infringement. Heim provided details of the events and timing surrounding conception of his one and two-wheel centerless grinders, as well as documents and other evidence to support his statements. He told of the hiring of William Weed and Weed's involvement with building Heim's first centerless grinder and the critical knowledge Weed had when he left the B&RB to work for Sanford. Heim even referred to Sanford's centerless grinder as "Mr. Weed's machine." Heim also testified about his meeting with Sanford to warn against infringement and the subsequent loss of customers and income as a result of sales by Sanford of his "A" and "B" machines.

Blair called many other witnesses to support the testimony of Pender and Heim. John Henry Roth and James Bennett, who were involved with the development of

Heim's centerless grinder, provided supporting testimony that Heim was the original inventor. Arthur Kirkland, who sold grinders for Detroit Machine Tool and Sanford, testified on the success of centerless grinder sales by Sanford as well as customer concerns with patent infringement.

Defendant's Case

In most trials, the plaintiff is first to present its case after which the defendant presents its case. In Heim v. Sanford, there was an unusual twist that allowed the defendant to present evidence out of order. On the first day of the trial *(May 16)*, Sanford's counsel, William Dodge, submitted an amendment *(No. XVI)* to its "Answer." In its amendment, Sanford claimed that the inventions of the patents in suit were not made by Heim but were made by L. S. Chadwick and reduced to practice in working machinery at the Ball Bearing Company of Boston, Massachusetts, around the year 1900. The amendment states that the inventions were described in an article titled "Among The Shops" published in the November 1900 issue of *Machinery*. The amendment further states that the machine was subsequently used at the Bantam Ball Bearing Company at Danbury, Connecticut, and that such use was known by Chadwick, W. S. Rogers, and others.

After submitting the evidence, there must have been an agreement between counsel because the first witness called was Winfield S. Rogers, a witness for the defendant. Rogers was the founder of the original Ball Bearing Company of Boston in 1894 and current chairman of the board of the Ball Bearing Company of Bantam, Connecticut. Rogers testified on the origins of his company and the roll-polishing machine cited in Amendment XVI. *(For a complete list of all people providing testimony, see Trial Summary in Appendix 3, A3-2.)*

Around 1896 or 1897, Rogers hired Lee Chadwick, a graduate mechanical engineer from Purdue University, to work at the Boston bearings plant. In 1899, Chadwick invented a two-wheel roll-polishing machine for use in manufacturing rolls used in roller bearings. Chadwick designed his machine with two wheels *(called discs)* mounted on parallel shafts and offset such that the overlapped wheel surfaces formed the grinding zone. The illustration *(Figure 8-6)* displayed in the *Machinery* article "Among The Shops" shows an inclined carrier for the rolls supported between the two wheels. The wheels were each about 16 inches in diameter and rotated in opposite directions and at different speeds. One wheel was faced with leather and rotated at 750 rpm and the other, the grinding wheel, was faced with leather coated with emery and rotated at 1,500 rpm. As seen in the illustration, the arrows indicate that "both wheels rotate downward" toward the carrier as well as the rolls in the grinding zone. The article states that the machine was able to polish a wide range of roll diameters and lengths with small, short rolls polished at a rate of 5,000 per hour.

Sanford's defense team discovered the Chadwick machine late in the process of preparing for the trial. As such, Chadwick's testimony was not taken until June 25, well after initial trial testimony was completed in May. During testimony, Chadwick was questioned about the design and operation of his machine compared to that described in the *Machinery* article. When asked about the direction of rotation of each

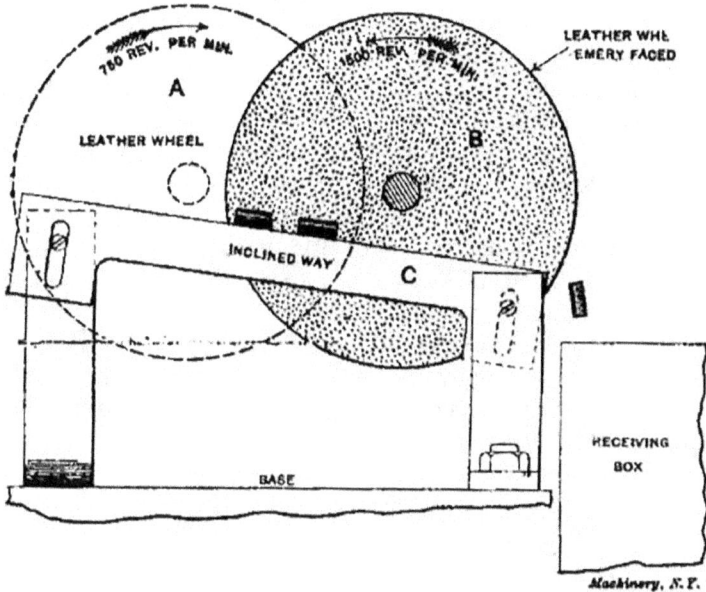

700 REV. PER MIN.

A

LEATHER WHEEL

600 REV PER MIN

B

LEATHER WHL
EMERY FACED

INCLINED WAY

C

RECEIVING
BOX

BASE

Machinery, N. Y.

Fig. 7. Roll Polishing Machine.

Fig. 8-6: Chadwick Roll-Polishing Machine Used at the Ball Bearing Company at Boston, MA, ca. 1900.

disc, Chadwick testified as follows:

> *The grinding-wheel drove downward against the inclined way. The leather-faced wheel drove in a direction tending to turn the roller toward the grinding-wheel. The earlier models were tested, driving the leather-laced wheel in different directions and it was found that grinding work could be done better when the wheels were as outlined.*

When asked to clarify his statement, Chadwick said:

> *The emery-faced wheel drove downward towards the center portion of the inclined way. The leather-faced wheel drove upward toward the center of the inclined way, both wheels tending to rotate the roller in the same direction.*[9]

Chadwick's testimony conflicted with the *Machinery* article, which stated both wheels rotated downward on the rolls. Under cross-examination, Chadwick explained that they tested different speeds and directions of the wheels and found that the grinder performed best with the slower wheel rotating upward. Chadwick also stated that the slower, leather-faced wheel controlled rotation and feeding of the work. So, except for the higher rotational speed of the leather-faced wheel, Chadwick's description was almost identical to the operation of the Heim centerless grinder and provided Sanford with a second machine, in addition to the French and Stephenson round bar grinder, to serve as prior art to invalidate Heim's patents.

In 1901, Rogers sold the Ball Bearing Company of Boston to the Standard Roller Bearing Company of Philadelphia that was founded by Samuel S. Eveland. In the sale, all machines, including the Chadwick roll-polishing machine, were also transferred to Standard Roller Bearing. Two years later in 1903, Rogers restarted the Ball Bearing Company in Bantam, Connecticut. At first, Bantam Ball Bearing manufactured both ball and roller bearings. Later, in 1914 or 1915, Bantam stopped manufacturing most roller bearings to focus on ball bearings that Rogers claimed made more money.

In his initial testimony on Monday, May 16, 1921, Rogers testified that he used a variant of the Chadwick machine at his company in Bantam to grind rolls for roller bearings. He built the variant by using a Gardner disc grinder as one wheel to which he mounted a carrier and second grinding wheel in an offset position. The Gardner disc grinder used by Rogers was the same type of machine modified by Heim in 1913 to construct his first double-ring wheel centerless grinder. Rogers, as shown in the following testimony excerpt, described the origins of the Chadwick variant *(in 1904)* and its nickname, "The Hog":

When they first bought the machine, when they first got the machine, we were doing our work in a slow way. We had not money to buy machines with and increase our plant. We had more orders than we could handle. The thing was to produce work in some other way faster. The Gardner disc grinder was practically new, so we purchased this machine for doing the work on brass parts that we had been doing on lathes, and after rigging up the machine, we put one of our lathe hands upon the machine. His first day's work was marvelous. He turned out more work on that machine than ten men would turn out on lathes and much easier. As the machine was new in its shape and everything else, to anything around the factory, that night at quitting time the men gathered around him to ask him what he thought of the new machine, and his remark (with the typical expression of a practical mechanic added to it) was that it was the hog, and the word "hog" has stuck to it ever since. We cannot eliminate it. After we found that the machine was doing that work so fast, then we decided to take the other side of it and put it right into service for grinding rolls, the same as we had in the Boston plant.[10]

Rogers testified that they used "The Hog" up to 1914 or 1915, after which they stopped manufacturing most roller bearings and no longer needed the machine. To back up his statements, Rogers offered to reassemble and demonstrate the Hog at his factory at Bantam, and Heim's attorneys accepted. At the same time, Heim's counsel *(Blair and Haynes)* requested pictures of the machine used at Bantam that Rogers agreed to submit.

Rogers returned to the trial on either Wednesday or Thursday *(May 18/19)*, and during testimony showed the court pictures of the Hog reassembled *(Figure 8-7)* by Bantam employees Monday night and Tuesday morning. During his second testimony,

Fig. 8-7: 1921 Reproduction of Bantam Two-Wheel Roll-Grinding Machine (Chadwick Variant) ***Used at the Ball Bearing Company, Bantam, CT.***

Rogers stated that after reassembling the machine they ground rolls successfully and promised that if Heim visited Bantam, a live demonstration would be performed.

On Thursday, May 19, a demonstration was held at the Bantam Ball Bearing Company where Rogers, who did not attend, assigned Henry Edwards, an engineer at the plant, to conduct the demonstration. When everyone arrived for the 5:30 p.m. event, they found that Edwards was ill prepared to conduct a proper demonstration. In fact, there was no demonstration at all. Edwards did not have hardened rolls available for grinding and the grinding wheel was too coarse to properly demonstrate its capabilities for fine grinding. When confronted with questions on the machine setup, operation, and history, many of Edwards's responses were either "I do not know" or "I am not sure" or "I do not remember." The only evidence provided to Heim were ten rolls that Edwards claimed were ground on the Hog prior to their arrival. The so-called ground rolls were rough in texture, which Edwards blamed on the course grit of the grinding wheel. All told, the demonstration was nothing more than a sham; that was fully apparent in the Edwards testimony.

Blair challenged the existence of the Chadwick machine at Bantam and its use at Boston. Under cross-examination, Rogers identified former employees who could back up his testimony. With names in hand, Blair tracked down seven former employees and on June 29 and 30, called them as witnesses for the plaintiff. Of the seven witnesses, five had worked at the Bantam plant while two were formerly employed at Boston. Under oath, all Bantam plant employees testified that they never saw the two-wheel Chadwick machine. They claimed that the only machines used at Bantam to make rolls for roller bearings were centered grinders and the Gardner single-wheel disc grinder with a grooved wood block that was manually held up against the grinding wheel.

The two former Boston plant employees confirmed the existence of the Chadwick grinder and that it was used to grind rolls for bearings. However, they conflicted with Chadwick's testimony and stated that both wheels rotated down on the roll as illustrated in the *Machinery* article. In addition, both said the Chadwick machine ruined a large portion of the rolls and that it was only used for two months. In its place, the Boston plant used the Gardner disc grinder and manually ground each roll with a grooved wood block.

The evidence surrounding the Chadwick machine gives rise to suspicions of a conspiracy between Rogers and Sanford. First of all, Chadwick was more than a former employee of Rogers as he was also his son-in-law. Thus Chadwick could have been coerced into modifying the true story of the machine. Second, Rogers testified that he knew and was a friend of William Weed who now worked for Sanford. Third, Rogers, as a manufacturer of bearings, was a competitor, so denying Heim patents to his machine would allow others to compete using similar machines, such as centerless grinders made by Sanford and Reeves. And fourth, Michael Dempsey, a witness for Heim, made an interesting comment during his testimony. He mentioned that Rogers told him he was having a lot of fun helping out a guy out who was in trouble and that Rogers told him that he should not testify or sign any affidavits.

Later in the first week, William Dodge presented the defendant's evidence against infringement. Just as Blair opened his case with an expert witness in Harold Pender, Dodge opened his case with an expert witness in patent litigation. Dodge called on Arthur Browne, a patent solicitor with 30 years experience and some expertise in the field of precision grinding. The purpose of Browne's testimony was to provide evidence to support the first two Sanford defenses, Non-infringement of the Sanford grinder and Anticipation of prior patents *(prior art)*.

Dodge first questioned Browne on the prior art machines and patents that the defendant relied on most to prove its case. The first patent cited by Browne was by French and Stephenson *(1,111,254)* for the two-wheel round bar grinding machine. In his testimony, Browne reviewed the structure of the F & S machine against Heim and made the case that both machines used a carrier placed between two rotating wheels, with one rotating down at a high speed and the other rotating up at a slower speed and that the difference in wheel rotation advanced the work through the machine. Browne testified that even though the F & S machine was designed for long round bars and not short rolls, there was no difference in either the mode of operation or the major components.

Browne went on to discuss other patents for machines that embodied some of the features patented by Heim and cited a machine invented by Jacob Reese *(U.S.65,832)* that was designed to straighten long round bars. The machine incorporated a flat carrier placed between two opposed rotating wheels. Browne stated that the Reese machine had demonstrated for more than 50 years the principle of rotating and advancing a round bar by a pair of opposed wheels.

Browne discussed the cork-grinding machine of Lowman *(BR 12,190 and U.S. 967,798)* that utilized a controlling wheel and a grinding wheel, with work advancing in between on a horizontal work rest *(carrier)*. Browne went on to discuss other grinding machines claimed as prior art for various components of Heim's centerless grinder, including those patented by Medart *(594,482)*, Eveland *(747,542)*, and Koerner *(951,401)*.

Following testimony on prior art, Browne addressed Non-infringement of the Sanford grinder. With questions choreographed by Dodge, Browne first addressed the double-ring wheel machine *(Reissue 15,035 and 1,210,937)*, claiming Non-infringement by reason of differences in design and operation. Browne claimed that Sanford's use of a flat carrier placed below the wheel centers and a tilted regulating wheel for roll feeding was fundamentally different from Heim's use of the wheel sides and an inclined carrier, and thus non-infringing.

Following testimony on Heim's double-ring wheel machine *(Re 15,035)*, Dodge attacked Heim's peripheral wheel machine *(1,264,930)* as a machine designed for a different purpose than Sanford's machine. Under questioning by Dodge, Browne stated that while Sanford's machine was designed to grind cylindrical work such as rolls, the Heim machine was designed to grind headed stock such as bolts and cited claim 8 of the patent as clearly stating this purpose:

8. A machine especially adapted for grinding the shanks of headed blanks, such as bolts, comprising a wheel having a grinding periphery driven at a relatively high speed, and oppositely moving regulating wheel driven at a relatively low speed, a blank-support between said wheels, shafts by which said wheels are carried, bearings for said shafts, and spring-pressed slides on which said bearings are mounted.

Browne tried to make a case that even though Heim's patent refers to grinding cylindrical blanks, it did not describe the method by which the two wheels advanced the work through the machine and hence, the machine was not designed for grinding rolls and was therefore, distinctly different from Sanford.

Under cross-examination, Blair attacked much of Browne's testimony regarding differences in machine design and operation. Blair structured his questions to force Browne to admit that the Sanford machine ground rolls using the same principles as Heim. But in many questions, Browne claimed that he did not know how the Sanford machine would operate, thus contributing to the defendant's weak evidence of Non-infringement.

After Browne, Dodge called on other witnesses to testify for the defendant. Outside of W. S. Rogers, Dodge called William Weed, the former head of grinding at the Ball and Roller Bearing Company, and James Daley, who installed and demonstrated Sanford centerless grinders. Weed testified that he did not provide details to Sanford on the Heim centerless grinder and that Sanford designed the machine without his help.

Daley, who was employed by Sanford as a demonstrator of Sanford centerless grinders, was another interesting witness. Daley was the only witness who had worked at all three companies—Sanford, Heim Machine, and the Ball Bearing Company in Bantam—and could confirm the machines they used. Regarding operation of the Bantam two-wheel *(Chadwick)* grinding machine, Daley was the only witness who testified to its existence and operation.

Dodge put Sanford on the stand as one of the final witnesses to counter the plaintiff's accusations that Weed provided the details of Heim's centerless grinder. Sanford testified that he conceived of the idea on his own using publicly available information on grinding and his own machine design skills with no input from Weed. Under cross-examination, Robert Blair, as he did with William Weed, surfaced conflicting answers from Sanford that punched holes in their testimonies. And Judge Thomas, as seen in the following excerpt from the opinion, did not believe one word of Sanford's story on the origins of his invention:

> *The claim respecting intervening rights hardly merits serious consideration and the defendant should be the last to raise this question, it having intervened through knowledge acquired from a former employee of the plaintiff.*[11]

On June 29, the trial was moved to the Ball and Roller Bearing Company in Danbury. During the period prior to this date, Heim and Pender prepared demonstrations of five grinding machines *(listed below)* to support the plaintiff's case for infringement and to negate the claims of *prior art*. Once again, Blair called upon Dr. Pender to demonstrate the machines and prove to the court that the defense's claims were without merit.

1. Heim X-tended Range Centerless Grinder
2. Grinding Machine Described in *Machinery* magazine
3. French and Stephenson Round Bar Grinding Machine
4. Sanford Centerless Grinder with Inclined Carrier
5. Heim Ring Wheel Centerless Grinder Equipped with 60M Grinding Wheel

The Heim X-tended Range Centerless Grinder was built to prove to the court that there was fundamentally no difference in operation between the double-ring wheel and peripheral wheel grinders. The machine was designed to allow the slides supporting the grinding wheel and regulating wheel to pivot from the ring wheel position to the peripheral wheel position *(Figure 8-8)*. In doing so, Heim demonstrated that it was the surface contact of the wheels on the work that made the invention and not wheel rotation as claimed by the defendant.

The next machine demonstrated by Pender was the two-wheel *(Chadwick)* roll-

Fig. 8-8: Photo of Heim X-tended Range Centerless Grinder Showing Pivoting between Ring and Peripheral Wheel Position, 1921.

polishing machine described in the November 1900 article in *Machinery* magazine. Heim and Pender reconstructed the machine according to the article description showing two offset ring wheels rotating in "opposite directions" and downward toward the rolls in the grinding zone. When Pender demonstrated its ability to grind rolls, the work would not feed through, even when forced. However, when one wheel was reversed and rotated in the opposite direction, the work shot through between the wheels at such a rapid speed, barely any grinding occurred. Pender described the action a bit more colorfully as follows:

> *... The operator has now made this adjustment so that the wheels are exactly in the same condition as they were before, except that the leather disc, the leather faced wheel, is rotating in the opposite direction. I will now attempt to feed some rolls through this machine and carefully warn everyone to get out of the way. I place a roll on the guide, having some difficulty to get one on, and it shoots through at bullet-like speed, suggesting that this machine might make an excellent machine gun. These rolls which came through have a slight degree of polishing, that is, a certain amount of the outside skin has been removed.[12]*

The third machine demonstrated was a replica of the French and Stephenson grinder for round bars. Heim modified one of his ring wheel machines to match the F & S design in the placement of the two wheels and the horizontal work rest *(carrier)*. In accordance with the patent, one wheel, the upwardly rotating wheel, was offset slightly ahead of the opposing, downwardly rotating wheel. The only change was the addition of vertical guides along the length of the carrier to prevent the rolls from falling off the sides. Wheel rotation and speeds were set according to the F & S patent at 575 rpm for the upward-rotating wheel and 1,400 rpm for the downward-rotating wheel.

Pender conducted the demonstration with both long and short rolls and commented that the machine, being designed for long rods, was favorable to longer rolls. The first test used rolls at 5/16 inches diameter and 3 inches long. Upon contact with the first, upwardly rotating wheel, the rolls flipped up and off the carrier and out of the grinding zone. Tests with shorter rolls at 5/16 inches diameter and 3/16 inches long produced the same results. Even with changes requested by the defendant, all tests resulted in the flipping of rolls up and off the carrier without being ground, thus proving that the F & S machine as patented could not grind the small rolls typically used in roller bearings.

The fourth demonstration was the Sanford centerless grinder equipped with an inclined carrier and parallel wheels. Sanford claimed that his machine, being distinctly different in operation, would not properly feed the work with an inclined carrier. To prove this claim as false, Pender modified the Sanford machine demonstrated at Norwalk Iron and Foundry Company *(Figure 8-9)*. Pender first changed the tilt of the regulating wheel to zero degrees such that its surface was parallel with the grinding wheel and then installed an inclined carrier similar to Heim. Upon operating the Sanford machine with the inclined carrier between parallel wheels, the rolls traveled through and rotated normally with no indication of anything abnormal in its performance, thus disproving a major claim by the defendant.

The fifth demonstration by Pender involved the two-wheel centerless grinder known as the Chadwick variant *(Figure 8-7)* that was rebuilt at the Bantam Ball Bearing Company and shown to but not operated for the court on May 19. Henry Edwards, the Bantam engineer conducting the non-demonstration, claimed that the

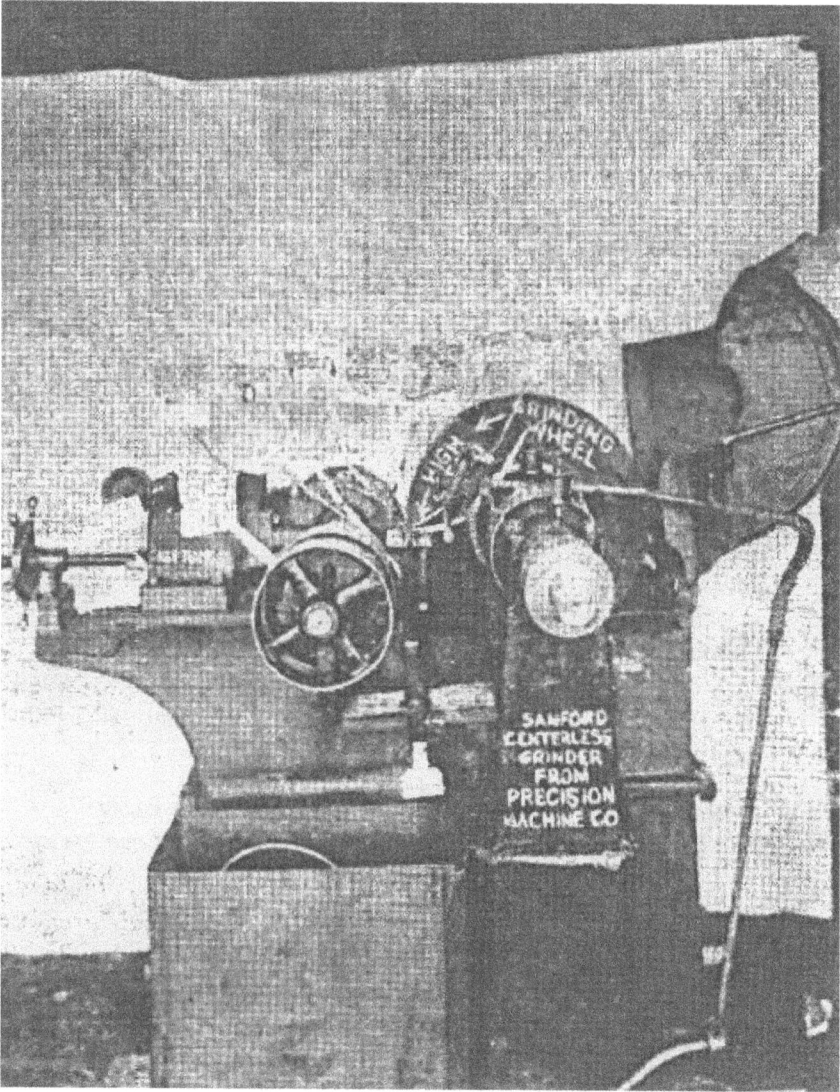

Fig. 8-9: Sanford Model B Centerless Grinder Installed at the Precision Machine Company, 1921.

rolls given to Heim were rough due to the coarse grit of the 60M grinding wheel. Upon demonstration of a Heim machine equipped with same 60M wheel, the rolls were ground to remove 3/1000 inches and produced a smooth and polished surface. Again, Pender proved that the defendant's evidence of prior art withered under testing against the Heim machine.

At the end of his testimony on June 30, Pender, commented on observations made at the Bantam demonstration and stated that the reassembled grinding machine showed evidence of recent modifications:

> In the Rogers factory at Bantam we found, when we stopped the machine, that the member of the machine to which the grinding head was attached showed evidence that it had been only recently adapted to the purpose, in that this member had been recently tapped. This was admitted by one of the Bantam Company's men, who was present. The work rest was also apparently newly constructed as evidenced by fresh saw cuts at the ends of the members. The flange collars holding the grinding wheel also showed freshly machined surfaces. The marks on the floor showed that the machine had been recently moved. (See the white spots on the floor of Figure 8-7), and the empty bolt holes which I mark B on the photograph. These evidences of recent work are a clear indication that the machine had been only recently assembled, and that the head holding the grinding wheel had never been used in that particular position before.[13]

Hence, Pender not so subtly accused the defendant of falsifying evidence and rigging the demonstration to claim prior art.

The defendant never proved that the Bantam *(Chadwick Variant)* two-wheel roll-polishing machine exhibited the inventive principles of the Heim patents. But that did not stop the defendant from claiming that testimony and evidence supported its claim of prior art. In its Brief for the Defendant *(pages 107–120)* submitted posttrial on September 19, William Dodge *(defendant's counsel)* made the following statements regarding the Bantam machine:

> . . . Without going further into the details of this disclosure, we submit that in and of itself, the evidence is conclusive as to the existence and the extended and successful use of said machine, long anterior to any invention or alleged invention of Lewis R. Heim set forth in any of the patents here sued upon. Even without the patents to French & Stephenson and to Lowman, this evidence is sufficient to establish prior knowledge and use of a grinding machine for metal rolls embodying the stated "essentials" of the Heim invention in substantially the same form and arrangement, and with the same vertical adjustments both bodily and pivotally, as are set forth in the Heim Patent 1,210,936 and in Reissue 15,035.

However, Judge Thomas must have seen through the charade or viewed the evidence differently, since the Bantam machine as prior art was never mentioned again.

In reviewing the testimony and evidence presented, it seems obvious that Dodge knew that his best chance of winning the case for Sanford was to attack the validity of the Heim patents through Anticipation *(prior art)*. The defendant submitted more than sixty domestic and foreign patents claiming inventions similar to Heim. Out of those placed in evidence by Sanford, two became the focus of the court. These were U.S. Patent No. 967,798 *(British Patent No. 12,190)* granted to Lowman and U.S. Patent No. 1,111,254 granted to French and Stephenson. The Lowman patent described a device for grinding corks using the periphery of one wheel and the radial side of the other wheel, and looked very similar to the Reeves centerless grinder. In Lowman, a cork was fed between the two wheels on a centerless support. The larger wheel, using the radial surface, performed both grinding and feeding while the other smaller wheel, using the periphery surface, held the corks up against the grinding wheel. In French and Stephenson, as previously described, a round metal bar was inserted between two oppositely rotating grinding wheels and automatically pulled through while being ground. And when first viewed, the French and Stephenson round bar grinder looks so similar to Heim's double-ring wheel machine that it is difficult to tell they are completely different machines. And these similarities are what Dodge used to make the case that the features in Lowman plus French and Stephenson were in fact the same features found in Heim's patents, thus proving prior art to Heim.

Dodge also attacked Heim patent 1,264,930 that used the peripheral surface of the wheels and claimed that Heim had abandoned the claims of that patent when it was not included in the application for the original patents for the double-ring wheel centerless grinder. At that time, inventors had a time limit of two years to file after an invention was reduced to practice. Patent 1,264,930 was filed July 13, 1917, almost four years after Heim first used a small prototype grinder to prove that his centerless concept could grind rolls used commercially in roller bearings.

Decision

The trial ran for eleven weeks and concluded on August 3, 1921. Both the plaintiff and defendant presented extensive arguments to support their positions with very few aspects agreed upon by both sides. To further support the evidence presented at trial, the defendant's counsel submitted to the court an all-inclusive "Brief for the Defendant" that was a 189-page document arguing all aspects of the evidence to support the defendant's many defenses. Given its level of detail and timely submittal on September 19, one wonders if others with a financial interest in the outcome, such as Reeves Pulley and Cincinnati Milling Machine, contributed to its development.

It would take Judge Thomas nine months to review the evidence, research precedents, and write an opinion, which he issued on May 6, 1922. In the opinion, Judge Thomas first ruled on the validity of Reissue Patent 15,035 that replaced the original machine patent 1,210,937. Thomas found the Reissue patent valid since he found no broadening and hence only narrowing of the claims that is allowed under patent law. Next, Thomas addressed the fifteen claims in suit to determine if the Sanford grinder used Heim's patented concepts. In this, Thomas found that the Sanford machine did use Heim's concepts and thus the pertinent claims in patent 1,210,937 *(Re 15,035)* were infringed.

Next, Thomas examined the two claims in suit from patent 1,264,930 covering the peripheral wheel centerless grinder. Thomas found that the design differed only in one respect and that is the use of wheel periphery, and that the claims should have been included under the original patent *(1,210,937)*. Thomas therefore held the claims to be aggregations *(same invention in a different design)* and hence invalid.

The last patent, 1,210,936, covered the feeding device for Heim's grinder. There were five claims in suit from this patent that Thomas examined against Sanford's machine. Thomas determined that none of Heim's features were found in Sanford's machine and concluded that no claims were infringed.

Thus remained the final defense, Anticipation *(prior art)* and in this area, Judge Thomas found it extremely difficult to sort out the arguments presented by both sides. In Thomas's view, a patent needed to embody something new and truly different. In this respect, the defense presented by William Dodge prevailed, convincing Thomas that the "general principles" described in both Lowman and French and Stephenson were also found in Heim and thus there was nothing new and inventive. With this finding, Judge Thomas concluded that the claims in suit were invalid due to prior art, and ruled against Heim in favor of the defendant, the F. C. Sanford Manufacturing Company.

Partial Rehearing

Heim had to be dismayed to lose the suit based on the prior development of machines with similar characteristics but totally incapable of precision grinding of bearing rolls. In most cases, the only course of action for the losing party is to file an appeal. This course of action was available but for this suit Blair believed it necessary to request a review of the decision that prior art invalidated Heim's patents. Blair needed some time to prepare and on June 30, 1922, petitioned the District Court for a partial rehearing limiting the review to fourteen claims in the Reissue patent that Judge Thomas had found to be infringed.

Sanford adamantly objected to the rehearing, but Judge Thomas believed the case was important enough to reconsider the issues leading to his decision. The rehearing focused on the invalidity of Reissue patent 15,035 in view of Lowman *(cork grinder, US 967,798/BR 12,190)* and French and Stephenson *(grinder for round bars, 1,111,254)*. Based on his opinion, it appears that Judge Thomas opened up the proverbial bees' nest. The rehearing turned into another trial, complete with both sides presenting extensive evidence in an effort to convince the court of its position. Blair developed a lengthy "memorandum" complete with seven charts and a printed brief containing multiple diagrams illustrating differences between Heim's machine and the machines of Lowman and French and Stephenson. The defendant replied with its own printed brief contesting every point made by the plaintiff. The evidence and arguments were so extensive and contradictory, the ensuing confusion caused Judge Thomas to make the following statement in his opinion:

> *With experts of standing in conflict, and able counsel contradicting one another upon almost every conceivable point, legal and mechanical, it results that, instead of presenting one or more clear-cut issues, the court is confronted with a veritable tangle of multiplied disputes. Therefore it becomes imperative to attack the problems presented as a matter of independent reasoning and this I shall proceed to do.[14]*

Judge Thomas first examined and commented in extensive detail the Lowman machine in comparison to Heim. In summary of his examination, Thomas concluded:

While Lowman clearly deprives Heim of any broad pioneership in the grinding of cylinders or rolls, using those terms generically, still there are obvious differences between the Heim machine and the Lowman machine. A machine organized precisely as disclosed by Lowman would not be suitable for grinding metallic rolls. It is unnecessary to speculate upon whether it could have been so modified as to be capable of grinding metal objects, without the exercise of the inventive faculty. Again, while Lowman suggest the use of a disc as a work-controlling device instead of his drum, he does not show any such disc; and further— perhaps most important—in the Lowman machine it is the grinding wheel that feeds the work longitudinally, and not the regulating or controlling wheel, a fact which plaintiff protests would militate against accurate and successful roll grinding, at least in the making of steel roller bearings. These differences, and some others more minute, sufficiently distinguish Lowman from Heim to make it at least likely that so far as concerns the Lowman patents alone there may be patentable invention beyond their disclosure in what Heim did; and any doubt in this regard should properly be solved in favor of Heim.

As seen, Judge Thomas, through his own analysis, concluded that Lowman by itself was not prior art. However, when French and Stephenson was examined in conjunction with Lowman, Thomas stood his ground regarding his original opinion. Thomas reinforced his original examination and in his second opinion made several condescending observations as follows:

But when one turns to the French and Stephenson patent, and assumes, as one must assume, that both French and Stephenson and Lowman were before Heim when he designed his machine, it becomes impossible to accredit Heim with anything more than the mere mechanical adaptation of what had already been invented or discovered.

The French and Stephenson disclosure is so similar to that of Heim that it fosters a moment of confusion between them. Compare Figure 2 of Heim with Figure 1 of French and Stephenson. The resemblance is striking.

The similarity between the Heim disclosure and that of French and Stephenson is so marked as to suggest the question of how Heim could have designed his machine or prepared his patent drawings without the French and Stephenson patent before him.

Therefore, in view of the Lowman and French and Stephenson patents, I am unable to see what is left to Heim beyond matters of adjustment, details which should have been obvious in purpose and effect to a mechanic skilled in this art.

In his final comment, Thomas stated:

The manifold excellencies accredited to the Heim machine as actually oper-
ated, marshaled at length and by no means minimized by counsel and experts,
appear upon careful study and in view of the statements of the witnesses to
be attributes of elements and operations common both to Heim and the prior
art, and not of the alleged differences between Heim and his predecessors.

In layman's terms, Judge Thomas concluded that even though Heim built a machine that grinds metal rolls suitable for bearings, something the machines of Lowman and French and Stephenson could not do, the improvements were nothing new
or inventive and therefore invalid due to prior art. The only recourse remaining for
Heim was to file an appeal, which was done shortly thereafter on September 6, 1923.

Judge Thomas's rehearing decision was issued on June 9, 1923, just short of three
years from the filing of the original complaint. During this period, in addition to Sanford and Reeves, other competitors emerged selling two-wheel centerless grinders using Heim's patented concepts. By June 1921, Detroit Machine Tool was selling its vertical two-wheel centerless grinder. Later that same year, Cincinnati Milling Machine
Company, through its subsidiary Cincinnati Grinders, began selling machines under a
license from Sanford. With the patent infringement trial requiring such a long time to
complete, Heim's four competitors were able to capture a majority of the market and
establish their brands for two-wheel centerless grinding.

As the infringement suit slowly progressed, Heim was very busy improving and
perfecting his centerless grinder. From 1920 through 1923, as described in Chapter 6,
Heim developed improvements focused on increasing precision and expanding grinding capabilities of the peripheral machine, including mechanisms to automate and increase production speeds. Along the way, Heim published articles periodically on his
machines and methods for centerless grinding. In 1921 Heim published an article on
his first peripheral machine in the February issue of *Machinery*, a trade magazine serv-
ing the machine tool and manufacturing industries *(see Appendix 2, A2-5-1)*. In May
of 1923, Heim published an article in the same magazine describing how centerless
grinding is conducted on his second-generation machine in which the regulating wheel
is tilted to induce automatic feeding through the grinding zone *(A2-5-3)*.

By 1924, Heim's centerless grinder, with optional attachments, could automatically
grind straight, tapered, and headed *(shoulder)* cylindrical work at high production rates.
Heim's latest machine was featured in the December 1924 issue of *American Machinist*
in an article titled "Heim Improved Centerless Grinding Machine" *(A2-5-4)*. The article
was in many respects an advertisement that described Heim's latest design for automatically grinding many types of round parts in a single machine, thus conveying to the
automobile and related industries the increased value to be found in centerless grinding.
None of the competitors offered machines with the capabilities of Heim. But many competitors, such as Cincinnati Milling and Detroit Grinder, seemed to have better business
connections than Heim, which allowed them to win orders that should have gone to the
Ball and Roller Bearing Company. After losing the rehearing trial, Heim must have felt
that the market for his machine was all but lost to his competitors.

Appeal: 1923

On September 6, Heim filed an appeal to the United States Court of Appeals. The appeal of The Ball and Roller Bearing Company v. F. C. Sanford Manufacturing Company was assigned to the United States Court of Appeals for the Second Circuit located in New York City. The Second Circuit is one of thirteen appeals courts in the United States and covers the states of New York, Connecticut, and Vermont. The Second Circuit became famous for one of its judges by the name of Learned Hand who was promoted to the court in 1924 by Calvin Coolidge and served until 1951. Hand was known for his decisions in fields such as patents, torts, admiralty law, and antitrust law that set lasting standards. Hand in one of his patent cases made the following statement on inventions:

The specification of a mechanical combination patent generally discloses a machine consisting of a large number of elements, most of them individually old in the art. The invention consists in the act of selecting some of these elements for a combination which constitutes an independent entity, serviceable to the art, and theretofore unknown. It is always this choice of the proper elements in combination which constitutes the invention.

On October 25, one and a half months after filing, the appeal trial began before Second Circuit judges Martin T. Manton, Charles M. Hough, and Julius M. Mayer. The court reviewed all case documents including testimony, evidence, and arguments without the need of a formal retrial. They also reviewed the laws and precedents considered by the lower court, and they issued their decision on February 18, 1924.

The judges approached the suit in the same manner as Judge Thomas by reviewing each patent separately on its own merits starting with the findings of Judge Thomas as follows:

1. Reissue Patent 15,035: Roll Grinding Machine *(double-ring wheel center less grinder)*. The District Court held that Claims 1, 2, 4, 6–10, 12, 13, 15, 18, 19, 20, and 21 were invalid due to prior art.

2. Patent 1,264,930: Roll Grinding Machine *(peripheral two-wheel center less grinder)* The District Court held that claims 1 and 4 were aggregations and therefore invalid.

3. Patent 1,210,936: Feeding Device for Grinding Machines. The District Court held that claims 2, 3, 4, 5, and 6 were not infringed.

The appeals court evaluated each patent for its inventive features, for its compliance with patent laws, and whether it was invalid due to inventions preceding Heim. First on the list was the primary patent, Reissue 15,035 *(1,210,937)* describing the double-ring wheel machine and its method for centerless grinding. The court first addressed compliance with patent laws, evaluating the reissue against expansion of claims, laches *(delay without a valid excuse)*, and the defendant's claim of intervening rights *(the changed claims in the Reissue patent caused Sanford to infringe)*. In this area the court agreed with Judge Thomas and found the reissue narrowed, valid, and not guilty of laches. With regard to intervening rights, the appeals court found that the Reissue patent embodied the language of the original seventeen claims and introduced nothing new to cause the defendant to infringe and hence denied the claim of intervening rights.

The appeals court next addressed the inventive features and validity against prior art. After defining the inventive features of the Heim patent, the question before the court became whether anybody before Heim devised a method of utilizing the principles employed in the Heim two-wheel centerless grinding machine. The court addressed the inventive principles and then judged their validity against prior art, specifically, the Lowman and French and Stephenson patents.

Upon examination of the Heim inventive principles and their use on ring wheel versus peripheral wheel machines, the court stated:

> *There is nothing in the patent confining invention to ring wheels, or those peripherally juxtaposed; the principle applies to both, and there is no difference at all between producing nonradial travel of the roll by directing its path between wheels parallel in vertical planes, and tilting the slower or regulating wheel to produce the same result, when the juxtaposition is peripheral. Defendant has been vindicated below (*by Judge Thomas*), not on noninfringement, not by a finding of anticipation, but by holding that in light of prior art, there was not invention in applying and utilizing the principle above stated. To this we are unable to consent.*

The court then addressed the evidence offered by the defendant as prior art. To get a feel for the temper of the court and the basis for its decision, this section of the opinion is presented in its entirety:

> *The voluminous record at bar is the best* (or worst) *example recently presented to us of useless and misleading references to earlier patents and publications. It seems necessary to apply to patent litigation from time to time the maxim that one cannot make omelettes of bad eggs—no matter how many are used. One good reference is better than 50 poor ones, and the 50 do not make the one any better. We decline to consider, as of any assistance, all references to single wheel machines, because they do not and cannot embody the principle stated. They cannot do this, because it is a vital part thereof that there shall be a regulating disc or wheel; the work rest or other appliance which brought the roll into operative touch with the grinder did not even suggest that idea.*

> *The case stands, as it did in the Patent Office, as an inquiry whether in the light of Lowman* (British 12,190 of 1905) *and French et al.* (1,111,254) *Heim showed invention as distinguished from mechanical skill. Lowman wished to clean and polish corks. Through a tube he fed between a sponge or brush disc slowly revolving in one direction and a wheel with a sandpaper face rapidly revolving in*

the other—the corks to be cleaned. The sandpaper wheel alone fed them along; the slow disc had no regulating power whatever, owing to construction. There was not, and could not be, any constant uniform rotation along a predetermined path, which is the feature of essential value in Heim's and defendant's machines.

French in 1911 (when he filed application) *undoubtedly had the idea of abrasive wheels beveled and oppositely rotating. But he expected both his wheels to grind, and both do. His purpose was not to grind rolls, but to remove from long rods 'the relatively thin surface or skin of more or less uncertain density or homogeneity' left on rods or bars after rolling or drawing. His wheels had to oppositely rotate at differing speeds, because, if they did not, no action would follow, and he arranged his wheels relative to each other, not to do what these parties are doing, but to enable him to get grinding or polishing action on a rod which was fed through in guides, not propelled by a regulating wheel. The whole machine was designed diverso intuitu from that of Heim, and it was demonstrated by experiment at trial that it could not be made to produce rolls. That there was not anticipation by all this is, we think, admitted. The question remains whether it was no more than the work of a skilled mechanic to advance from the 'grinding and polishing machine' of French et al. to the 'roll-grinding machine' of Heim.*

The question would be more difficult, if the principle of operation characterizing Heim and the defendant were even dimly present in French; but his inventive concept did not contain any thought either of producing their result or obtaining his result by and through a regulating wheel. The practical business comment on this invention is conclusive. The French machine was used and known, but it moved no one to try it for rolls; we see no reason why it should do so; and rolls continued to be made by the centering method, or on single wheel machines, until Heim appeared.

The decree below is reversed as to this patent, and the claims in suit held valid and infringed.[15]

"Reversed!" The appeals court identified Heim's regulating wheel and his grinding method the novel inventive features not found on any other machine and that no one prior to Heim developed a machine that could produce precision-ground rolls equal to Heim's two-wheel centerless grinder. Whereas the lower *(District)* court found that similar features on earlier machines constituted prior art, the appeals court, in view of Learned Hand, looked at the intent of the invention, its mechanical features and capabilities compared to perceived prior art, and found that none existed before Heim.

The judges next examined patent No. 1,264,930 for Heim's peripheral wheel centerless grinder. The appeals court not only agreed with the lower court but also made some additional findings defined in the following statement:

As a matter of law we are of opinion that the disclosure unaccompanied by any claim in Heim's earlier patent constituted a dedication to the public of all other devices, combinations, and improvements apparent from the specification of that earlier patent but not claimed.... Further, we incline to think as matter of fact that plaintiff had used the substance of this improvement in public for more than two years prior to filing application for this patent. We express no opinion as to whether the substance of the claims under consideration could have been inserted in the reissue 15,035. The point at bar must be decided as it stands, viz: It presents an original application, and no question of reissue, renewal or amendment. As to this patent, the result below [Decision by Judge Thomas] *is affirmed.*

In its statement, the appeals court agreed with Judge Thomas's decision that 1,264,930 was an aggregation in relation to grinding rolls and hence void. Second, the appeals court found that since there was no reference to it in Heim's original double-ring wheel patent *(1,210,937)*, the peripheral design, outside of using a regulating wheel, was abandoned as an invention. It should be noted that patent 1,264,930, describing Heim's first peripheral wheel centerless grinder, has been erroneously credited as the original patent for centerless grinding.

The last patent was 1,210,936 for the feeding device. The court concluded that the device would occur to any ordinary mechanic and there was no invention and thus affirmed the lower court decision that the claims were not infringed.

Upon completion of its examination, the appeals court made a final statement regarding the appellant *(Heim and the Ball and Roller Bearing Company):*

The appellant, having been victorious in the major portion of this appeal, is allowed one-half costs in this court; the costs below are left to the discretion of the District Court.

Decree reversed, and cause remanded, with directions to enter a new decree in conformity with this opinion.

In what had to be an emotional victory for Heim, the Second Circuit Court of Appeals reversed the lower court decision and remanded the case to the Connecticut District Court for settlement of costs and lost profits from Sanford. But the legal process took three and a half years after filing the original complaint, and during that period Heim and his partners in the Ball and Roller Bearing Company lost a tremendous amount of business to Sanford Manufacturing, Reeves Pulley, Detroit Machine Tool, and Cincinnati Milling Machine Company. Winning his patent suit gave Heim not only a large settlement but also a legal basis for suing others who made, sold, or used infringing machines and that is exactly what Heim did.

Chapter 9

Heim Grinder Versus Fafnir

The Heim Grinder Company

U p to 1921 there were only three companies selling two-wheel type centerless grinders to externally grind cylindrical metal parts. These were the Ball and Roller Bearing Company of Danbury, Connecticut, selling the Heim machine, the Gardner Machine Company of Beloit, Wisconsin, selling the Reeves hybrid machine, and Russell, Holbrook and Henderson of New York City selling the Sanford Models A and B machines. Even though Heim had two machines for centerless grinding, he was mainly selling his second design that used the peripheral wheel configuration. One of these machines, installed in the Crosby Street factory, is shown in Figure 9-1. As discussed previously, the peripheral design had much more flexibility than the original double-ring wheel machine in that it had the ability to grind all types of cylindrical work including work of uniform and varying diameter *(tapered, concave, convex)*, short rolls and long rods, plus headed work such as bolts. Early machines sold by Heim, such as the machine shown in Bulletin 110 *(A2-6-1)* published in 1920, were designed to grind work of only one diameter. These machines had the ability to grind a variety of automotive parts including wrist pins, camshafts, valve lifters, shackle bolts, and other parts up to 3 inches in diameter and 15 inches long. Later, machines such as that shown in Bulletin 140 *(A2-6-2)* were able to grind work of varying diameters.

Reeves advertised his hybrid two-wheel machine to grind rolls without centers up to 3 inches in diameter and 5 inches long. And Sanford sold his Models A and B peripheral wheel machines to grind straight cylindrical work up to 6 inches in diameter and 20 inches long. Typical advertisements for various manufacturers citing features of their centerless grinders can be seen in Appendix 2 *(A2-7)*.

On February 18, 1921, Heim filed his second application for a peripheral wheel centerless grinder *(described in Chapter 6),* that incorporated three major improvements. The first was tilting the regulating wheel shaft downward 2 to 3 degrees that induced feeding of the work and allowed use of various types of fixed carriers. The second and third improvements related to a flat carrier with a slanted *(side-to-side)* work blade combined with positioning the grinding wheel axis below the regulating wheel axis. Together, these two improvements produced the "automatic corrective action" for grinding out imperfections to achieve highly accurate roundness and dimensions. The inventive features of Heim's centerless grinding technology made his machines the best among his competitors.

By the end of 1921, Detroit Machine Tool and Cincinnati Milling Machine had entered the market for two-wheel centerless grinders. Ford Motor Company, one of the largest users of machine tools, was one of the first big manufacturers to embrace two-

Fig. 9-1: Heim Centerless Grinder in Crosby Street Factory, ca. 1923.

wheel centerless grinders. Ford employed more than 200 distinct grinding operations for the parts used in its automobiles that required the use of several thousand grinding machines.[1] Ford bought its first two-wheel centerless grinders from Sanford to manufacture bearing rolls, and later from Frederick Geier, after he acquired the Cincinnati Grinder Company.

By 1923, as the Sanford trial continued, two-wheel centerless grinders using Heim's regulating wheel invention were fast replacing centered and single-wheel centerless grinding machines for the manufacture of small cylindrical parts used in automobiles and other machinery. In March 1923, *American Machinist* published an article titled "The Production of Small Parts by Centerless Grinding." The article, which can be found in Appendix 2 *(A2-5-2)* provided examples of automobile parts made, as well as production rates achieved on machines made by Heim, Cincinnati Milling Machine, Reeves Pulley, and Detroit Machine Tool. Parts included piston pins, gear shifter rods, flange yoke pins, motorcycle pistons, bearing rolls, and cam rollers. What was interesting about this article was the introductory statement made by the author in which he summarized the tremendous impact that changes in grinding methods such as centerless grinding were having on the automobile industry:

The art of grinding metals, which for many years was considered so standard that there appeared to be only one way of handling a grinding job, has undergone so many changes in the last three or four years, that no one can say with certainty what can be done or what can be expected next. In the old days, grinding constituted a trade in itself, but now the latest types of machines have reduced what was almost a science to an operation. The human element which entered into the grinding of a piece of work by an operator who placed one piece of work at a time into the machine and turned the feed wheel as his conscience dictated, has been eliminated by a machine into which the work is fed through a trough. And there will undoubtedly be other changes, just as radical, in the next few years.

One job, which, because of its simplicity of design is being done almost universally on centerless grinding machines, is the automotive piston pin. Some shops use only the centerless machines while others, such as the Standard Gear Co., Detroit, Mich., use a center machine for the first operation and centerless machines for the subsequent operations.

Another interesting aspect of the article was the inclusion of all three methods for grinding small cylindrical parts. There was the Norton machine employing "centered" grinding *(Figure 9-2)*, the Detroit Machine Tool "single-wheel centerless grinder" as well as the four two-wheel centerless grinders employing Heim's regulating wheel. With all the precision and reliability problems of the single-wheel centerless grinder, this may have been one of the last articles published that featured the machine.

Between 1921 and 1923, Heim filed patent applications for 14 inventions related to centerless grinding and the largest number of applications filed in any three-year period of Heim's career. Regardless of the competition and the ongoing patent infringement trial, Heim dedicated a significant amount of time advancing his centerless grinding technology.

On January 18, 1924, Cincinnati Milling Machine Company filed it first patent application for a centerless grinder of its own design. The invention that became patent 1,575,558 was designed to automatically grind various types of cylindrical work including uniform and tapered diameters and articles with a head or shoulder. The primary invention of the machine was a mechanism for automatically ejecting headed or shoulder work after grinding. This was Cincinnati's first attempt to differentiate itself from the competition and create its own brand of two-wheel centerless grinders. Later that month on January 28, Cincinnati filed its second patent application *(1,524,969)* for a centerless grinder with variable speed drive systems for the grinding and regulating wheels.

On February 18, three weeks after Cincinnati Milling filed its second patent application for centerless grinding, the Second Circuit Court of Appeals in New York City issued its opinion that reversed the lower court decision of the Ball and Roller

Fig. 9-2: Rough Grinding Piston Pins on Norton Centered Grinding Machine, American Machinist, *March 1923.*

Bearing Company versus F. C. Sanford Manufacturing Company. Whereas Heim was jubilant, Frank Sanford, Milton Reeves, Frederick Geier, and Detroit Machine Tool had to be shocked and disappointed. The reversal meant that Heim's key patents were valid and legally created a monopoly for Heim to sell two-wheel centerless grinders that employed a regulating wheel.

With the decision reversed, Sanford was immediately out of the centerless grinder business. But not so for the Cincinnati Milling Machine Company, as Frederick Geier apparently made contingency plans should Heim win his appeal. Geier knew that if the Appeals Court reversed the lower court decision, Cincinnati Milling would be barred from the centerless grinder business until Heim's patents expired. Being a seasoned businessman and seeing an opportunity to have a monopoly on a machine high in demand led Geier to make Heim an offer for his centerless grinder patents. Another influence may have been Henry Ford. Since Geier knew that Ford was very much interested in centerless grinding, it is very possible that Ford urged Geier to make Heim an offer for his patents.[2]

The date of the offer is not known but appears to have been made in January or early February 1924. The offer by Frederick Geier to buy Heim's patents must have been very good since Heim accepted very quickly. But the sale did not close right away. Apparently, there were conditions made by Geier requiring Heim to shut down all remaining competitors to secure a complete monopoly. The agreement struck between Geier and Heim involved the following conditions:

1. *Heim would spin off the grinder business from the bearing business of the Ball and Roller Bearing Company. A new company (the Heim Grinder Company) would be organized and dedicated to manufacturing and selling centerless grinders.*

2. *All patents for centerless grinding owned by Heim and other parties would be assigned to the Heim Grinder Company.*

3. *Heim would issue a non-exclusive license to Cincinnati Milling to manufacture and sell centerless grinders* (of its own design) *and operate its business separately and independently from Heim Grinder.*

4. *Heim, if necessary, would initiate legal action against all infringing companies that refused to heed the Appeals Court decision and shut down their operations.*

5. *Once all infringing companies had been shut down and Heim established a legal monopoly, Geier and Cincinnati Milling would close the sale and Heim would receive final payment for the patents and business for centerless grinding.*

Outside of Sanford, there were three others—Reeves Pulley, the Norton Company, and Detroit Machine Tool—that were still producing two-wheel centerless grinders. In February 1924, the Norton Company, owned by Charles Norton, had five patent applications pending for centerless grinding. But since there is little historical information available, Norton must not have been in the business very long and all five patents were eventually assigned to Cincinnati Milling.

Heim Grinder v. Fafnir Bearing Company

Heim originally manufactured the centerless grinder at the Ball and Roller Bearing Company at the corner of Maple Avenue and Crosby Street in Danbury, Connecticut. After the third expansion of the building in 1917 there was no space left for increased grinder production. In order to accommodate increasing demand, Heim evaluated several new sites around Danbury and finally constructed a new factory directly opposite the B&RB at 18 Crosby Street sometime around 1921.

After the Appeals Court found for Heim on February 18, Geier and Heim quickly arranged to enact their agreement. First on the list was to separate the centerless grinder business from the B&RB. On February 25, 1924, the Heim Grinder Company was incorporated in and under the laws of both New York and Connecticut.[3] On March 1, 1924, the Heim Grinder Company began operations as an independent company in the Crosby Street factory. A little over six weeks later on April 9, Reeves Pulley assigned its six patents for centerless grinding to the Cincinnati Milling Machine Company.[4] The assignments effectively signaled that Reeves was out of the business of making its hybrid two-wheel centerless grinder. Since Cincinnati Milling facilitated the patent assignments, it is probable that Geier knew Milton Reeves personally and convinced him that he would lose any protracted legal fight.

On April 10, Heim and the Ball and Roller Bearing Company transferred 28 patents for centerless grinding to the Heim Grinder Company.[5] The B&RB transferred the six patents originally assigned by Heim and Heim transferred the remaining 22 patents for improvements made after 1920. Whereas Heim owned 30 percent of the B&RB, his ownership in Heim Grinder would have been the same plus the value of the 22 transferred patents. But Heim was cashing out, as he had decided to exit the bearing business and leave Connecticut. To do this, Heim sold a portion of his investment in the Ball and Roller Bearing Company to his partners to gain 100 percent ownership of Heim Grinder. Heim retained some interest in the B&RB but far less than his previous 30 percent.

Around the same time that the patents were transferred, Heim issued a license to Cincinnati to make and sell its centerless grinder. The Digest of Assignments, maintained by the U.S. Patent and Trademark Office, contained the following statement on the license:

> *This deed states that Heim owns all right in certain patents and applications herein referred to except that he has heretofore granted a non-exclusive license to the Cincinnati Milling Machine Co. and that the Ball and Roller Bearing Co. owns all right in certain other patents and applications herein referred to except that it has heretofore granted a non-exclusive license thereunder to the Cincinnati Milling Machine Company and that all right in all said patents and applications are with the exception of said license owned either by Heim or the Ball and Roller Bearing Co.*

On May 17, Cincinnati Milling assigned the six Reeves Patents to the Heim Grinder Company.[6] In addition to the six patents for centerless grinding, there were three additional patents by Reeves that were assigned to Heim Grinder at some later date that related to truing and balancing grinding wheels. Also assigned at a later date were five patents with one by Heim for a feeding device for end grinding and four patents for centerless grinding granted to Clement Booth. With these assignments, Heim Grinder owned a total of 42 patents related to centerless grinding.[7]

After receiving the Reeves patents from Cincinnati Milling, Heim Grinder subsequently advertised ownership of the Reeves Patents as well as the validity of Reissue Patent 15,035 in Circular HG1 *(A2-6-4)* published in September 1924 as follows:

> *The Heim Centerless Grinder is being built under the following patents issued in the name of L. R. Heim.*
>
> *No. 1,210,936   No. 1,264,930   No. 1,278,463*
> *No. 1,281,366   Reissue No. 15,035*
>
> *The Heim Grinder Company has acquired the ownership of the centerless grinder patents issued and pending, formerly the property of Lewis R. Heim*

and The Ball and Roller Bearing Co., and is the successor to the centerless grinding business of the latter company.

Reissue Patent No. 15,035 was recently sustained by the decision of the U. S. Circuit Court of Appeals, Second Circuit.

We also won the following patents covering features in centerless grinding machine design and issued in the name of M. O. Reeves.

| *No. 1,264,129* | *No. 1,410,956* | *No. 1,430,754* |
| *No. 1,440,796* | *No. 1,440,795* | *No. 1,456,462* |

We will fully enforce our rights under these patents.

Cincinnati also conveyed to the market that it was a legal manufacturer of centerless grinders. In its advertising media, Cincinnati included a notice on patents that stated it was licensed to use Heim's patents and that its clients were protected against infringement. An example of the notice can be seen in Appendix 2 *(A2-7-18)*, which was published in *Machinery* in August 1925.

With the Heim Grinder Company established, Cincinnati Milling under license, and Reeves Pulley and the Norton Company out of the business, the next action was to shut down the one remaining company still producing centerless grinders.[8] On November 11, 1924, Heim filed his second complaint for patent infringement against the Fafnir Bearing Company of New Britain, Connecticut.

In the Complaint for infringement *(see Appendix 3, A3-4)*, Heim cited two patents. The first was Reissue 15,035 that was the machine patent found valid in the first trial against Sanford. The second was patent 1,281,366 that was granted on October 15, 1919, with the title "Method of Grinding Hardened Rolls." Patent 1,281,366 covered the "Method" for centerless grinding and was divided out of the original "Machine" patent *(1,210,937)*. The other two patents for the Carrier Feed System *(1,210,936)* and the Peripheral Centerless Grinder *(1,264,930)* were omitted since they were nullified by the Second Circuit Court of Appeals in Sanford.

Fafnir manufactured ball bearings whose components required manufacture to precise dimensions *(see Figure 9-3)*. Over time, Fafnir bought seven two-wheel centerless grinding machines from Detroit Machine Tool to manufacture its bearing rings *(races)*. In an affidavit, Raymond Searles, vice president and factory manager at Fafnir, explained why they needed centerless grinders:

The entire business of the Fafnir Company is the manufacture of ball bearings. These ball bearings are composed of two rings and balls between them. The outside of the outer ring of these bearings is ground to an exact circle with great accuracy. In our plant we employ centerless grinding machines made by the Detroit Machine Tool Company for grinding the outer surface of these outer ball bearing rings. Fully 90% of our product, that is the outer rings of our ball bearings, are ground on these Detroit machines. Before we

FAFNIR

DOUBLE ROW
RADIAL

SINGLE
ROW
RADIAL

RADIAL THRUST

MAGNETO TYPE

ADAPTER

TRANSMISSION TYPE

THRUST

SELF-ALIGNING THRUST

There's a FAFNIR for Every Purpose

Fafnir Ball Bearings are made in *all* standard types and sizes.
All are manufactured with the utmost accuracy from *thorough-
ly heat-treated* (not casehardened) chrome-carbon alloy steel.
There is a Fafnir Ball Bearing for every bearing purpose.

THE FAFNIR BEARING COMPANY
NEW BRITAIN, CONN.

CHICAGO, ILL., 537 South Dearborn St.   CLEVELAND, OHIO, 1016-1017 Swetland Bldg.
DETROIT, MICH., 120 Madison Ave., Room 511   NEW YORK, N. Y., 5 Columbus Circle
NEWARK, N. J., 271 Central Ave.   PHILADELPHIA, PA., 1427 Fairmount Ave.

Fig. 9-3: Fafnir advertisement in Machinery *magazine, August 1923.*

used these Detroit machines we were grinding the outside of our ball bearing rings on center grinding machines.[9]

As discussed, many buyers of new types of machines required the seller to indemnify them from patent infringement. Since Fafnir bought and used Detroit centerless grinders, by suing Fafnir, Heim was also suing Detroit Machine Tool. For this suit, Detroit Machine Tool hired and paid for the legal representation necessary to defend Fafnir and itself.

Prior to development of its own machine, Detroit Machine Tool had approached Heim in an attempt to obtain rights to manufacture his centerless grinder. Heim, in the following testimony made during the Sanford trial, described the offer made by Detroit Machine Tool *(referred to as Detroit Grinder)*:

Q.318. Did the Detroit Company, who, as I believe you have formerly stated, manufactured a single wheel grinder, ever make an offer to you involving your machine?
A. Yes, sir.

Q.319. Please state in general terms what that was.
A. I cannot say right offhand, but we have it as a matter of record. One of the officers of the—I believe their name is the Detroit Grinder company— called on us and wanted an interview with me. It was granted to him, and he informed me that they had in the neighborhood of $100,000 worth of stock which they wanted to dispose of and was interested in our centerless cylindrical grinder and made overtures to me to the end of wanting me to interest myself in their company and give them the right to manufacture our centerless grinder. This man wanted me to come to Detroit and see their president and talk over details with him; in fact, we have several letters bearing on that subject. They are not altogether clear so that anyone not familiar with the subject which we had under discussion at that time could tell what they were trying to drive at, but the sum and substance was that they wanted me to take some stock in their company, and in return for that stock let them use our machine.[10]

When Detroit failed to get an agreement with Heim to use his centerless grinder technology, it developed and patented its own design in which the grinding wheel was positioned directly over the regulating wheel that also served as the carrier supporting the work. Detroit probably believed that its vertical design was inventive enough to stand on its own. Detroit Machine Tool, as did Sanford, knew the value of centerless grinding and invested heavily in its unique design. In a testament to its investment, Detroit published a two-page advertisement *(see Appendix 2, A2-7-14 and 15)* in the April 1925 edition of *American Machinist* citing improvements to its latest machine, the Model 4C, which incorporated a heavier, more rugged, and enduring construction with guarantees that work is ground free from waves, flats, and chattering marks. The advertisement included not only a picture of the machine but the many types of straight, tapered, and shoulder work that could be ground.

On January 11, 1926, fourteen months after filing the Complaint, the patent infringement trial between Heim Grinder and Fafnir began in the United States District Court of Connecticut in South Norwalk before Judge Edwin S. Thomas, the same judge who presided in the first trial against Sanford Manufacturing. Heim retained Robert Blair from the Sanford trial as his legal counsel and Detroit Machine Tool hired the Boston law firm of Fish, Stackpole and Holmes.

Since the Complaint named Fafnir as the Defendant, Fafnir submitted the answer that was prepared by the legal team hired by Detroit Machine Tool. But, for all intents and purposes, the patent infringement suit was actually between Heim Grinder and Detroit Machine Tool. The defenses claimed by Detroit Machine Tool boiled down to two points of patent law: invalidity due to prior art and non-infringement due to the unique differences in the design of the Detroit Centerless Grinder. In its answer to the Complaint, Detroit made many of the same claims as Sanford, as follows:

1. Both patents *(Reissue 15,035 and 1,281,366)* were not properly or legally issued.
2. The Heim inventions and improvements were not capable of practical use.
3. Defendant's *(Detroit)* grinder did not contain any of the alleged Heim inventions.
4. Heim was not the first and original inventor.
5. The alleged Heim inventions were known and used by others prior to Heim's patent applications.

Detroit cited nine patents as evidence of prior art, far fewer than the fifty-plus patents cited in Sanford. Included were the Lowman and French and Stephenson patents plus one new British patent *(Edwards, BR 1,292)*. As did Sanford, Detroit relied primarily on two machines as evidence of prior art. Whereas Sanford cited Lowman and French, Detroit cited French and a two-wheel grinding machine once used at the Newark, New Jersey, plant of the Hyatt Roller Bearing Company.

Detroit claimed that it did not infringe on the Heim invention since its machine did not use a separate carrier to support the rolls. Detroit used vertical construction with the larger grinding wheel placed directly above the smaller and a slowly rotating work-driving *(regulating)* wheel. The lower wheel performed three functions: roll support, roll rotation, and roll feeding. To keep the work from rolling off the work-driving wheel, Detroit used two work-holding members placed on either side of the work. This gave the Detroit machine a four-point system for centerless grinding versus the Heim three-point system *(grinding wheel, carrier and regulating wheel)* and allowed Detroit to claim its invention was novel and distinctly different from the Heim invention.

In its arsenal of prior art evidence, Detroit cited a grinder once used by the Hyatt Roller Bearing Company that was never patented. Hyatt had operated a number of machines to grind rolls for its spiral-wound roller bearing in the years prior to 1913. These machines used two grinding wheels with a carrier between them to support continuous feed of the spiral-wound rolls used by Hyatt. Detroit stated that in the Hyatt machine, the wheels rotated in opposite directions with one wheel beveled, rotating upward on the work, and its axis swung out of alignment to cause feeding action. Detroit claimed that the Hyatt grinding machine, with its automatic feed system, embodied the Heim invention. But Detroit could not provide physical evidence of the grinder since Hyatt no longer used the machines. To prove its claim of prior use, Detroit built a replica of the Hyatt grinder and operated it for the court against the performance of one of Heim's older machines. Not only did the Hyatt machine fail to replicate the action of the Heim regulating wheel, it failed to grind rolls to the same precision as Heim.

Detroit also tried to prove that the French and Stephenson machine used at Crucible Steel in Syracuse embodied the Heim patents, despite the prior ruling in Sanford. Detroit claimed in its testimony that Crucible had operated its machine not only at the patent-specified 1,500 and 750 rpm speeds but at slower speeds of 1,100 and 160 rpm, with the lower number approaching the range used by the Heim regulating wheel. During the trial, Detroit attacked Heim, claiming he knew about the slower wheel speeds but did not admit it, resulting in false testimony as was cited by Judge Thomas in his opinion:

> *The attacks upon the testimony of the witness Heim on the Crucible Steel wheel speed are unjustified. He is corroborated by Roth and contradicted only by Pardee, who, however, admitted that, even at a later date, he was experimenting with higher speeds. Hayes testified merely that the machine "operated" as usual. This is a frail support for a charge of false swearing against witnesses who gave their testimony in open court and apparently stated the facts to the best of their recollection.*

To prove that the F&S machine could grind bearing rolls, Detroit built a machine based on the French specifications but modified to grind small rolls and demonstrated it before the court. During the demonstration the machine performed poorly with a distinct lack of regulating action and to worsen matters, it virtually ruined every roll it attempted to grind.

After both machines *(Hyatt and F&S)* failed to demonstrate Heim's regulating wheel action and grinding performance, Heim pointed out that both machines had been abandoned. Hyatt replaced its machine with another grinder different from Heim, and Crucible replaced its French machine with a Heim centerless grinder. In addition to Heim's evidence, the court *(Judge Thomas)* pointed out that while both machines were abandoned, Heim's machine was a sweeping commercial success.

In his opinion issued on May 17, 1926, Judge Thomas made the following summary on the defense offered by Detroit:

> *In the foregoing discussion the invention in suit is taken as the broad Heim combination of a grinding wheel, a regulating wheel, and an interposed carrier, and it is against this combination that defendant's attack is leveled. Although additional features, such as the invention of the process patent in suit, the endwise feed of the work, and the variation of the latter, apparently involve patentable advances beyond the broad invention, it is unnecessary to dwell upon them, for, if the broad invention be novel, then a fortiori this*

invention plus the additional features is not anticipated.

The question of infringement requires no lengthy discussion. If the Heim invention be made use of by the defendant, it matters not that defendant may have added to it, or improved it. By way of missing elements of the Heim combinations in its machines, defendant names only the carrier or work support. We are told that by the transposition of the Heim machine, and throwing the weight of the work entirely, instead of partially, on the regulating wheel, the latter becomes both carrier and regulating wheel, and the former, in so far as the Heim invention is concerned, ceases to exist. Obviously something far more sound and simple than such a theory is required to exonerate one who has bodily taken a meritorious invention.

Both patents in suit are valid, and all claims thereof are infringed. There may be a decree for the plaintiff, with costs to abide the event; and it is so ordered.[11]

The suit was really never about the question of Detroit using Heim's invention. Detroit used the regulating wheel and even though its machine embodied many differences, it could not hide that fact using its vertical orientation. But the case was about prior invention and this time Judge Thomas, heeding the findings of the Second Circuit in the Sanford appeal as well as the reasoning of Learned Hand, realigned his requirements for identifying prior art. Given that both machines cited by the defendant were abandoned and that Heim's machine was not only a sweeping commercial success, but copied in various forms by many other makers of machine tools, was undeniable proof that prior art did not exist.

With the opinion rendered, Judge Thomas issued his decree on July 2, 1926. In Sanford, the court found that only a few claims were infringed. In Fafnir, the court found that "all claims in both patents" were valid and infringed. The decree also stated that Detroit Machine Tool, being the party that hired and paid for the defense, was the real defendant and therefore, was bound by all results of the suit. The decree ordered Fafnir to pay costs and lost profits to Heim Grinder and ordered both Fafnir and Detroit Machine Tool to cease all use, manufacture, and sale of Detroit's centerless grinder. On July 3, Judge Thomas issued a permanent injunction to Fafnir and Detroit that sent a strong message industry-wide to heed the patent rights issued to Heim for centerless grinding.

Even with the decision for Heim and the reversal in the Sanford suit in favor of Heim, Fafnir and Detroit did not give up, for on July 30 Fafnir submitted an appeal to the Second Circuit Court of Appeals in New York City. But something happened in the months after the appeal was filed that caused the defendants to change their minds. As documented in the Court Order issued December 20, 1926, Fafnir and Detroit Machine Tool withdrew their appeal and requested dismissal, with costs paid to Heim Grinder. With the defendant's withdrawal, the Second Circuit Court of Appeals on December 29, 1926, and in the name of the Chief Justice of the Supreme Court, William Howard Taft, dismissed the appeal that ended Heim Grinder v. Fafnir Bearing Company.

Fig. 1—Heim Centerless Grinding Machine Improved. Fig. 2—Rear view of the machine

Fig. 9-4: Final Heim Centerless Grinder Design: American Machinist, September 1926.

Consolidation of the Heim Grinder Company

In May 1926 when Heim won his suit against Fafnir, he was still competing against Cincinnati and Detroit Machine Tool for sales of centerless grinders. It is not known who was selling more machines but Heim had an advantage over Cincinnati and Detroit Machine Tool. During the trial years as discussed in Chapter 6, Heim developed and patented many improvements that put Cincinnati and Detroit at a disadvantage in the market. Shortly after the trial, Heim introduced his most advanced, and what became his final design for a centerless grinding machine. The grinder was profiled in both *American Machinist* and *Machinery* magazines in their September 1926 publications, with the *Machinery* article shown in Appendix 2 *(A2-5-5)*. The machine *(Figure 9-4)* was designed for the full range of work feed arrangements consisting of in-feed, through-feed, and spot feed. The machine was larger, more automated, and the first by Heim to use hydraulics, as described by the following article excerpt:

> *The chief feature of the improved machine is the use of an hydraulic control for the regulating-wheel slide. This control is used in spot or in-feed grinding, and it is said to eliminate all waste time in the movements of the regulating-wheel slide. There is also a hydraulic control for the truing diamonds by means of which the diamonds may be fed uniformly at a fast rate across the wheel or with an almost imperceptible movement. Other improvements embodied in the machine included "Texrope" drives to both the regulating wheel and the grinding wheel, a quick-change gear box, visible settings, and simple adjustments. The machine is of greater capacity than previous models, and it will accommodate work ranging from 1/32 to 6 in. in diameter. In through-pass grinding, bars of any length can be handled and in spot grinding, work up to 8 in. under the head can be ground.*

In a testimonial to Heim's centerless grinder design, Louis Henes, a seller of machine tools in San Francisco, described Heim's advantages over the Cincinnati machine in a May 1926 sales letter to the Mann Manufacturing Company *(see Appendix*

Fig. 9-5: Modern CNC Centerless Grinder Manufactured by Cincinnati Machines, 2015.

2, A2-6-6). In the last paragraph of his letter to Mann, Henes sums up his opinion of the Heim machine:

> *With the photographs that I will leave with you you will find detailed spec-ifications and comparison between the HEIM and Cincinnati machines but please do not misunderstand the motive that prompts me to place this data before you. It is not done with the idea of "knocking" our competitor's ma-chine because they make a very good grinder and they have many of them in use but I really believe that if you study the new HEIM machine thoroughly, you will agree with me that it is superior to the present Cincinnati machine.*

If the Henes letter was any indication of industry opinion, Cincinnati Milling, two years after signing the agreement with Heim, was selling a second-class machine. For-tunately for Geier, with the Fafnir suit won and Detroit Machine Tool out of the center-less grinder business, he had the agreement to buy Heim Grinder. In early September 1926, Geier completed the deal and paid Heim for Heim Grinder along with 42 patents for centerless grinding. The patents included 29 by Heim, nine by Reeves, and four by Booth.[12] On September 10, 1926, Geier formed Cincinnati Grinders Incorporated with an authorized capital of $1,500,000. The next day, September 11, Cincinnati Grind-ers took over the grinding business of Cincinnati Milling Machine and the centerless grinder business of Heim Grinder Company.[13]

The price paid by Geier for Heim Grinder is not known. One obituary in 1964 stated that Heim sold his patents for $1,000,000, but there were also buildings and equipment involved, so the price may have been much higher.[14] There is also a story in which Frederick Geier paid Heim in gold coins. In an article published in 1993 about Frederick Geier and the Cincinnati Mill, the author stated that Heim insisted on being paid an undisclosed amount in gold coins.[15] According to some Heim family members, there was some type of transaction; however, any payment in coins, gold, or even silver could only be a fraction of the buyout price. Since the price of gold in 1926 was around $20 per troy ounce, one million's worth would weigh around 3,400 pounds. So, possibilities include Geier offering gold to convince Heim to sell or Heim asked for some type of initial payment to seal the deal.

Cincinnati Grinder v. LMV

Cincinnati Milling Machine Company, with its newly created grinder division, Cincinnati Grinders, quickly surpassed Brown and Sharpe in sales and became the largest manufacturer of machine tools in the United States. With uncontested ownership of the American market for centerless grinders, Geier set about securing the market in Europe. In 1927, there was only one manufacturer of two-wheel centerless grinders. This was Lidkoping Mekaniska Verkstad AB (LMV) of Sweden. As discussed in Chapter 7, Carl Ekholm, an engineer employed by LMV, obtained the Heim concepts for centerless grinding on his visit to the United States in 1921. By 1927, LMV was selling two-wheel centerless grinders in many European countries.

On November 2, 1927, Cincinnati Grinders notified LMV that it infringed on patents it owned for centerless grinding. Three months later on February 23, 1928, Cincinnati filed a Complaint against LMV with the Landgericht Berlin, a regional court in Germany. LMV, as a result of the Complaint, was forced to restrict production and cease all sales in Germany.[16]

Before 1978, patent applications in Europe had to be filed in each country where protection was desired. Since the cost of filing in all countries can be very expensive, companies generally selected the larger countries with developed markets for its products. In the 1920s, Germany was a growing manufacturer of automobiles and rolling bearings and a large and important market for machine tools. Given that the Complaint was filed in Berlin, it follows that Heim had previously obtained patent protection for Germany and not for Sweden.

The Cincinnati complaint appeared to be on the same basis as Sanford and Fafnir in which the suit alleged infringement of the machine and method of centerless grinding. LMV answered the Complaint with a request for abrogation of the Cincinnati patents on the basis of prior art with similar machines already known. On November 8, 1928, the Landgericht ruled in favor of LMV, agreeing that similar machines did exist.

Cincinnati appealed the decision to the Reichsgericht in Leipzig (Germany's Supreme Court), which required both companies to produce their respective machines for a comparative demonstration. LMV, to prove its claim, had to construct a replica of the machine they cited as prior art. In the following excerpt from his letter, Ekholm describes how both companies tried to cheat the test by supplying pre-ground rolls. Reference to the "professor" is Professor Schwerd, who was appointed by the court to monitor the test:

One episode occurring during test grinding should perhaps be rescued from oblivion. During grinding in front of the professor and his assistant, Cincinnati's grinder suddenly said to our grinder; "Your rollers are ground beforehand and then blackened so that they look unground." The response of our man was not left wanting. He replied: "Yes, that is correct; and so are yours. Shall we swap?" The professor overhead this exchange, and as a result he suggested that they swapped (sic) rollers. This took place, and it turned out that the other party's rollers were better in our machine than our rollers were in theirs.

In this statement, with the Cincinnati ground rolls performing better in the LMV machine, Ekholm confirms that the Cincinnati machine was the superior grinder. According to Ekholm, the result of the test led to the ruling by the Reichsgericht in favor of LMV.

But Cincinnati had a backup plan if they lost. There was another patent previously approved in Germany by Heim that made it impossible for LMV to build its centerless grinder for the more precise grinding performance of the latest Cincinnati machines. Even though it was not identified, the patent was No. 1,579,933 that included in the claims an inclined carrier *(slanted downward from the grinding wheel to the regulating wheel)* for producing an "automatic corrective action" and perfectly round rolls and later cited by Woodbury as the second most important critical invention for centerless grinding.[17]

Again, LMV decided to fight Cincinnati and filed a nullity suit with the Berlin Patent Office. But after losing two decisions, Cincinnati apparently decided that settling with LMV was the more prudent course of action. Cincinnati made an offer to LMV for free use of all of Cincinnati's patents for centerless grinding in exchange for withdrawing the nullity suit and suspending all sales activity in the United States for ten years *(approximately until 1938)*.

This is the story told by Ekholm, but it seems out of character for Geier to offer free use of the centerless grinding patents for which he just paid a large sum. Before 1995, the life of a patent was 20 years from filing or 17 years from grant, whichever period is greater. In the United States, the Heim machine patent *(1,210,937)* was valid until March 1935 and the patent for the slanted carrier *(1,579,933)* was valid until April 1943. With those and other Heim patents, LMV was locked out of the U.S. market until they expired. Thus, there may have been much more to the story than recalled by Ekholm.

However, Lidkoping Mekaniska Verkstad *(LMV)*, with the information obtained by Carl Ekholm, did what Reeves, Sanford, Detroit Machine Tool, and others failed to do. With the help of the German courts, LMV skirted Heim's patents to introduce modern centerless grinding to Europe. As of 2015, LMV continues to be a major manufacturer in Europe selling centerless grinders and other machine tools under the name of KMT Precision Grinding. Cincinnati Grinders eventually became Cincinnati Machines and also continues to be a major manufacturer of modern computer-controlled centerless grinders *(Figure 9-5)*.

Recognition

From 1918 through mid-1926 Heim sold about 250 centerless grinding machines. Sales would have been much greater had not so many infringing machines entered the

Fig. 9-6: Grinding History Display from the "Hall of Tools" Exhibit: 1964–1986, Smithsonian Institute of American History.

market so quickly, and prolonged by the glacial pace of patent infringement litigation. Unlike Sanford, who had some centerless grinders returned, not one customer returned a Heim machine.[18] Heim's centerless grinder revolutionized manufacture of cylindrical metal parts and was later adopted for thread grinding and thread rolling. The fact that so many companies quickly copied Heim's invention supports the notion that centerless grinding was a disruptive machine tool technology that significantly contributed to the rapid growth of the automobile, aerospace, and metal manufacturing industries.

More than 100 years later, centerless grinding continues to be widely used for the grinding and mass production of precision cylindrical work. Grinding accuracies have improved from Heim's best of 0.0001 inches *(2.54 microns)* in the mid-1920s to under 0.00002 inches *(0.5 microns)* in 2017.[19] Centerless grinders have advanced in size, accuracy, and production rates, with most machines equipped with many sensors and controlled by computers. Even with all the advancements, modern centerless grinders use the basic mechanical design and methods invented by Lewis Heim.

The Smithsonian Institute

In 1964, the Smithsonian Institute National Museum of American History opened the Hall of Tools exhibit in Washington, D.C. One of the technologies presented was

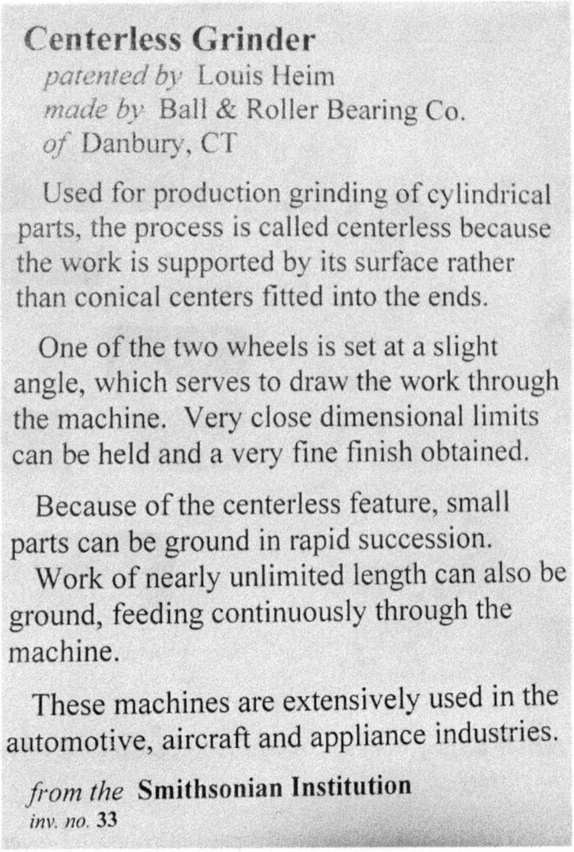

Centerless Grinder

patented by Louis Heim
made by Ball & Roller Bearing Co.
of Danbury, CT

Used for production grinding of cylindrical parts, the process is called centerless because the work is supported by its surface rather than conical centers fitted into the ends.

One of the two wheels is set at a slight angle, which serves to draw the work through the machine. Very close dimensional limits can be held and a very fine finish obtained.

Because of the centerless feature, small parts can be ground in rapid succession.

Work of nearly unlimited length can also be ground, feeding continuously through the machine.

These machines are extensively used in the automotive, aircraft and appliance industries.

from the **Smithsonian Institution**
inv. no. **33**

Fig. 9-7: Description of Heim Centerless Grinder Presented by the Smithsonian Institute at the 1964 Hall of Tools Exhibit.

the history of modern grinding, which included the Heim centerless grinding process. The Hall of Tools exhibit was open from 1964 through 1986 and included a display titled "Grinding" *(Figure 9-6)* that described the five grinding technologies developed between 1881 through 1918. The text inscribed in the display is as follows:

Grinding

The familiar grinding wheel of natural sandstone, which removes metal by heating and tearing off small particles from a surface, has been used to sharpen edge tools for more than 1000 years.

Development of the grinder as a machine tool began about 1800. The transition from the old idea of grinding as a wearing-away action to the present concept of grinding as a cutting process occurred during the last quarter of the 19th Century.

Growth Of An Idea: 1881–1918

Centerless Grinding
Was originally applied to external grinding of round steel rods, of roller-chain rolls, and of roller-bearing rolls. It has been widely adapted to grinding engine parts and other mass-produced items. Through this process, concentricity of the workpiece can be held within less than 0.001 inch.

Samuel Tretheway: 1881
External centerless grinding in which the workpiece is guided against the grinding wheel by means other than mounting it on centers was done on a machine patented by Samuel Tretheway of Pittsburg, Pa. This machine for grinding round steel rods and tubes used slightly skewed opposing cylinders to advance the work past the wheel.

Hans Renold: England, 1906
Similar in principle to Tretheway's machine, Renold's machine—used for grinding rods for roller-chain pins—employed slightly skewed opposing disks to advance the work past the wheel.

Robert H. Grant: 1914
Robert H. Grant of Detroit, Michigan patented a machine for grinding roller-bearing rolls in which the rolls were guided in an undercut groove and were advanced past the wheel by the wheel's slightly conical shape.

French & Stephenson: 1914
Applied to the finishing of cold rolled steel rods by E. L. French and G. W. Stephenson of Syracuse N.Y., the feed disks of the Renold machine of 1906 were replaced by grinding wheels turning in opposite directions and at different speeds.

Louis R. Heim: 1918
Substitution of two peripheral abrasive wheels for the disks of the French & Stephenson machine was a principle adopted by Lewis R. Heim of Danbury, Conn. and later used in virtually all subsequent machines.

In current machines, the grinding wheel rotates as shown—with a surface velocity of about 6500 ft. per minute (nearly 75 miles per hour). The surface velocity of the regulating wheel ranges from 36 to 900 ft. per minute. The work blade and regulating wheel control the rotation of the workpiece.
The regulating wheel is slightly skewed to move the workpiece past the grinding wheel and to discharge it from the machine.

(Author's Note: The exhibit used two spellings of Heim's first name with Lewis being correct.)

In addition to the display on "Grinding" history, the Smithsonian displayed one of Heim's original centerless grinding machines. The significance of Heim's invention was briefly described on a card placed on the machine shown in Figure 9-7. The machine was donated to the Smithsonian in 1962 by the Danbury Centerless Grinder Company and subsequently restored for the Hall of Tools exhibit. The machine is one of Heim's peripheral centerless grinders manufactured at the Ball and Roller Bearing Company after 1921, with the nameplate indicating it is No. 209. Both the machine and nameplate can be seen in Appendix 2 *(A2-8-10)*.

There are two other original Heim centerless grinders with the nameplates showing numbers 158 and 225. Both are peripheral models with No. 158 manufactured at the Ball and Roller Bearing Company and No. 225 manufactured at the Heim Grinder Company. No. 158 is an unrestored machine manufactured after 1921 that resides at the American Precision Museum in Windsor, Vermont. A picture of this machine and its nameplate can be seen in Appendix 2 *(A2-8-9)*. No. 225 resides at the Heim Company in Fairfield, Connecticut, and was still being used in 2014 to make bearings. The machine was made after the formation of the Heim Grinder Company in April 1924 and designed to allow the grinding wheel to oscillate back and forth over the work.

National Register of Historic Places

On July 7, 1989, the Connecticut State Historic Preservation Office nominated the buildings and property of the former Ball and Roller Bearing Company at 20–22 Maple Avenue in Danbury, Connecticut, to be entered into the National Register of Historic Places. The nomination claimed that "the Ball and Roller Bearing Company was significant for its association with inventor Lewis Heim. In the factory, Heim developed the modern centerless grinding machine, an important advance in machine tool technology." The nomination was quickly accepted and on August 25, 1989, the site of the former Ball and Roller Bearing Company was entered in the National Register of Historic Places administered by the National Park Service.[20]

Chapter 10

The Florida Years

Mount Dora

After World War I ended, the United States experienced two recessions, with the second being more of a short depression that ended around July 1921. Afterward, the United States experienced a strong recovery with unprecedented industrial growth and prosperity, which period became known as the "Roaring Twenties." One of the beneficiaries of this prosperity was Florida, which was experiencing a land boom resulting from a statewide strategy to encourage tourism and retiree migration. The land boom resulted in the formation of new cities, with many being created out of the Everglades region near Miami. The Miami area was advertised as a tropical paradise that fueled the development boom that attracted investors from outside Florida. But the land boom was not just associated with Miami; it stretched far and wide across Florida. One small Florida town that encouraged land development was Mount Dora, which was located in the picturesque lake country northwest of Orlando.

In 1919, Connecticut native Bob White moved to Mount Dora and became a real estate developer. White was a former student of Lewis Heim, who taught Sunday school at a local Methodist Church in Danbury. Outside of his church, one of Heim's other activities was dabbling in Danbury real estate, which was, in some respects, a side business. Danbury land records from that period contain many entries for houses, commercial buildings, and land bought and sold by Heim. Descendants said Heim helped his family in some of these transactions to own houses or other properties for their businesses.

Bob White knew that Heim was interested in real estate development opportunities and encouraged him to move to Mount Dora.[1] Florida obviously also offered a much warmer climate than Connecticut, which was an attraction to Lewis, who was tired of the cold Connecticut winters. Whereas real estate development was the primary reason for selling his centerless grinder patents and most of his shares in the Ball and Roller Bearing Company, there were other events that may have contributed to Heim's decision. These included the deterioration in his health due to a two-year *(1917–1919)* battle with exhaustion, the stress associated with patent infringement litigation, and the death of his younger brother, William.

William Heim, who started his career as a machinist in the hatting industry and then as a bicycle maker, opened Heim's Music Store in 1905 at the age of 27. In 1912, William married Ethel Sperry *(27)* of Lawyersville, New York. Their first child, Jeanette Adelaide Heim, was born in August 1914. On January 4, 1919, at the age of 40, William died of complications from pneumonia and influenza. A little over five months later in June, Ethel gave birth to their second child, Willa Sperry Heim. Heim's

Music Store was sold the same year to L. Jackson and F. Hanson, who kept the music store name unchanged for many years.

In 1924, after establishing the Heim Grinder Company and signing the centerless grinder deal with the Cincinnati Milling Machine Company, Heim and his wife, Anna, moved to Mount Dora. Once in Mount Dora, he initiated construction of a large house at 347 East Third Avenue *(Figures 10-1 and 2)* that was completed in 1926. In 1924, Heim's brother Alfred also left the Ball and Roller Bearing Company and bought Peffer's warehouse at 14 Crosby Street, just down the road from the Heim Grinder Company. Alfred operated the warehouse until 1928, when he sold the business and his house at 40 North Street and with his wife, Clara, also moved to Florida to be near Lewis and Anna.

After establishing himself in Mount Dora, Heim began to plan his real estate development project that he named Sylvan Shores. Together with partners, Heim formed the Sylvan Shores Company and purchased land south and west of Lake Gertrude. The portion of the highway bordering the south side of the property, that ran west to Tavares and east to Mount Dora, was renamed "Heim Road". Sylvan Shores was a high-density, planned community designed for a mixture of luxury and modest-sized homes. The property was divided into 48 sections containing 931 building lots and some open spaces. To help sell lots, Heim built a small hotel on the property to cater to prospective buyers and published a colorful brochure *(Figure 10-3)* that contained a map of all streets and home sites.

To complement his investment in land, Heim also invested in a local bank with the purpose of providing mortgages and other funding for Sylvan Shores. In September 1925, the Mount Dora Bank and Trust Company opened and, by the following March, occupied an attractive Georgian-style building at the corner of Fifth Avenue and Donnelly Street. Heim, along with another retired businessman, provided much of the operating capital and each became a vice president. In 1927, the bank applied to join the Federal Reserve System and after acceptance changed its name to the First National Bank of Mount Dora.

Starting in 1925, the Florida real estate boom began to show signs of stress. Claims of inflated land prices and negative press started to appear throughout the state, warning investors that the land boom was starting to fizzle out. Adding to the real estate problems were a pair of hurricanes, one in 1926 and the other in 1928, that devastated southern Florida, forced many developers into bankruptcy, and severely deflated land values.

Up to the summer of 1929, Heim built nearly three-dozen expensive homes around the south end of Lake Gertrude and on Morningside Drive and Sheridan Road to the west. Further west of the lake, the lots in Sylvan Shores were slow to sell. When the Great Depression hit in October 1929, sales completely stopped. By March 1932, Heim needed cash to shore up Sylvan Shores' finances and put 233 lots up for auction. Nearly 100 of them found buyers but this was not enough, eventually forcing Heim to sell out in 1934. The new owner of Sylvan Shores had little success as 780 lots remained unsold in 1942. In the late 1950s, there was a rush into Mount Dora real estate and the development finally sold out.

In 1925, in addition to his investments in Sylvan Shores and the bank, Heim bought a small weekly newspaper called the *Mount Dora Topic*, but after four years sold

Fig. 10-1: Heim House at 347 East Third Avenue in Mount Dora, front view after construction was completed in 1926.

Fig. 10-2: Anna, Lewis, Charles, and Florence Heim in Mt. Dora, May 1927.

Fig. 10-3: Title and Summary Pages from Sylvan Shores Brochure, ca. 1926.

it back to the original owners. About the same time he bought the newspaper, Heim contributed to the development of Mount Dora with a gift of land on Liberty Avenue on which he built a baseball field complete with a grandstand. "Heim Field" became the home of the Mount Dora town baseball team, the Vikings. In addition to Heim Field, in 1932 he also gave funds to the town to erect a line of decorative streetlights along the highway from downtown to Sylvan Shores. These lights were often called the "white way" and continue to light up Mount Dora each night. In remembrance of Heim, the town designated his house on Third Avenue as a historic landmark.[2] The failure of the Sylvan Shores project cost Heim most of his savings, ended his career in real estate development, and, in 1932 at age 58, forced him to return to his previous profession of making bearings.

Experimental Bearing Designs

After moving to Florida and up through 1927, Heim focused his attention on real estate and his bank, with no activity on new inventions. In 1927, Heim became active again in the development of bearings and machines. In his quest to develop a lower-cost bearing without compromising durability and reliability, Heim experimented with two new concepts for making bearings. The first concept involved the use of flat bar stock made from sheet metal to make metal shells that form the bearing housing and races. Whereas most bearings at that time were fabricated from milled metal bar stock, Heim's concept consisted of forming cylindrical or tubular-shaped shells in a suitable metal press using preformed dies, hardening said components, then fitting them together with cylindrical rollers for a complete bearing. The concept of "Stamped Sheet Metal Components" allowed bearing parts to be made from standard-size, readily available, and lower-cost raw metal stock that significantly reduced machining and assembly costs. Additionally, Heim's new concept facilitated designs using a minimum of parts that were ready for assembly by unskilled labor, with negligible risk of creating defective bearings through careless workmanship. These concepts produced durable, reliable, and lower-cost bearings that allowed Heim to compete against seasoned manufacturers.

Fig. 10-4: Centerless Grinder for Flat Bar Stock. Patent 1,958,001, filed Aug. 9, 1927.

Even though Heim switched from milled metal to sheet metal for his new bearing construction, some components still required grinding to achieve precision dimensions with a very smooth surface finish. At that time precision grinding of flat material was labor intensive and expensive since the work surface, generally large in area, had to be traversed relative to the grinding wheel. This was achieved by shifting *(manually or mechanically)* either the grinding wheel against the fixed work or the work under the fixed grinding wheel. Either method involved a great loss of time handling the work as well as slow production speeds.

The lack of a commercially available high-speed grinder led Heim to develop a machine that could continuously and rapidly grind flat bar stock such as sheet metal in succession with minimal operator input. The machine was Heim's fourth invention of a precision grinder and incorporated the concepts of his centerless grinder. The machine *(Figure 10-4)* consisted of a table on which the grinding and regulating wheels were

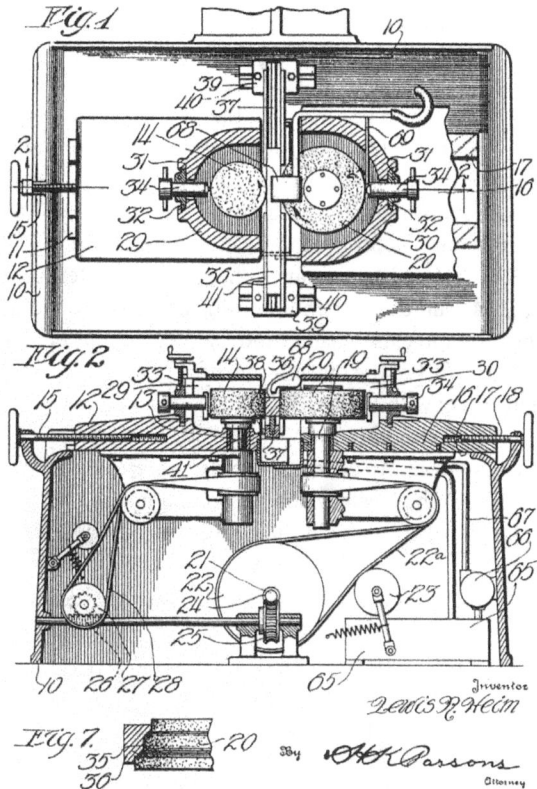

mounted horizontally such that the work to be ground traveled along the flat surface and passed between the rotating periphery surfaces. The machine operated similar to Heim's peripheral wheel centerless grinder except that the surface of both wheels contacting the work rotated in the "same" direction. Whereas the high-speed wheel performed grinding and pulled the work forward, the slow-speed wheel held the work back and regulated its forward speed. Heim's grinder continuously ground not only work with a simple flat face but also work with curved or contoured shapes (*see Figure 7 in Figure 10-4*). This feature made Heim's machine both novel and valuable as it lowered the cost of machining complex shapes. Heim filed the patent application (*1,958,001*) on August 9, 1927, with the U.S. Patent and Trademark Office granting letters patent almost seven years later on May 8, 1934. But ten days prior to filing the application, on July 30 Heim assigned all rights to the patent to Cincinnati Grinders.[3] The reasons behind the assignment and hence sale are not known, but the sale may have been due to financial need by Heim or due to a previous agreement with Frederick Geier.

Five years later, on May 6, 1932, Cincinnati Grinders sold Heim's patent (*1,958,001*) along with many other patents to the Heald Machine Company of Worcester, Massachusetts.[4] The Heald Machine Company was a very successful manufacturer of machine tools with a specialty in machines to grind the inside surfaces of cylindrical parts, such as the cylinders of internal combustion engines. The company was acquired by Cincinnati Milacron in 1974 and subsequently closed in 1992. The sale to Cincinnati Grinders and Heald demonstrated recognition by others of the value of Heim's invention.

Based on his new concept of making roller bearings from low-cost stamped sheet metal, Heim developed three designs for shafts and pulleys that were filed in October (*1,943,864*), November (*1,885,914*), and December (*1,976,019*) 1928. The first bearing was designed for leaf springs typically installed in the suspension systems of automobiles and trucks in which the bearing experienced oscillation and hence limited rotation. In these applications, limited rotation

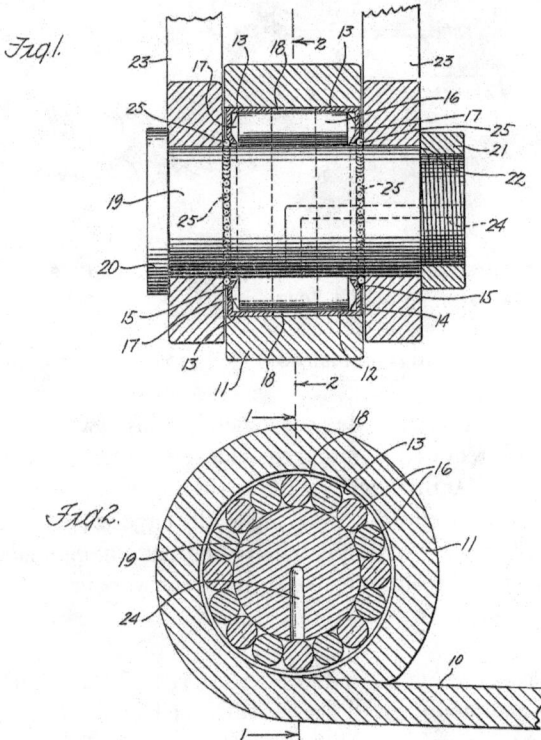

Fig. 10-5: Stamped Sheet Metal Roller Bearing for Limited Rotation. Patent 1,943,864, filed Oct. 11, 1928.

tends to squeeze out lubricant, resulting in dry surfaces, accelerated bearing wear, and eventual failure. In his patent *(1,934,864)*, Heim designed his bearing with two cup-like members that contained two features to retain lubricant. The first was a space *(see item 18 in Figure 1 of Figure 10-5)* between the cup-like members that acted as a reservoir that allowed lubricant to disperse circumferentially to all the rollers. The second feature was to crimp the ends of the flanges on the cup-like members that provided not only a track for the rollers but to retain lubricant. Another feature of the bearing was its ability to accommodate friction generated from axial thrust induced by the end shackles supporting the bearing. Heim provided a space between the bearing shell and the end shackle for insertion of small balls that absorbed these loads while eliminating friction generated from rotation.

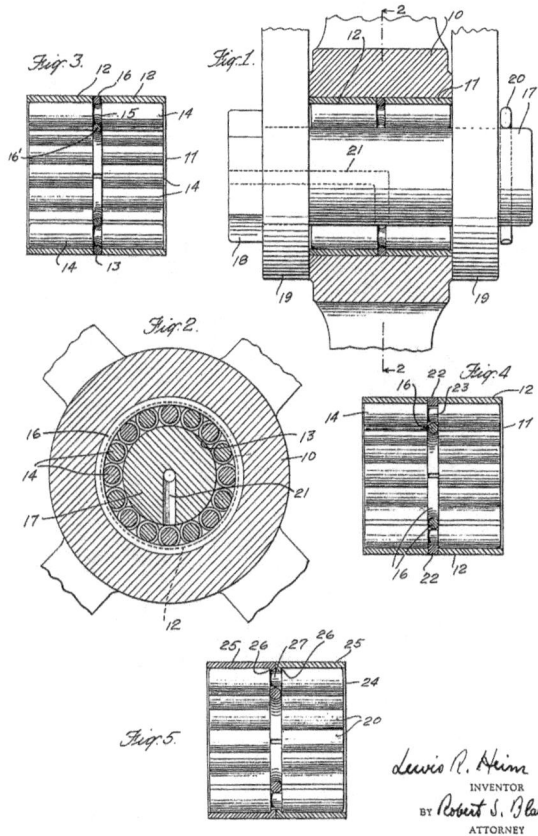

Fig. 10-6: Stamped Sheet Metal Roller Bearing for Wheel Axles. Patent 1,885,914, filed Nov. 2, 1928. Retention Ring: #16′ in Fig. 3 and #16 in Fig. 4.

The second bearing developed by Heim using stamped sheet metal construction was a single-shell roller bearing *(1,885,914)* designed for shafts such as wheel axles. The bearing *(Figure 10-6)* consisted of a single outer race fitted with a raised ring at the center, onto which were inserted a full complement of rollers, each machined with an annular *(ring-shaped)* groove. The novel feature of this bearing was the use of a retention ring inserted on the inside that allowed the bearing to be shipped and stored without dislodging the rollers. When installed onto an axle, the retaining ring was automatically dislodged, allowing the axle to rotate freely against the bearing rolls.

The third invention *(1,976,019)* was not a bearing, but a method for making from stamped sheet metal, a bearing designed for use in wheels, rollers and similar components. In this invention, the single race bearing was assembled from two cup-like sections that form the outer race. When mounted on a wheel, a shaft passes through the center of the bearing and makes contact with the rollers within the bearing.

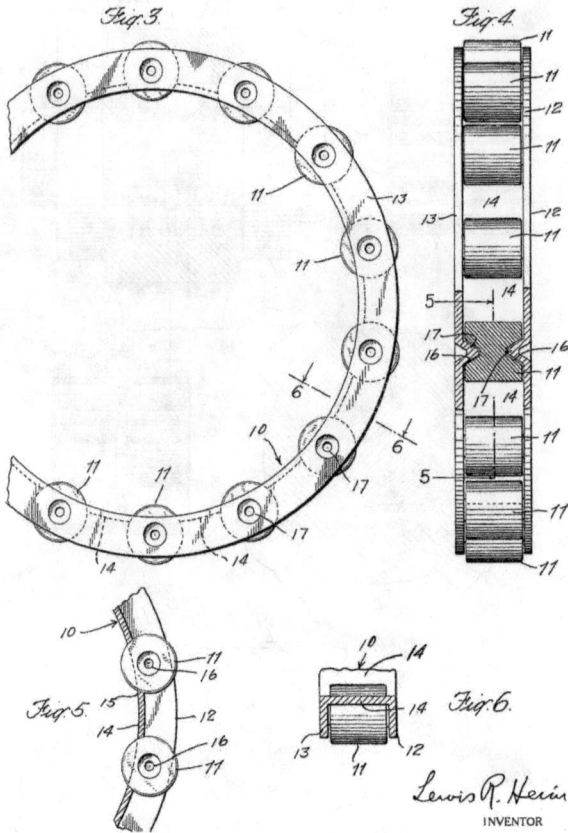

Fig. 3.   Fig. 4.

Fig. 5.   Fig. 6.

Lewis R. Heim
INVENTOR

Fig. 10-7: Stamped Sheet Metal Roller Bearing for Relative Motion. Patent 1,922,805, filed Oct. 4, 1930.

The novel feature of this invention is the relatively quick and easy method for securely mating the two cup-like sections that involve one or more conical bosses *(protrusions punched into the metal housing)* to create a force-fit locking system that lowered the total cost to manufacture this type of bearing.

Following his three applications in 1928, Heim filed two more in 1930 that incorporated his new construction method using stamped sheet metal components. Both of these roller bearings were composed of a one-piece shell that was formed into a circular tube then crimped on both sides to form flanges. Loosely fitting rollers were inserted in between the flanges and retained using the two methods described below. The bearing was designed to allow relative motion between two rotating members, which was accomplished with rollers greater in diameter than the flange height.

Heim's first design *(1,984,213)* for this application was filed in August and incorporated slots punched out of the shell into which the rollers were inserted. To retain the rollers, the flanges were pressed between suitable dies to form projecting portions at regular intervals. The second bearing *(1,922,805)*, filed in October, incorporated a different method to retain the rollers. Instead of pressed portions, Heim used cone-shaped projections punched inward on both sides that can be seen in Figure 10-7. Both of these designs permitted the use of more rolls in a bearing of this type, thus easing the load on each roll. Heim's designs provided a simple and reliable means for maintaining the rolls in proper alignment with reduced friction that increased durability and bearing life.

190

Telescopically Interlocked Construction

The second concept Heim developed while living in Florida involved a new design and assembly method that he called "Telescopically Interlocked Construction," in which a pair of shells made from stamped sheet metal "overlapped and tightly fit" to form a structurally sound bearing. This method eliminated the need for soldering, welding, or other methods of connecting bearing components and further reduced manufacturing time and labor costs while minimizing the risk of fabricating a defective bearing through poor workmanship or careless assembly.

Heim developed three rolling bearing designs incorporating telescopically interlocked construction and filed separate patent applications for each with the first *(2,074,182)* filed on December 3, 1930. All bearings were made for shafts and formed from various

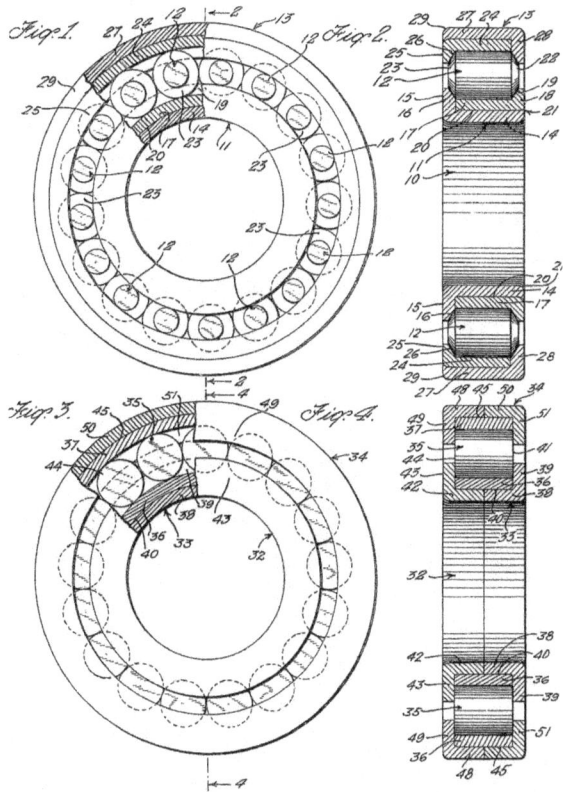

Fig. 10-8: Roller Bearings with Inner and Outer Races Formed from a Pair of Telescopically Interlocked Cylindrical Shells. Patent 2,074,182, filed Dec. 3, 1930.

combinations of either two or four cylindrical shells, with each shell flanged to form the sidewalls on each side of the bearing to retain the rollers endwise. Two designs were assembled using inner and outer races, while the third design comprised only an outer race with the shaft acting as the inner race. One design incorporated a cage to space the rollers while the other two retained a full complement of rollers to maximize its load-carrying ability. Figures 10-8, 10-9, and 10-10 illustrate the various designs and Heim's telescopically interlocked construction.

There was also one other patent application filed while Heim was living in Florida. The application related to a new type of centerless grinder equipped with two regulating wheels and a stationary grinding element that ground over a large area of the work instead of along a thin line, as done on conventional peripheral wheel machines. The machine consisted of two long rolls placed side by side that acted as the carrier to support and rotate the work against a stationary grinding element positioned above the

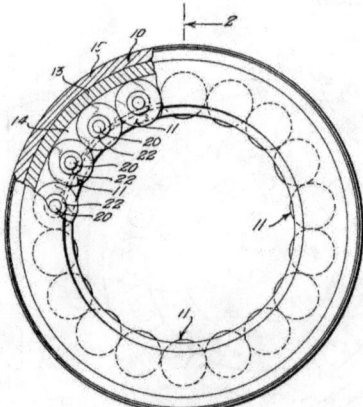

Fig. 1.

Fig. 2.

$\mathcal{L}ewis\ R.\ Heim$
INVENTOR

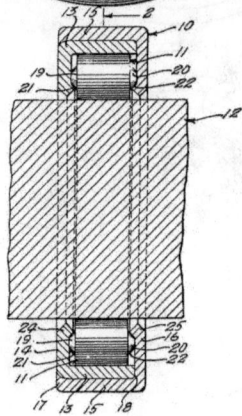

gap between the rolls. One roll was inclined slightly to feed the work along the stationary grinding element. The grinding element consisted of an abrasive block with a width a little more than the diameter of the work. The grinding element at the point of contact with the work was concave, allowing contact over a large area of the cylindrical work.

The purpose of the machine was to grind uniform diameter cylindrical objects to achieve a very fine finish at high production rates and adapted to grind not only work made of hardened steel, but also other materials including cast iron, brass, aluminum, glass, fiber, and wood. The two primary features were the use of a concave grinding element and the ability to adjust grinding pressure on the work surface. The use of

Fig. 10-9: Roller Bearing with Outer Race Formed from a Pair of Telescopically Interlocked Cylindrical Shells. Patent 2,080,609, filed Jan. 2, 1931.

a stationary grinding element, that contacted work over a large area, was better able to grind out imperfections such as high spots *(bumps)*, low spots, and uneven diameters along its length. The concave profile inherently applied greater pressure to imperfections producing perfectly round and straight work. And the ability to control grinding pressure over the entire surface allowed fine grinding or lapping of other types of materials that could not be performed in conventional centerless grinders. Details of the machine showing the rolls and grinding element are illustrated in Figure 10-11 taken from the patent.

The application for this machine that became patent 1,987,850 was filed on May 14, 1931, with the inventor named as Charles R. Heim of Mount Dora, Florida. Charles was the son of Lewis Heim and at the time the invention was developed was attending Yale University in Connecticut. Given the complexity of the machine, it is obvious that his father, Lewis, was deeply involved in its development.

Based on the description in the patent it appears that Heim made a serious attempt to develop another type of centerless machine for special-purpose grinding. The extent to which Heim used this invention is not known and since it never became widely commercialized it is presumed that either its performance was no better than other methods or there was no market for this type of machine.

From 1928 until the end of May 1931, while living in Mount Dora, Heim developed and filed eight patent applications for his experimental bearing designs and two applications for grinding machines. The bearings described above became the springboard for Heim to re-enter the manufacturing business and penetrate a competitive market with lower-cost products well adapted to achieve reliability and durability standards demanded by customers.

Return to Connecticut

The city of Danbury published a directory of incorpo-

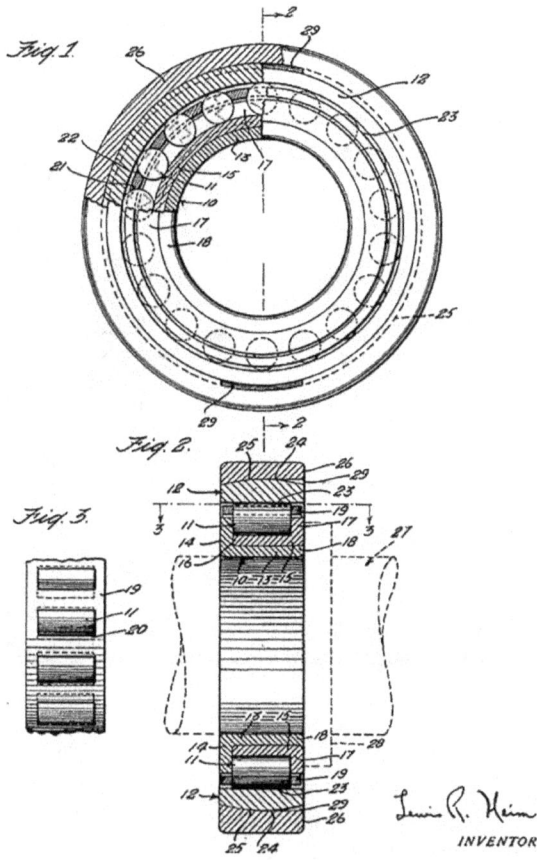

Fig. 10-10: Roller Bearing with Inner and Outer Races Formed from a Pair of Telescopically Interlocked Cylindrical Shells and Cage for Rollers. Patent 2,044,168, filed Jan. 24, 1931.

rated businesses that listed the names of the top officers and their positions. When the Ball and Roller Bearing Company *(B&RB)* incorporated in 1914, Lewis Heim was listed as holding the positions of president and treasurer with brother-in-law, John Henry Roth, holding the position of secretary. In 1917, Roth took the position of treasurer and a young engineer, Howard I. Beard, around 24 years old, became secretary. In 1928, both Beard and Roth were replaced by George Rockwell as secretary and Charles Waters *(Heim's cousin)* as treasurer. The origins of Rockwell and his ownership relationship with the B&RB are not known, but he was listed as secretary up to 1934, after which he changed positions with Charles Waters to become treasurer.

In 1928, the B&RB added the position of vice president and appointed William Barrett who, in addition to Roth, was one of the original partners with Heim when the B&RB incorporated in 1914. Heim, being the founder of the company as well as a minority shareholder, retained the title of president up through 1933, even though he

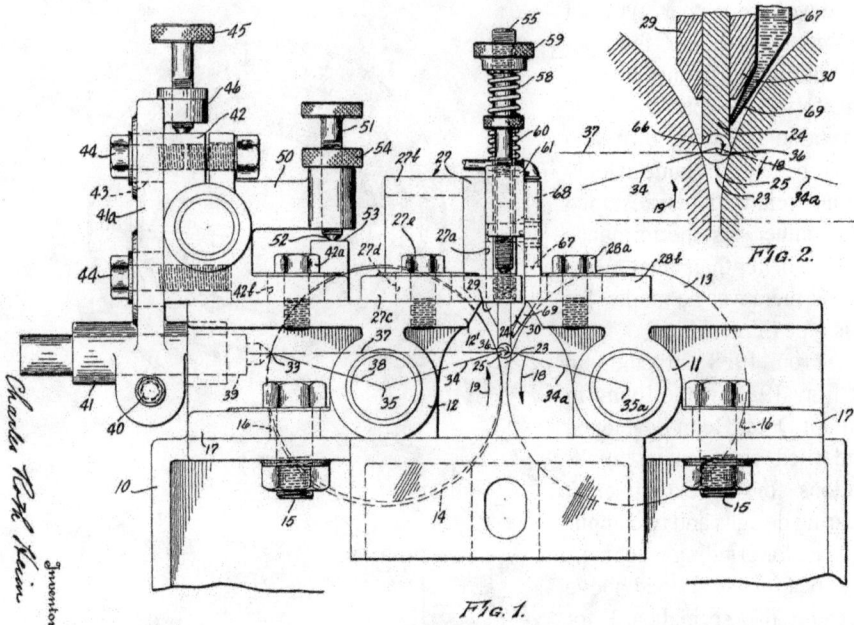

Fig. 10-11: Charles R. Heim Grinding and Lapping Centerless Grinder. Patent 1,987,850, filed May 14, 1931.

moved to Florida in 1924 to focus on real estate development. With Heim in Florida and inactive in its operation, Barrett as vice president most likely managed the B&RB.

When Heim returned to Danbury in 1932, he attempted to rejoin the B&RB and since he was still technically president and held a minority interest, tried to regain his former responsibilities. However, the other shareholders reportedly rebuffed Heim and instead temporarily promoted William Barrett to president until 1934, when the shareholders promoted Howard I. Beard to the position of president.[5] Beard, who was an engineer and also received letters patent for inventions of rolling bearings, apparently won the battle against Heim for control of the Ball and Roller Bearing Company.

With Beard as president, the Ball and Roller Bearing Company continued in business, reaching a production peak during World War II when it manufactured tank bearings for the U.S. Army and employed 115 people. The B&RB remained at 20–22 Maple Avenue until 1974 when, nearing bankruptcy, the factory was sold and the company moved to New Milford, where it continues in business on a modest scale, selling many of the same bearings developed by Heim.[6]

Chapter 11

The Heim Company

A. H. Nilson

In 1932, the United States continued to experience the effects of the Great Depression, with industrial production falling 46 percent from 1929 before beginning to rebound in 1933. By the end of 1932, unemployment was still rising and approaching 20 percent nationwide. Being part of the industrial sector, the machine tool industry was suffering to the extent that its annual trade shows in 1931 and 1933 were cancelled. The hard times did have a positive side in that they forced many companies, in order to compete and survive, to develop new technologies and innovative lower-cost products. One example was the introduction of the metal-cutting contour bandsaw invented by Leighton Wilkie in 1933, which became one of the most widely used tools in manufacturing. Another example was the development of a V-8 engine by Ford Motor Company that was the first engine cast in one piece. The casting process for producing the one-piece engine not only cut down on machining time but also reduced the risk of machining errors that forced the scrapping of the entire engine. In the bearings industry, sometime around 1933 the Torrington Company developed the needle bearing designed for constraining shafts that became the company's primary product for many years.

After failing to return to the Ball and Roller Bearing Company and with no factory to make his new bearing designs, Heim was left with few options. With the savings that remained from his Florida real estate project and the sale of his B&RB stock, Heim decided to start a new company in the already depressed bearings industry and formed the Heim Company in Danbury.[1] The Heim Company began in a small rented shop, with five employees manufacturing Heim's novel bearings made from stamped sheet metal and interlocked construction. One of the employees was Heim's son, Charles *(Figure 11-1)*, who was finishing up his bachelor of science degree at Yale. In addition to manufacturing bearings, the Heim Company performed grinding work on contract for other companies.[2]

Between 1932 and 1935, Heim conducted experimental work to perfect his bearing ideas at the A. H. Nilson Machine Company in Bridgeport, Connecticut.[3] Nilson Machine was owned by Axel Hilmer Nilson, who was born in Wermland, Sweden, in 1849 and immigrated to Bridgeport in 1880 at the age of 31. Nilson first worked at the Bridgeport Organ Company as a cabinetmaker and in 1896 founded the A. H. Nilson Company, which manufactured corset-making machines of his own invention. Around 1905, after selling the business, Nilson founded the A. H. Nilson Machine Company located at 1525 Railroad Avenue, which first manufactured metal and wire-forming machines and later expanded into automatic stapling machines, chain-making ma-

Fig. 11-1: Charles Roth Heim (Age: 24), ca. 1932.

chines, wire reels, wire straighteners, and dipping machines.

Like Heim, Nilson was an inventor, but specialized in machines that automated the handling of wire. Inventions include a wire-straightening and cutting machine *(701,375)*, a wire-working machine *(804,029)*, and a wire-bending machine *(969,488)*, all of which were filed between 1901 and 1909. Nilson was well known in the trade for his "Four-Slide" automatic wire and metal-forming machine that was capable of producing a wide range of formed parts to close tolerances at high speeds. Per Nilson Bulletin #61, the machine performed four operations that took metal directly from the coil, straightened it, fed a predetermined length, cut it off, and then formed the part while eliminating any secondary handling of the work. With this machine, many different forms of wire could be produced automatically in large quantities, in sizes ranging from 3 to 30 inches.

In 1910, Heim filed two applications for inventions similar to the wire-bending and forming machines sold by Nilson. The inventions related to a simple and inexpensive stapling machine and the wire-feeding mechanism incorporated within the stapling machine *(Figures 11-2 and 11-3)*. The patent applications for each were filed in January and February 1910, with the stapling machine *(1,036,841)* granted in 1912 and the wire-feeding mechanism *(1,119,510)* granted in 1914. Patent 1,036,841 described an invention for a low-cost, hand-operated, and foolproof stapling machine that automatically feeds wire from a coil, cuts off a blank, forms a staple, and drives it downward to staple the object in one stroke of the plunger. Patent 1,119,510 described the wire-feeding mechanism that used a system of geared feed rolls activated through the plunger. The pinions *(gears)* were designed with smooth sections that interrupted feeding of the wire after a predetermined length had been fed and was patented separately since the mechanism could be used in other devices. There are many similarities between Heim's stapling machine patents and Nilson's wire-handling machines. However, no records were found describing what Heim did with his invention. He may have manufactured it at the Heim Machine Company prior to founding the Ball and Roller Bearing Company or he may have offered it to Nilson. Regardless, this was the last device of its type invented by Heim.

Both Heim and Nilson were inventors of machines and operated manufacturing plants in Connecticut. They even used the services of the same patent attorney, A. M.

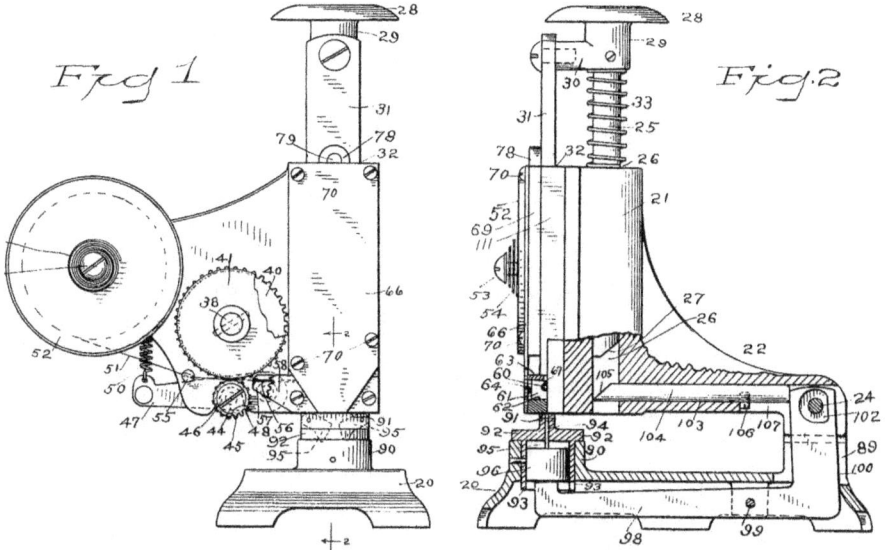

Fig. 11-2: Heim Automatic Stapling Machine. Patents 1,036,841, filed Jan. 14, 1910 and 1,119,510, filed Feb. 3, 1910.

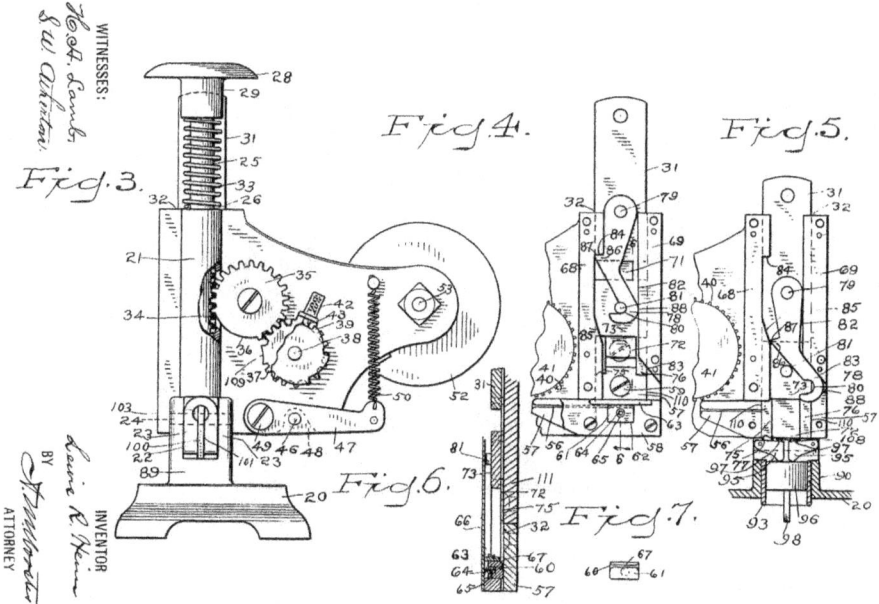

Fig. 11-3: Heim Automatic Stapling Machine Showing Wire Feeding (Fig. 3) and Stapling Mechanisms (Figs. 4 and 5). Patents 1,036,841, filed Jan. 14, 1910 and 1,119,510, filed Feb. 3, 1910.

Wooster, of Bridgeport, Connecticut. Given that Heim cited Nilson as the host for perfecting his new bearing designs created in Florida, it seems logical that they were friends through business and possibly social circles. Their friendship also explains the origins of the two stapling machine patents that had no relation to any of Heim's hat-making, collar-ironing, and bearing-making machines.

Fairfield

In 1935, after three years in Danbury, Heim relocated his company to 46 Sanford Street in Fairfield, Connecticut. Heim moved into his house on Fairfield Beach and used his home as a second office. He converted one of the rooms overlooking Long Island Sound into a glassed-in study in which he installed a drawing board so that he could continue his work at home on designs for new bearings and machines.[4]

Between 1934 and 1936, Heim filed three additional patent applications related to bearings made from stamped sheet metal. The first two applications were filed on July 18, 1934, and related to two inventions for bearings using the telescopically in-terlocked construction developed in Florida. The objectives of the inventions were to fabricate a low-cost bearing in large quantities by unskilled labor that was simple in design, durable in construction, reliable, and built for heavy loads relative to its size. Also, the bearings were to be fabricated from inexpensive materials, require a minimum of machining operations and labor, and be rapidly assembled without com-promising quality.

The bearing was formed from a pair of shells and was similar to the two-piece bearing described in 2,080,609, but designed with much wider shells and rollers for heavier shaft loads. Outside of the bearing design, what made this patent different was that it combined two inventions into one patent. The first was an invention for the bear-ing "Construction" and the second was the invention for the "Machine and Method" for making the shells that formed the bearing housing.

The construction invention was divided out from the method invention and filed as a separate application that became patent 2,301,399. This patent describes a roller bearing formed from a pair of cylindrical sheet metal shells consisting of an inner sleeve and an outer sleeve and "Telescopically Interlocked" by force-fitting the outer sleeve over the inner sleeve, creating a rigid bearing housing. The inner sleeve was designed to provide a tracking surface for the rollers that was continuous and unbroken axially and circumferentially, providing for smooth and unobstructed movement of the rollers. There was no inner race for this bearing, so Heim sized the rollers to fit tightly against the inner sleeve *(outer race)* such that each roller acted as a keystone to the other rollers to prevent radial dislocation. Heim developed three bearing configurations using various combinations of flanged and unflanged shells that are shown in Figure 11-4 *(Figs. 1, 12, and 13)*. A short article describing one of the bearings was published in *American Machinist* on June 1, 1938. The article, which includes a picture identical to the bearings in Figure 11-4, describes the bear-ing with self-contained rollers manufactured to within 0.0001 inch *(2.54 microns)* and designed to retain lubrication when used with shaft diameters up to 1.5 inches.

The method application that became patent 2,102,460 described a machine and

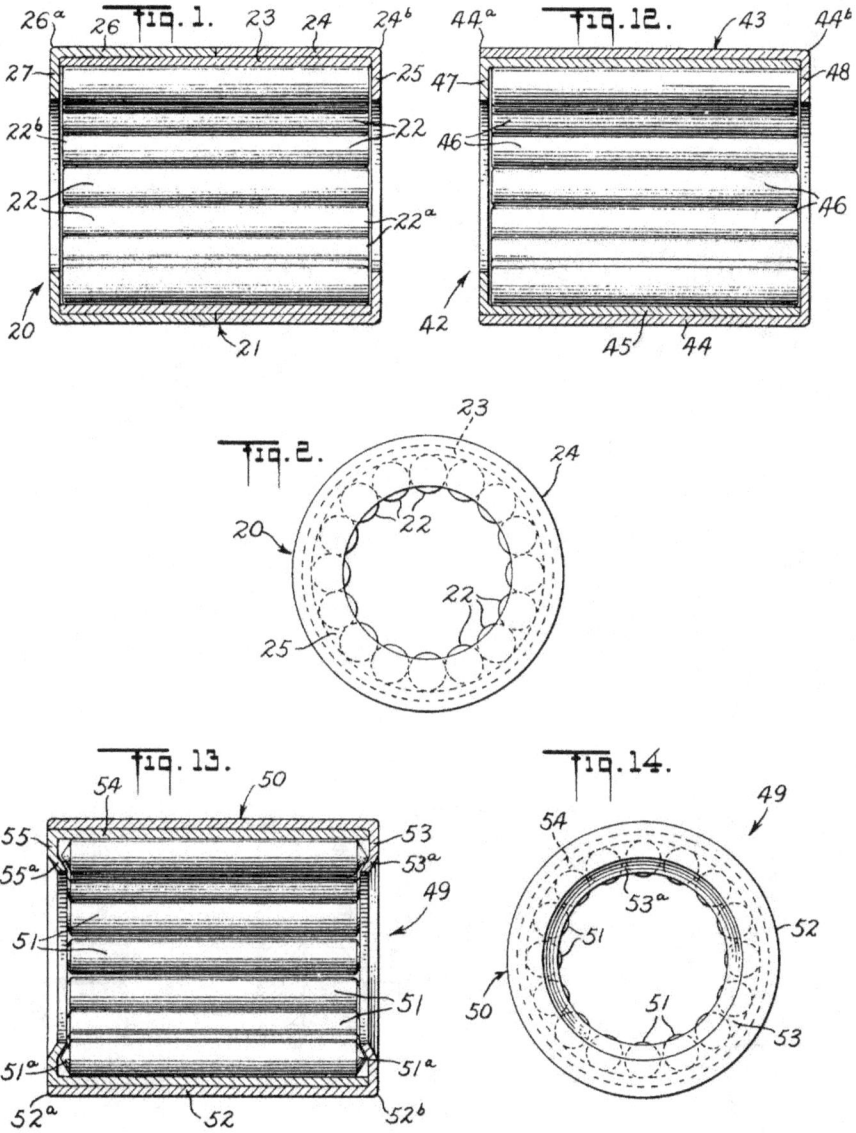

Fig. 11-4: Roller Bearing Designs Using Telescopically Interlocked Construction. Patents 2,102,460, filed July 16, 1934 and 2,301,399, filed Nov. 19, 1937. Figures (1, 12, and 13) showing three shell designs.

procedures for fabricating bearing shells from stamped sheet metal. The machine consisted of a plunger fitted with a male die that rammed flat sheet metal into a female die to form flanged cup-like shells. The method called for stamping pre-cut sheet metal into a rough cup-like shape, followed by several drawing and forming operations to

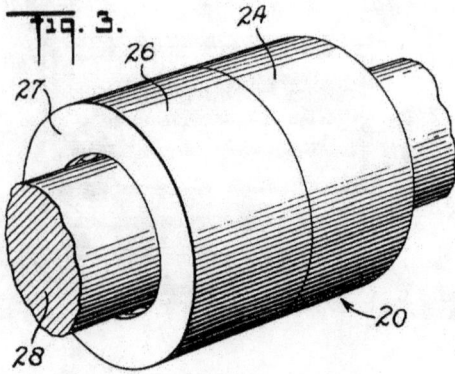

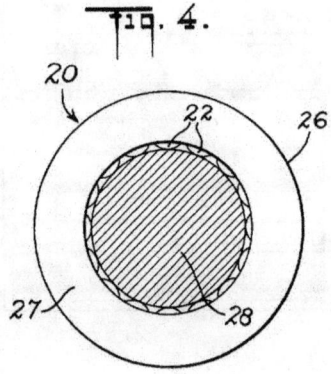

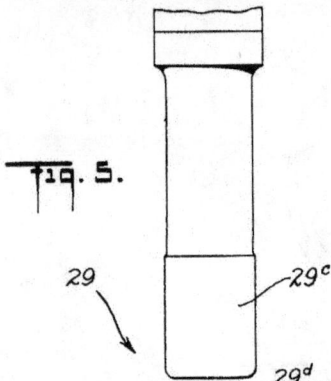

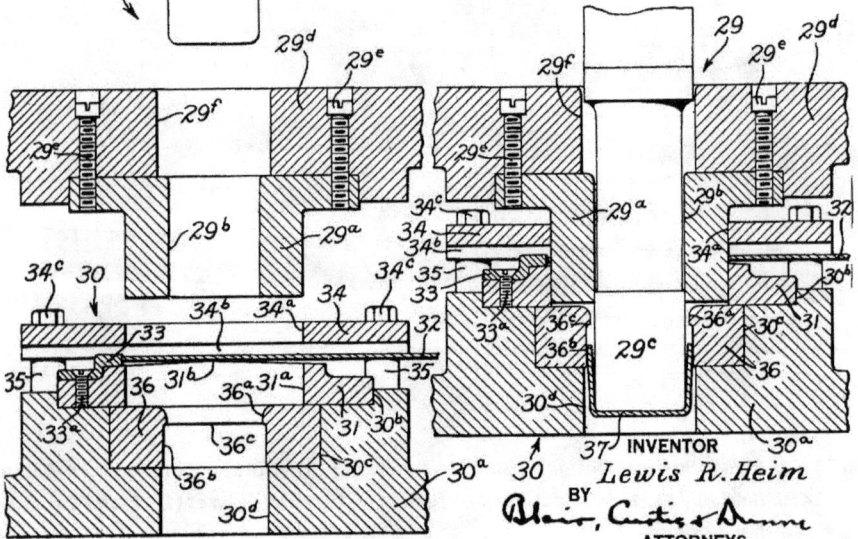

Fig. 11-5: Machine for Fabricating Cylindrical Shells from Stamped Sheet Metal. Patent 2,102,460, filed July 16, 1934. Figures 5 and 6 showing stamping of flat sheet metal (32) into a cylindrical shell (37).

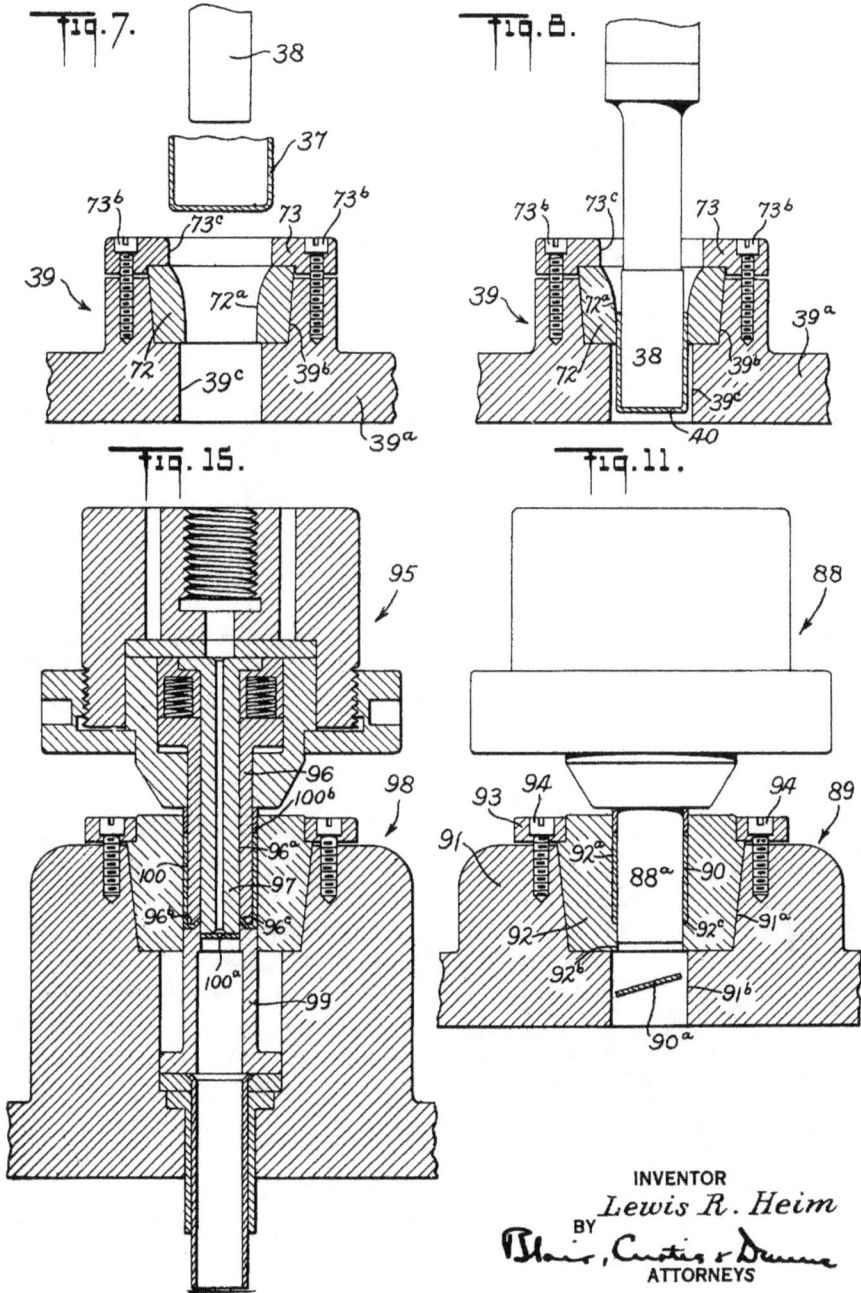

Fig. 11-6: Machine for Fabricating Cylindrical Shells from Stamped Sheet Metal. Patent 2,102,460, filed July 16, 1934. Figures 7 and 8 showing drawing and forming dies (72) to shape shells and form flanges.

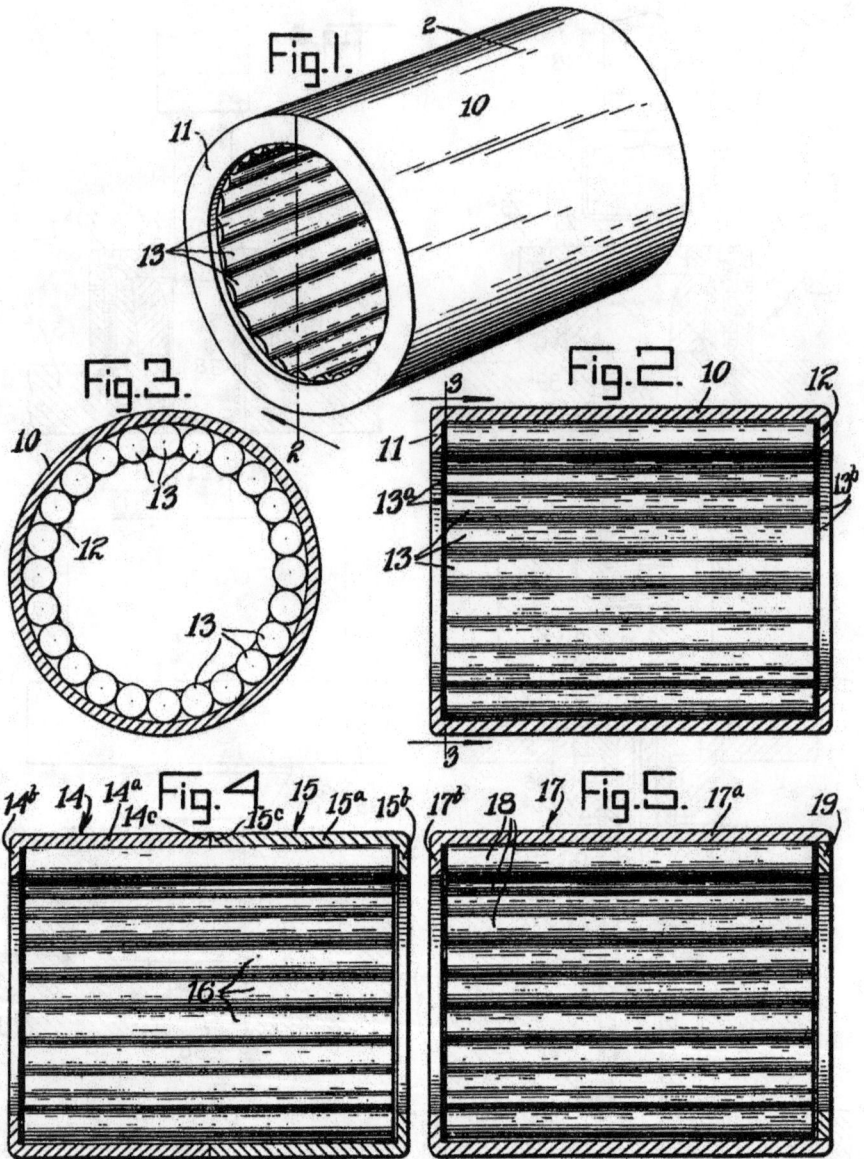

Fig. 11-7: Single Shell Roller Bearing Assembled Using Heating and Cooling of Components. Patent 2,160,362, filed Aug. 1, 1936. Figures 2, 4, and 5 show different shell configurations.

increase the cup depth while reducing its diameter, then perforating the bottom of the cup to create flanges in the cylindrical shell. Using various dies and a succession of shell-reduction operations yielded a process that Heim called a "Flowing of Metal," where any irregularities in the wall of the shell were automatically ironed out to produce a very smooth surface that did not require supplemental grinding, thus eliminating post-forming machining operations. Figures 11-5 and 11-6 show the machine, dies, and steps used to form the bearing shells and flanges.

In 1936, Heim developed a new method for assembling roller bearings that required less material and labor to fabricate, resulting in a lower-cost bearing. The method involved heating and chilling components to cause expansion or shrinkage that allowed assembly of precision-made and tightly fitting parts. Applying heat to the shell and significantly raising its temperature results in expansion of the metal and a subsequent increase in diameter and length. For a bearing designed with a full complement of precision-sized self-locking rollers, expansion increases clearances to allow insertion of all the rollers. Upon cooling, the shell diameter and length shrink to their original dimensions, compressing the rollers into their self-locking position. Also, if required to obtain the necessary clearances for assembly, chilling of the rollers or other components can be performed to shrink their size.

Heim filed a patent application for this method on August 1, 1936, that described the fabrication process as well as three roller bearing designs assembled from a "single" outer shell *(Figure 11-7)*. In the application that became patent 2,160,382, Heim described the advantages of his new design as follows:

> *This invention is to provide a roller bearing which requires a minimum amount of room for mounting and which, while sturdy and durable is relatively light in weight. Another object of the invention is to provide a roller bearing well adapted to effect an anti-friction mount for small high speed shafts and which is easily lubricated under practically any operating conditions.*

As described above, Heim used drawing and shaping dies to form the shells of his roller bearings. The inside surfaces of the dies were precision-machined to the extent that the shells formed in the dies did not require fine grinding to meet specified tolerances and surface finish. Throughout the 1930s Heim used various machining and grinding processes to manufacture the dies. Apparently, the processes for finish-grinding the dies were expensive and led Heim to invent a new type of machine that rapidly polished the inside surfaces of dies.

Polishing metal is a form of fine grinding in which an abrasive is used on metal to create a very smooth, defect-free, and mirror-like surface. To polish the dies used to make bearing shells, Heim developed a machine *(Figure 11-8)* similar to a band saw that used an endless flexible *(polishing)* belt coated with a very fine abrasive that was threaded through the die then rubbed up against the die at high speed. The polishing belt was circulated on two large motor-driven pulleys and guided through the work by four smaller pulleys. The work was placed on a circular table that rotated while the belt passed through to polish the entire inner surface evenly.

What made Heim's machine innovative was its ability to polish circular dies formed with irregular shapes. Some dies used to make shells were formed of constant diameters but many dies required irregular or complex shapes such as tapers and bends. In order to make irregular-shaped dies, Heim designed his machine to use a tool supported from above the work that dropped down inside the work to press the moving belt against the entire surface. The tool contained an arbor shaped to match the con-

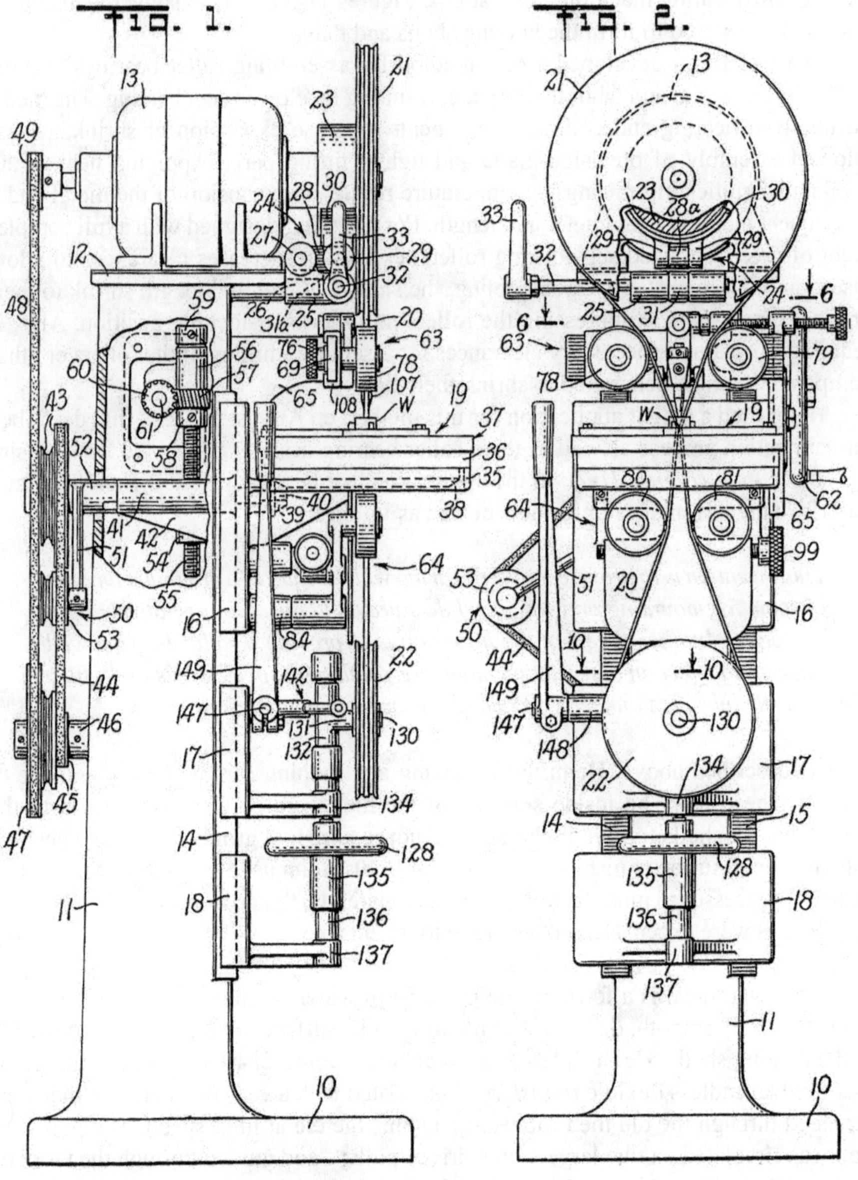

Fig. 11-8: Machine for Polishing Dies Used to Form Bearing Shells. Patent 2,372,722, filed July 28, 1942.

tour of the die surface such that pressure of the belt was distributed evenly across the work surface *(Figure 11-9)*.

The machine for polishing dies used to make bearing shells was essentially the fifth grinder that Heim invented. Heim designed his machine to be relatively simple for unskilled labor to set up and operate, as well as to rapidly and efficiently polish the dies used to make his bearings formed from stamped sheet metal. Heim filed for patent protection of his die-polishing machine on July 28, 1942, which was granted No. 2,372,722 in April 1945.

From his new company in Fairfield, Heim manufactured roller bearings and supplemented sales with other metal products requiring machining and heat-treating. One such product was the dowel pin made from alloy steel that was heat-treated and ground to a tolerance of 0.0002 inches. Heim stocked dowel pins in sizes ranging in diameter from 1/8" to 7/8" with lengths between 7/8" to 4".[5]

But Heim's primarily business was the manufacture of precision-made and low-cost ball and roller bearings and throughout the 1930s built up his product catalog. To make his bearings competitive, Heim invented new methods and machines that allowed him to compete in a crowded market. Even though the United States was still experiencing the effects of the Great Depression, Heim succeeded in profitably selling his innovative bearings fabricated from stamped sheet metal and his novel construction methods. It was also during the early years of the Heim Company that Heim's son, Charles, learned the trade of making and selling bearings.

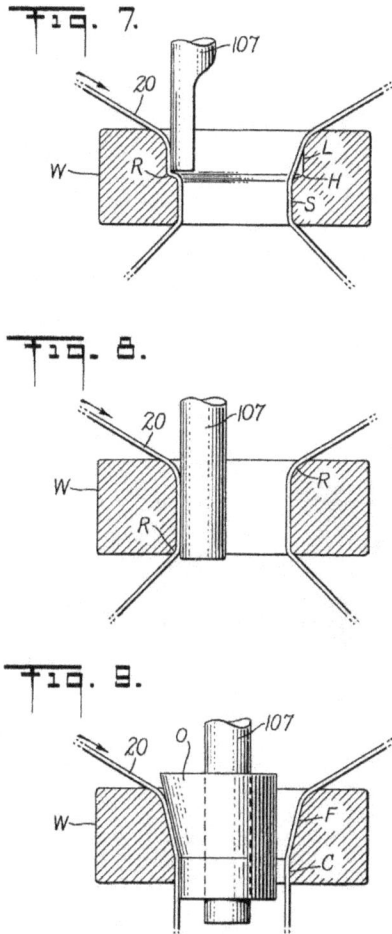

Fig. 11-9: Typical Arbors Used in Polishing Dies of Different Shapes. Patent 2,372,722, filed July 28, 1942.

Chapter 12

The Rod End Bearing

Messerschmitt Secret

In the 1930s, as the Great Depression ravaged Europe, Germany started rebuilding its military and in 1933 established the Reich Air Ministry *(Reichsluftfahrministe-rium/RLM)* to develop and manufacture all aircraft built in Germany. Aircraft for Germany's air force, the Luftwaffe, had top priority and by 1935 had introduced one of the most successful and lethal fighters of World War II, the Messerschmitt Bf 109. Developed primarily by Wilhelm Messerschmitt, the Bf 109 was a lightweight, fast, and highly maneuverable aircraft that could outfly most aircraft at that time.

The original Bf 109 could fly 290 mph and was about 100 mph faster than comparable fighter aircraft. By 1938, the company that made the plane, Bayerishe Flugzeugwerke *(Bf)*, introduced the fifth variant, the 109E, which had a newly developed Daimler-Benz DB 601A liquid-cooled inverted 12-cylinder engine equipped with direct fuel injection that produced a maximum of 1085 HP. With a top speed around 345 mph, it was the primary single-engine German fighter used during the Battle of Britain, where it outperformed the Hawker Hurricane but was only evenly matched against the Supermarine Spitfire. In late 1940, in an effort to dominate the skies, Germany introduced the sixth variant, the 109F, with an improved DB 601E engine that produced 1330 HP and much improved aerodynamics. The plane *(Figure 12-1)* had a top speed around 410 mph, an improved climb rate and maneuverability, and was at the time superior to the two British fighters. The 109F was first deployed in late 1940 in the Battle of Britain and later to Africa in the fall of 1941 to support Rommel's Africa campaign against the British.

Sometime in late 1941 or early 1942, the British obtained a downed Bf 109 and sent it to Vultee Field in Downey, California, for dismantling and analysis by the U.S. Army Air Force Matériel Command. The model sent to Vultee Field is not known but given its description in an article published by the *Bridgeport Post* in 1944, it was most likely a 109F.[1] One of the features attributed to the plane's maneuverability was a new type of bearing that was used to connect the control system of the plane to its wing flaps, ailerons, and rudder. The bearing installed in the Messerschmitt 109 was a less versatile version of the modern-day rod end bearing. Whereas modern rod end bearings consist of a drilled-through ball inserted into the head of a threaded shaft *(Figure 12-2)*, the Messerschmitt bearing consisted of a shank ending in a slotted housing with a drilled-through spherical slice that looks more like a rounded ring than a ball. The ring is inserted along its narrow width into the slot, turned at a 90-degree angle to lock it into the housing, then flipped upward

Fig. 12-1: Messerschmitt Bf 109F.

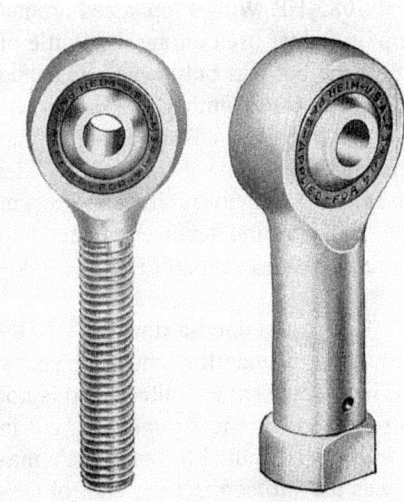

Fig. 12-2: Modern Spherical Rod End Bearing: 2016 Components: 1) Drilled-Through One-Piece Ball Bearing, 2) Circular Housing for Ball Bearing, and 3) Internally or Externally Threaded Shaft Attached to Housing. (Photo courtesy of the Heim Company).

or downward 90 degrees to expose the drilled-through hole. A shaft is inserted through the hole in the ring in which it would swivel. Within the Messerschmitt Bf 109, the slotted bearing design was used not only in rod ends but also in other applications such as actuating levers *(Figure 12-3)*. A modern example of the German rod end bearing with the drilled-through spherical slice *(ring)* is shown in Figure 12-4. After completing their evaluation of the Messerschmitt rod end bearing, the AAF Material Command set out to duplicate and test similar bearings with the intent of installing them in U.S. fighters to improve maneuverability. The AAF Material Command contacted a local bearings broker in Los Angeles by the name of Edward Maltby who subsequently contacted various bearings manufacturers, one being the Heim Company in Fairfield, Connecticut. The story surrounding the AAF Material Command and how it found Heim was

Fig. 12-3: Slotted Bearings Inserted into Radiator Door Actuating Lever of Messerschmitt Bf 109G: ca. 1944. Photo courtesy of Flugmuseum-Messerschmitt, Manching, Germany, 2017.

told in a *Bridgeport Post* article that can be found in Appendix 4 *(A4-1)* and summarizes the events and coincidences that led the AAF to the Heim Company.

When Maltby contacted Heim in early 1942, the Heim Company was still recovering from a fire that destroyed the assembly and inspection departments in June 1941.

Maltby showed Heim the Messerschmitt bearing, explaining the advantages of the German design versus the rod end bearings used in American planes, which were generally dependent on a circlet of balls in the housing. According to the article, Heim studied the German rod end bearing and within five minutes began sketching an improved design. Shortly thereafter, he was able to produce a bearing similar to the Messerschmitt's but with the added advantages of being easier to manufacture, using less critical material, and working more efficiently. Heim's bearing omitted the slotting and slicing by using an unbroken outer race *(head)* into which a drilled-through perfect sphere *(ball bearing)* was inserted and allowed to swivel with 360 degrees of freedom compared to the Messerschmitt's 90 degrees. The ball swivel was securely fixed into

Fig. 12-4: Modern Slotted Rod End Bearing. Photo courtesy of the Heim Company, 2016.

the head using a pair of stamped-in brass inserts, resulting in a simple four-piece design. Steel against brass created less friction than steel on steel, thus reducing wear that allowed Heim to use lower-cost and more readily available materials. The head and rod were formed from a single piece of bar stock that made Heim's bearing the first "integral" rod end bearing used in commercial applications.

Heim submitted samples of his four-piece rod end bearing to the AAF Material Command, which put them through rigid engineering tests at Wright Field in Dayton, Ohio. Testing found Heim's bearing performed better than competing designs, leading to the AAF's first purchase order on September 4, 1942. Within 67 days, the Heim Company began delivery of Heim rod end bearings to aircraft manufacturers. Monitoring of rod end production became a priority of the Area Office of the AAF Material Command. From their location in New Haven, AAF representatives visited the Heim Company constantly to test finished work, iron out production snags, and aid in raw material procurement. By November 1944, the Heim Company was producing more than 20 times that of its first order. All bearings were shipped to the AAF Material Command center at Wright Field for allocation to aircraft manufacturers, with many returning to Connecticut and delivered to Sikorsky *(for their R-4 helicopters)* in Stratford and Chance-Vought *(for their F4U Corsair)* in Stratford and Bridgeport.[2]

A rod end bearing is constructed from a spherical plain bearing connected to a threaded rod *(shaft)*. With a drilled-through ball mounted into the head of the rod, these bearings facilitate multidirectional rotation as well as high relative load-carrying capacities and are ideally suited for applications requiring linear motion at odd or varying angles.

One very important feature of the rod end bearing is its ability to accommodate misalignment of the shaft or other hardware passing through the ball swivel. A second feature, due to its threaded shaft, is the ease in mounting and adjustment of its position that cannot be achieved in a fixed position spherical bearing. And a third feature is its ability to carry higher relative radial and thrust loads. This is achieved with the use of a single ball revolving within spherical inserts that provides a larger bearing surface compared to other bearing types and was referred to as the Heim Unibal principle in many advertisements *(see Appendix 4, A4-4)*.

With its ability to handle misalignment and linear motion at varying angles as well as relatively high loads, rod end bearings are well suited for use in mechanical linkages that have tremendous appeal in aircraft, specifically in control applications such as rudders, ailerons, elevators, and carburetors. For these reasons, by the end of World War II, the United States installed Heim spherical rod end bearings in nearly every American warplane.[3]

Patents for Rod End Bearings

On September 29, 1942, Heim filed a patent application for his rod end bearing approved by the Army Air Force that was for both the "device" *(rod end bearing)* and the "method" for making and assembling the bearing. The device portion of the invention consisted of the "four-piece spherical plain bearing" that Heim stated was simple and sturdy in construction, capable of extended rigorous use and inexpensive to manufacture. The reason that Heim was able to quickly design and produce his rod end bearing was due to the many years developing roller bearings made from stamped sheet metal. The application of two brass bushings to retain the ball within the enclosure was derived from Heim's telescopically interlocked construction that was previ-

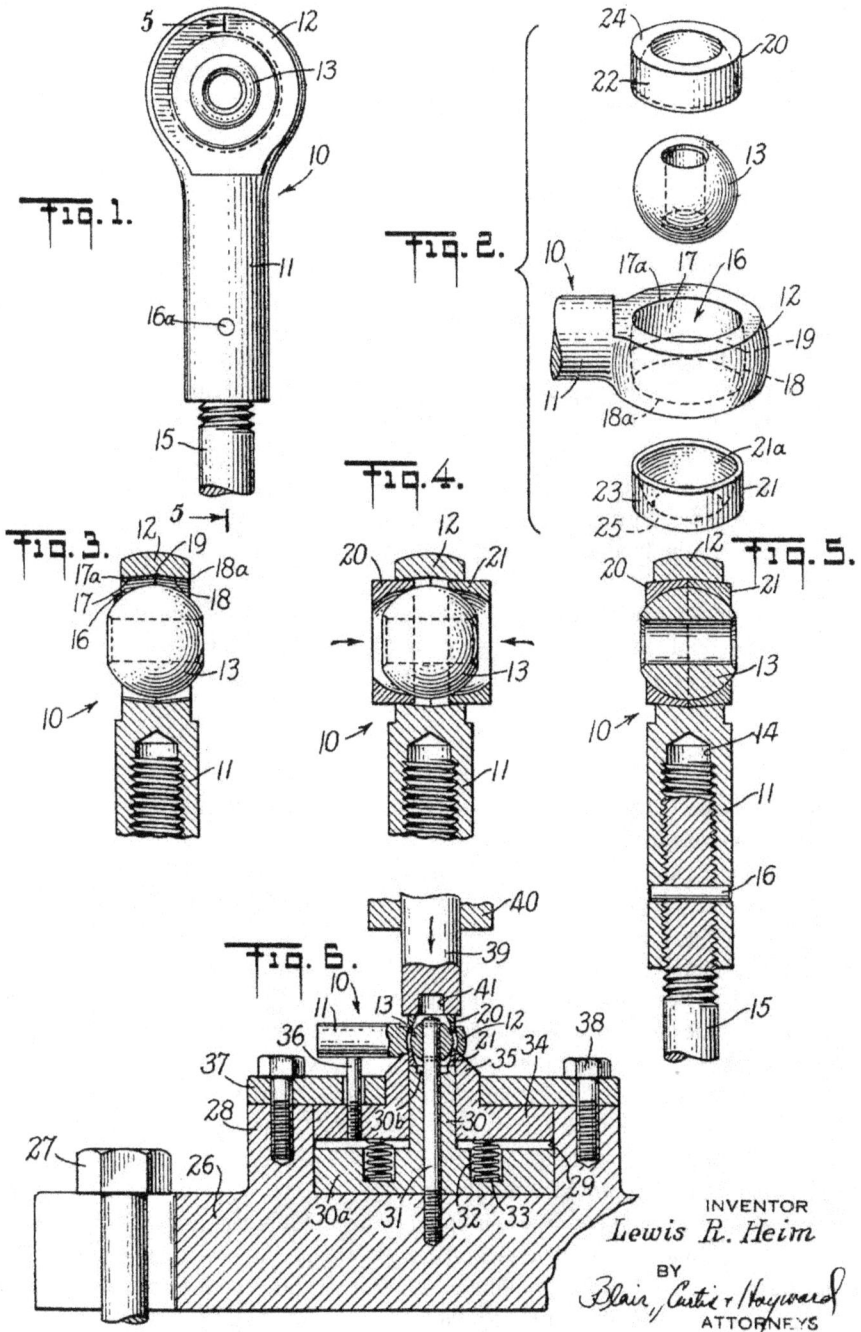

Fig. 12-5: Original Heim Four-Piece Rod End Bearing: "Heim Joint" Patent 2,366,668, filed Sept. 29, 1942.

ously proven through many years of use.

The method portion of the invention was divided into two parts, one for making the bearing components and the other for assembling the components. Using standard round bar stock, Heim fabricated a one-piece "integral" body comprised of a shank *(rod)* and a circular head sized to retain a drilled-through ball. The head was machined with a pair of tapered openings *(one on each side)* bored to expand from the outer openings toward the center of the head. The shank was drilled and threaded as required for a typical female connection. The head, shank, and drilled-through ball were made of hard steel while the two bushings were made of a soft metal alloy such as bronze. Soft, malleable metal bushings were necessary to facilitate deformation *(molding)* around the ball and the inner surface of the head when pressed into place during assembly. Details of the bearing components are shown in Figure 12-5.

For assembly of his rod end bearing, Heim developed a machine called a "Die Set" consisting of a punch press equipped with dies that conformed to the shape and sizes of the bearing. The bearing components were placed in the dies in proper order such that when the punch press actuated downward, the bushings were forced into the head of the bearing. During the press operation, the soft metal bushings readily distort and expand into the tapered head due to pressure from the ball. When the operation is complete, the ball is locked in place within the head for an assembled bearing. The punch press with a Heim rod end bearing ready for assembly is shown in Figure 6 of Figure 12-5. Heim was granted letters patent for his invention on January 2, 1945, which gave him exclusive production rights into the early 1960s.

Rod End Bearing Designed with Lubrication

Rod ends are members of the family of plain bearings in which one bearing surface slides over the other. In such bearings, without proper lubrication, friction is generated especially under high loadings that can lead to wear and failure. In Heim's original rod end design, there were no provisions for lubrication. This led Heim to quickly develop a second design and within two months on November 5, 1942, he filed a second application to patent a four-piece rod end bearing modified to accommodate lubrication. This bearing added features that allowed injection of graphite or machine oil to lubricate the ball swivel. Heim redesigned the bearing head to include an annular groove that held a lubricant and allowed it to flow around the ball swivel. To prevent the two bushings from closing off the annular groove when pressed into the head, the head was machined with a rib at the center of the inside surface.

With modifications made to the bearing head, Heim developed two methods for lubrication. One design provided for a reservoir in the shaft opening to the base of the ball swivel. Inserted into the reservoir could be a solid lubricant such as graphite or a wick containing machine oil. The second design provided a hole drilled into the top of the head for insertion of solid graphite or other lubricants. The respective designs are shown in Figs. 3 and 4 of Figure 12-6 taken from patent 2,400,506. Heim developed two additional improvements to enhance bearing lubrication in his rod end bearings and filed applications on July 24, 1943, and February 24, 1945. The first invention *(2,488,775)* embeds a wick containing oil in the annular groove surrounding the ball

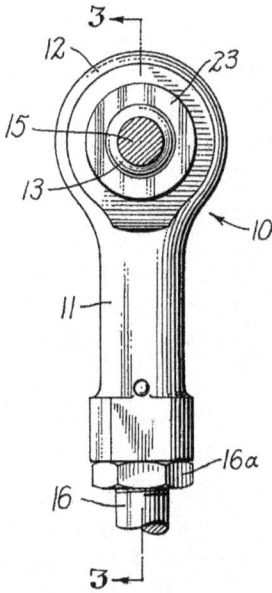

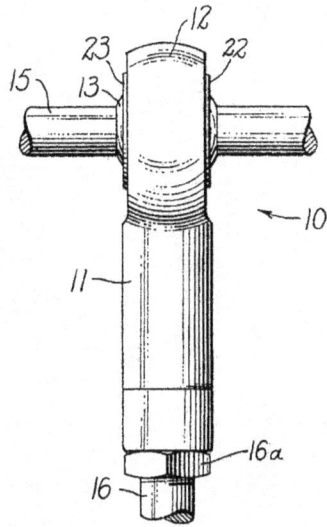

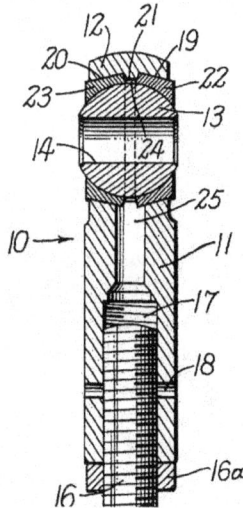

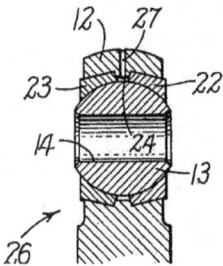

INVENTOR.

Lewis R. Heim

Fig. 12-6: Heim Four-Piece Rod End Bearing Designed with Rib and Annular Groove for Ball Swivel Lubrication. Patent 2,400,506, filed Nov. 5, 1942.

swivel. The second invention *(2,454,252)* incorporates a lubricator head system employing a spring-loaded ball check valve installed into the threaded shaft.

Method to Loosen Ball Swivel

As described above, Heim's assembly process for spherical rod end bearings used the method of pressing two soft metal bushings over the steel ball. Under high pressure, the bushings deform to fit the shape of the inside surface of the head and the outside surface of the ball. After pressing, the bushing was interlocked into the bearing head and tightly gripped around the ball. The pressure of the bushing against the ball prevented free movement and required post-assembly loosening.

Heim's first solution for loosening of the ball was to invent a device and method for freeing mechanical joints. The method involved suspending the bearing on supports and using a punch lowered from above to press down on the head to induce pressure on the bushings, causing additional deformation that created a very small gap that freed up the ball to swivel. This device *(Figure 12-7)* was developed almost immediately after receiving his first purchase order from the U.S. Army Air Force and on December 14, 1942, he filed an application that became patent 2,476,728.

After using the ball-loosing device for two years, Heim developed a second method that was much simpler and cost effective. The second method involved heating the ball to expand its diameter prior to assembly. After assembly, the ball cooled and its diameter shrunk sufficiently to create a small gap against the bushings and completely free movement. This method for assembly reduced both time and cost and became the preferred method for bearing assembly. Patents for the method and bearing were filed on December 30, 1944, and granted as 2,541,160 in 1951 and 2,675,279 in 1954.

Rod End Bearings by Other Inventors

Heim was not the first inventor of rod end bearings in the United States. There were several inventions using ball joints developed in the 1930s that had characteristics similar to rod end bearings used in tie rods and drag links of automobiles and trucks. One such bearing *(Figure 12-8)* was developed by William Flumerfelt in 1936 *(2,108,814)*, which incorporated a ball joint connected to a threaded shaft.

In what may be the first true rod end bearing patented in the United States, Horace Steele invented a bearing where the spherical element within the housing *(called a "retaining ring" in the patent)* is composed of two parts: a bearing ring and a slotted cylindrical collar. The components of Steele's rod end bearing *(2,309,281)* are illustrated in Figure 12-9 and show the rotatable spherical element and the threaded shaft.

In May 1942, Anthony Venditty filed to patent *(2,350,482)* what he called a ball joint to be used in airplane control rod assemblies. Venditty's invention *(Figure 12-10)* contains the essential components of a rod end bearing with a drilled-through rotatable ball enclosed in a round housing attached to an internally threaded shaft. In order to assemble the ball within the head *(race)*, Venditty used precision-made parts including two ring bearings to hold a two-piece segmented ball and spring washers seated into each side of the enclosure to retain the ring bearings. Details of the ball and enclosure can be seen in Figs. 3, 4, and 5 in Figure 12-10.

Shortly after Heim filed his second patent application, Frank Hill of Cherry Valley, Illinois, filed an application to patent a rod end bearing for use on control rod ends

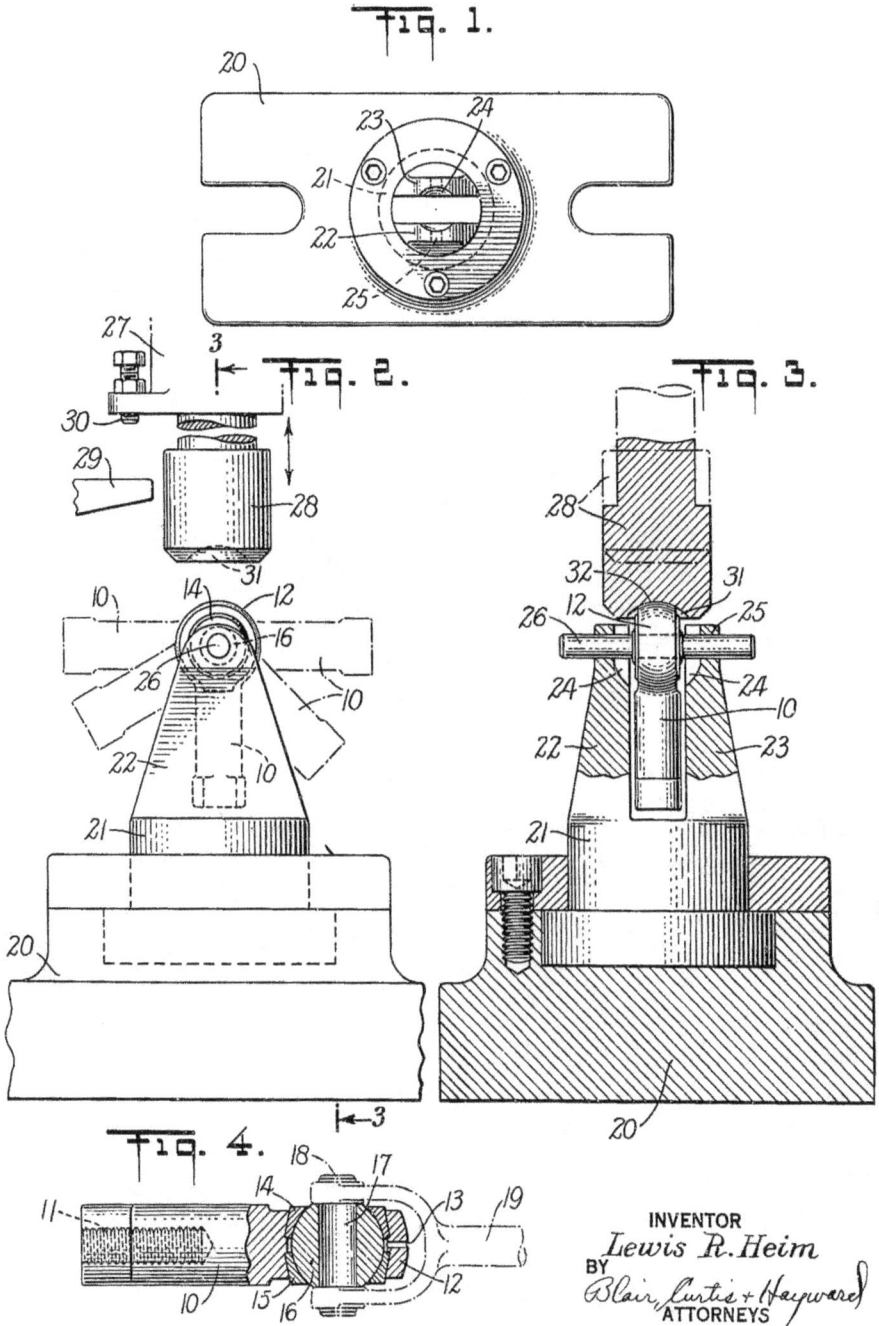

Fig. 12-7: Device for Freeing Balls in Rod End Bearings. Patent 2,476,728, filed Dec. 14, 1942.

and other applications on aircraft where frequent lubrication is impractical. Hill's rod end bearing looks similar to Heim's bearing but contains six components versus four for Heim's bearing. To retain the ball swivel, Hill uses two races made from powdered metal and adapted to retain a lubricant. To retain the races in place, Hill uses two corrugated spring washers, one on each side and made from a softer metal such as brass. The washers are inserted in between the races and the flanged sections of the bearing head that prevent the races from dislodging.

A second purpose of the washers was to prevent foreign material from penetrating the space between the ball and races. The washers, being corrugated, were designed to snuggly fit up against the ball swivel that acts as a barrier to dust and dirt that Hill claimed made his bearing indestructible and suited for use in airplanes. Hill's bearing is shown in Figure 12-11, taken from patent 2,365,552.

Heim Joint and Heim Rod Ends

Heim's inventions from the 1940s became the standard for modern rod end bearings. The popularity of the Heim bearing eventually led to its two nicknames, the "Heim Joint" and "Heim Rod Ends," which are widely used in aerospace, military, industrial, agriculture, and consumer applications. Heim's invention of the four-piece rod end bearing led him to develop the two-piece design as well as advancements in the general class of spherical bearings that are discussed in the next chapter.

Rose Joint

In the United Kingdom and Europe, rod end bearings of Heim's design are referred to as "Rose Joints," with credit given for their origin to the Rose Bearings Company of Saxilby, United Kingdom. For the record, the Rose Bearings Company did not invent or otherwise originate its Rose Joint rod end bearing. During the years between 1942 and 1945, all rod end bearings produced by the Heim Company were allocated to American aircraft. In order to support the needs of British aircraft manufacturers during World War II, in 1943 or early 1944, the Rose Bearings Company was provided with the technology and a license from the Heim Company to fabricate Heim rod end bearings.[4, 5, 6]

Internal Grinding Machine

In 1945, Heim developed his sixth grinding machine that he used to make the bushings and balls for his rod end bearings. The machine *(Figure 12-12)* was designed to grind both straight and tapered holes in spherical *(ball bearings)* and circular *(bushings)* work in a quick and accurate manner. The grinding portion of the machine is divided into two assemblies, with one holding and rotating the work and the other supporting and positioning the grinding wheel. The work holder consists of a chuck that supports a collet into which the work is inserted and a device ancillary to the chuck for manually loading the work into the collet. The collet serves as a type

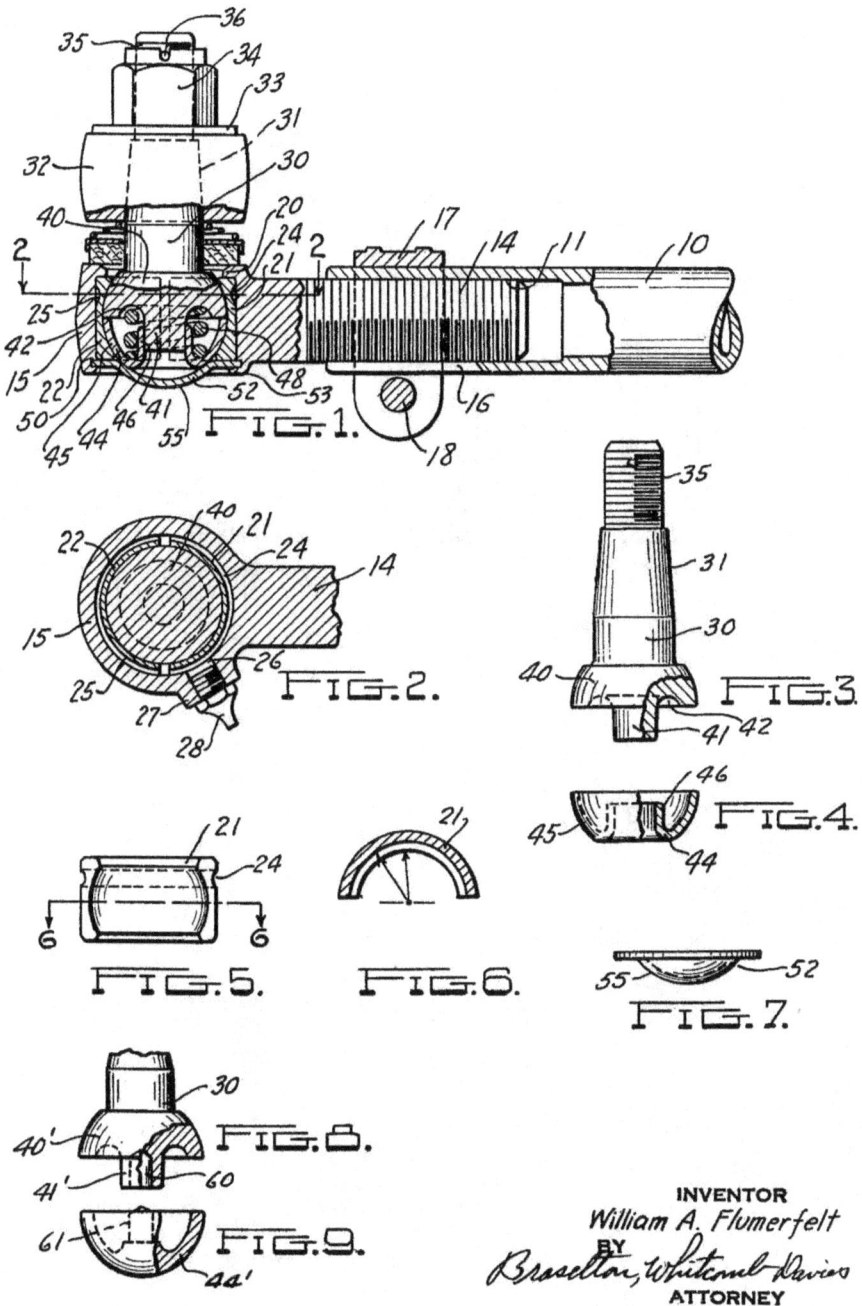

Fig. 12-8: W. A. Flumerfelt Ball Joint Rod End Bearing. Patent 2,108,814, filed June 22, 1936.

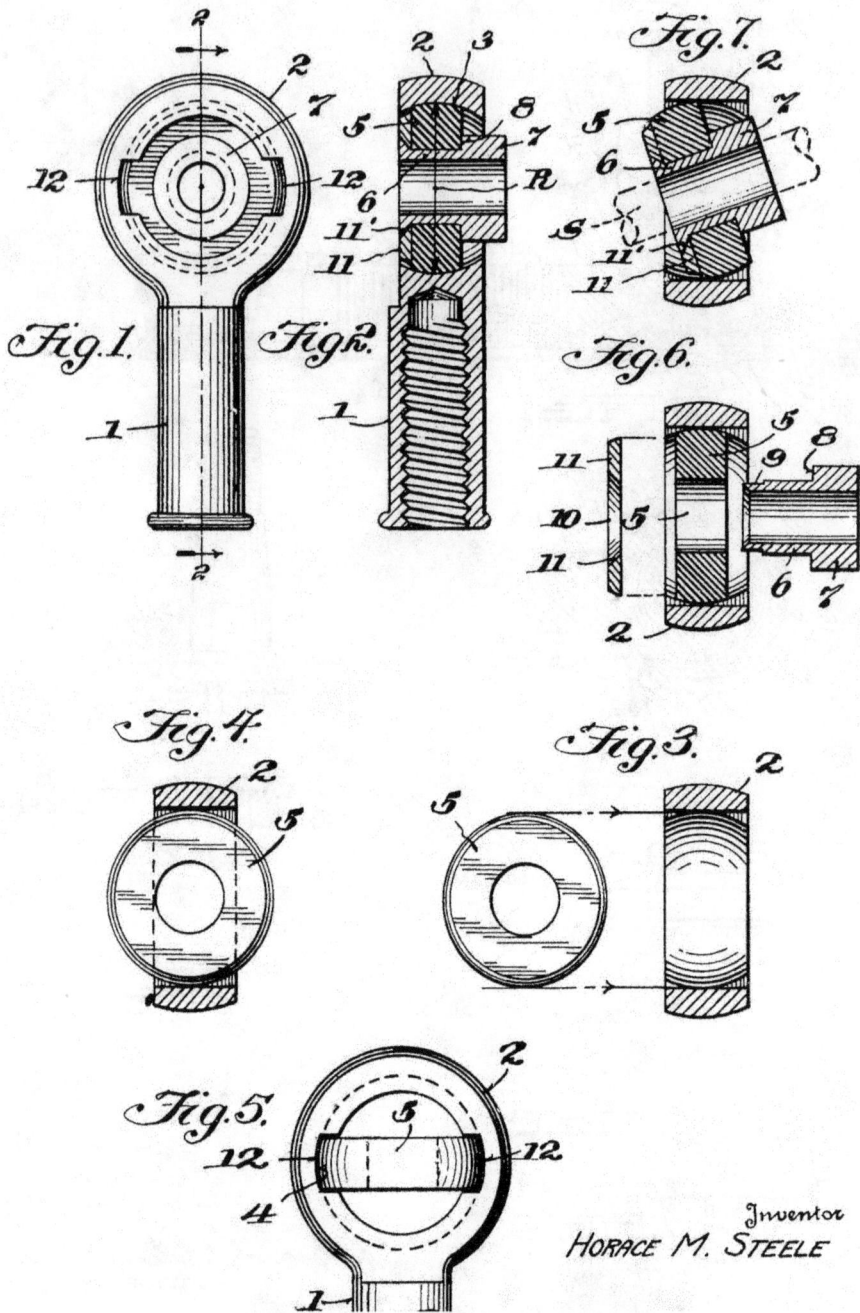

Fig. 12-9: H. M. Steele Rod End Bearing. Patent 2,309,281, filed Jan. 27, 1942.

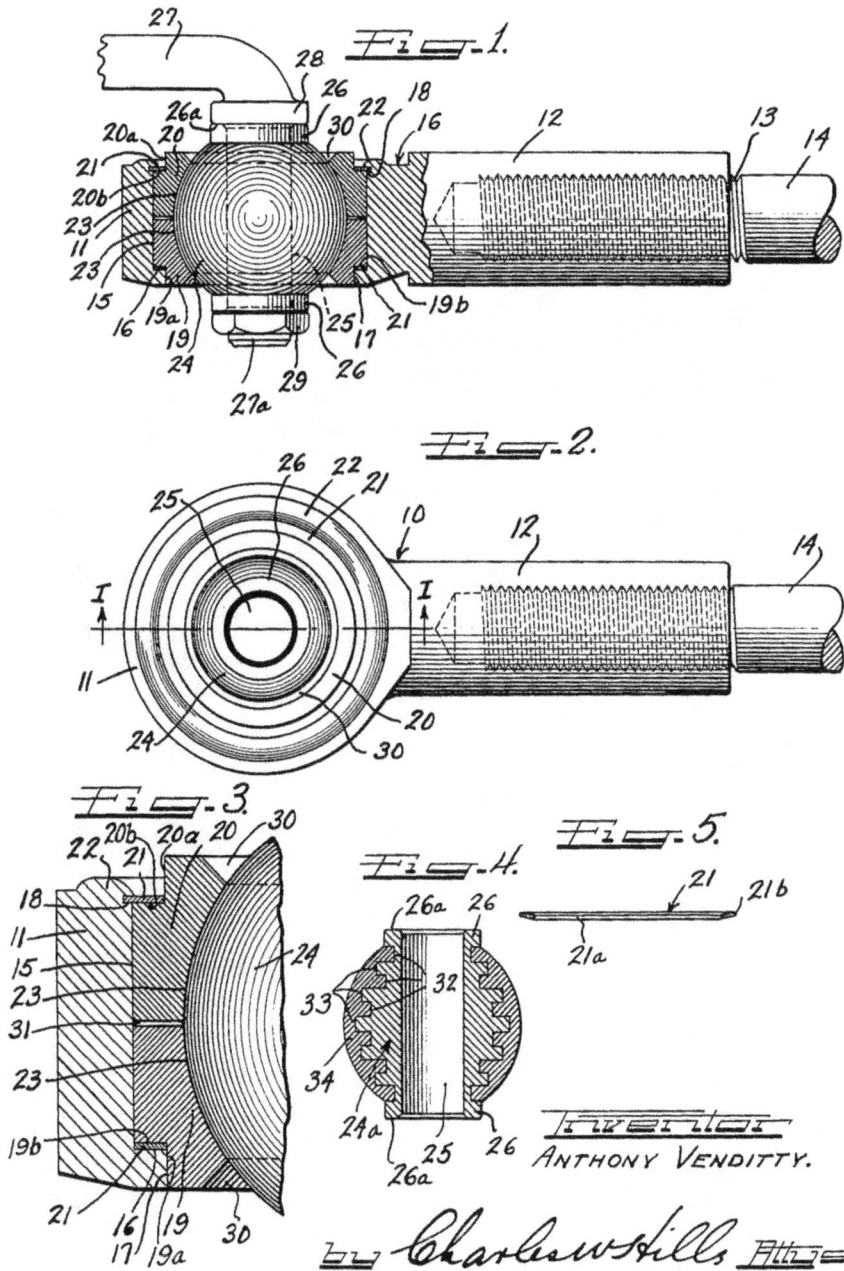

Fig. 12-10: Venditty Ball Joint Rod End Bearing. Patent 2,350,482, filed May 2, 1942.

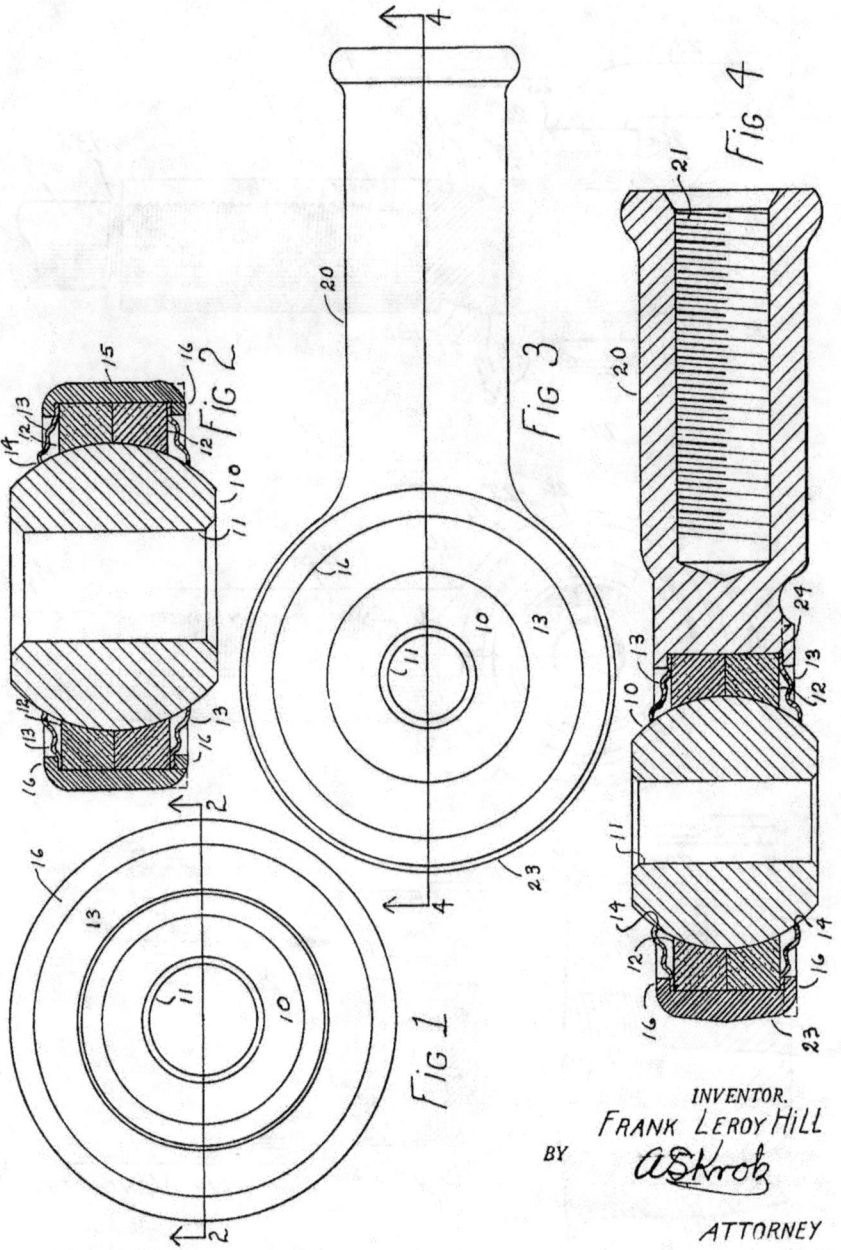

Fig. 12-11: Hill Mono-Ball Rod End Bearing. Patent 2,365,552, filed Nov. 9, 1942.

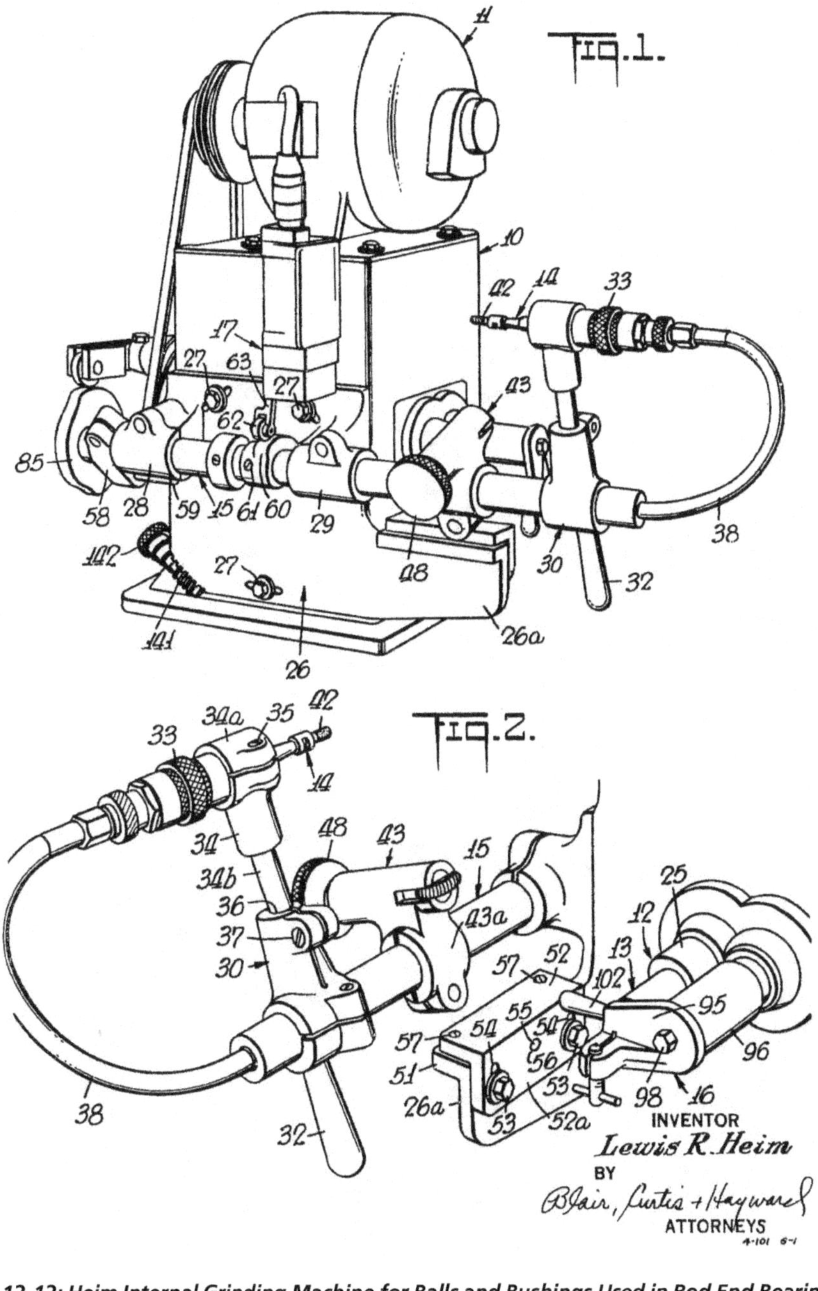

Fig. 12-12: Heim Internal Grinding Machine for Balls and Bushings Used in Rod End Bearings. Patent 2,430,423, filed Jan. 8, 1945. Figs. 1 and 2 show the machine and the grinding assembly in the retracted position.

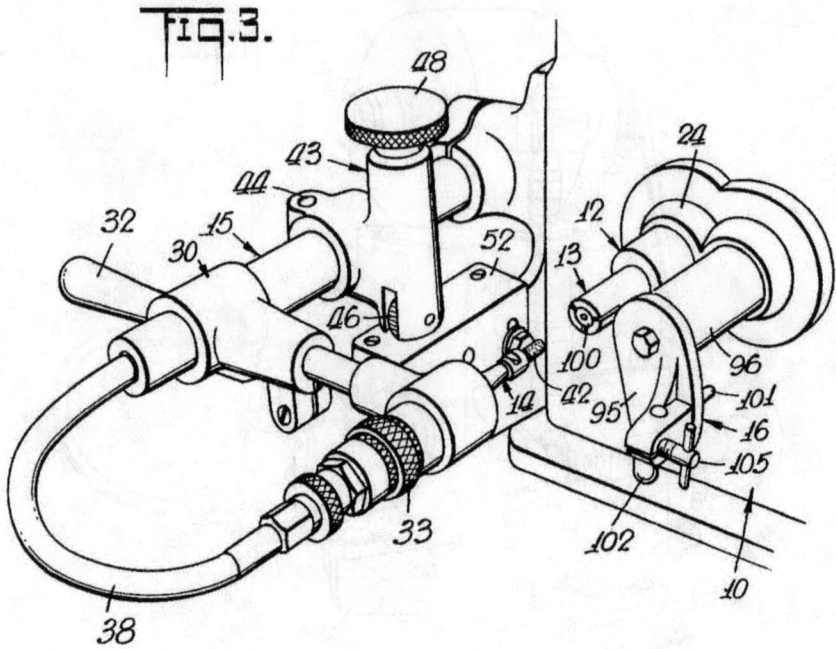

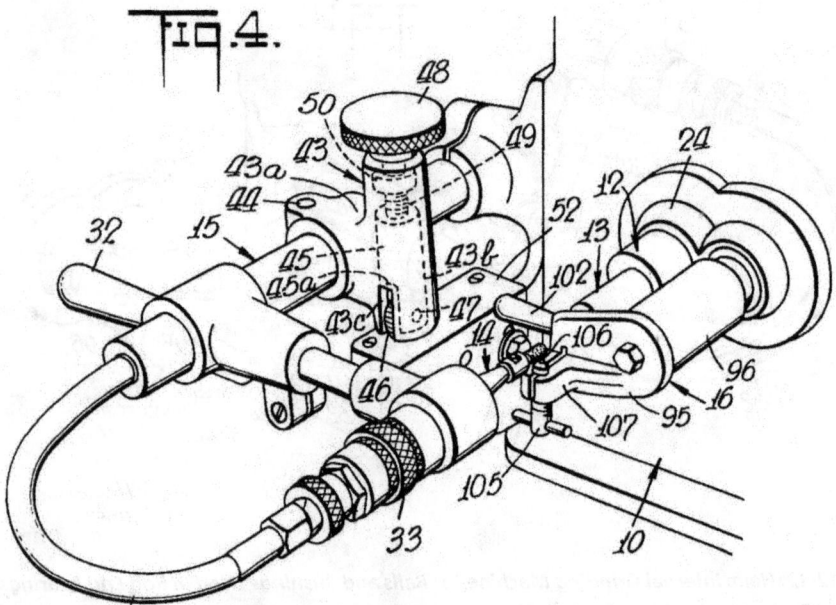

Fig. 12-13: Heim Internal Grinding Machine for Balls and Bushings Used in Rod End Bearings.
Patent 2,430,423, filed Jan. 8, 1945. Figs. 3 and 4 show the grinding assembly in the grinding
position against the work holder.

of chuck that forms a collar around the work being held and exerts a strong clamping force to keep the work secure during grinding.

The grinding assembly is mounted to the side of the work holder and designed to rotate about 90 degrees to position the grinding wheel quickly against the work. The retracted position can be seen in Figure 2 of Figure 12-12. The grinding wheel, generally in the form of a cylinder, is rotated independently of the work using a separate electric or air-driven motor. The grinding assembly is equipped with various mechanisms to position the grinding wheel accurately relative to the work, to limit its movement, and to control the grinding action. The grinding position can be seen in Figs. 3 and 4 of Figure 12-13. A cam and brake system is employed to automatically stop chuck rotation when the grinding wheel is retracted so that the finished work can be removed and new work inserted into the collet. The invention became patent 2,430,423 in November 1947.

Chapter 13

Unibal and the Art of Making Industrial Bearings

New Markets for Rod End Bearings

After World War II ended in September 1945, the U.S. economy continued to experience growth and, instead of a depressed economy with a reduction in military spending, pent-up consumer demand powered the economy forward. Industries such as transportation and housing provided the foundation for continued growth. Available and affordable mortgages for former servicemen stimulated housing, while the automobile industry quickly and successfully converted back to non-military car and truck production. In addition, expansion of the relatively new aviation and electronics industries that introduced new products during the war added fuel to economic growth.

However, some companies were stung with the sudden reduction in military spending and the Heim Company was one such casualty. With 100 percent of its production of rod end bearings sold to the government, the Heim Company lost most of its only source of revenue. On top of that, all customers that bought Heim ball and roller bearings before the war were now buying from other sources. Complicating the matter was that rod end bearings were newly developed and therefore prewar customers to replace lost government orders did not exist. As with most products, introducing something new and generating orders requires time and thus, Heim had to find a way to fill the void.

During the war years, Heim increased employment more than 70 percent to around 350 employees, with many of them experienced machinists whom he did not want to lose. He knew it would take several years to develop new customers for his rod end bearings and therefore needed to find sources of revenue to keep his factory busy. One tactic readily available was to restart sales of ball and roller bearings that were sold before the rod end bearing. However, reestablishing customers lost during the war would also take some time. Another tactic employed was to temporarily manufacture a simple commodity item that could be quickly manufactured in large numbers and readily sold. Heim did this with the mass production of salt and pepper shakers made from machined steel. Heim designed a set that consisted of a closed cylinder *(shaker)* that was drilled and tapped at one end, into which was placed a smaller cylinder tapped on the outside so that it could be screwed into the metal shaker. The smaller cylinder was open at one end and capped at the other, with holes drilled through to allow salt or pepper to pass through during shaking. The smaller cylinder extended into the shaker far enough to prevent leakage when standing upright. However, in order to sell the shakers, they were priced at a loss, with Heim funding the balance of operations out of his own pocket until bearings orders picked up. Even with this strategy, downsizing the company was still necessary.[1]

AMF Pinspotter

In 1946, the American Machine and Foundry Company, better known as AMF, introduced the first fully automatic pinspotter *(pinsetter)* machine for bowling alleys. In bowling, the function of spotting or setting of pins was originally a manual task performed by a person stationed behind the pins. The first mechanical pinspotter appeared around 1936 but still required monitoring by a person to ensure continuous operation. Even though the AMF machine in 1946 revolutionized bowling, it employed suction mechanisms to pick up and hold pins, and these mechanisms were not reliable in operation. Over the next two years, AMF made significant changes to the mechanical lifting systems and developed an entirely different machine. In 1947, AMF approached the Heim Company for help in designing the linkages used to guide and stabilize the spotting table as well the drive-and-sweep mechanism.[2] With Heim's guidance, AMF replaced many of the linkage bearings with Heim rod ends that had the ability to accommodate misalignment and shock forces, with a resulting improvement in operation and reliability. An advertisement from the late 1950s describing the advantages of Heim rod ends in AMF pinspotters can be seen in Figure 13-1.

The ability of Heim's rod end bearings to accommodate misalignment of moving components in machines was a great advantage. Many of the applications were in linkages but there were also applications in supports, shafts, and other machine components. As the 1940s wound down, many companies started buying various types of rod end bearings designed and manufactured by Heim. Applications included general machinery, machine tools, textile looms, construction and mining equipment, industrial boilers, and transportation systems. A sample of applications and the advantages of rod end bearings are shown in a collection of 1950s-era Heim Company advertisements found in Appendix 4 *(A4-4)*.

Also adopting the use of rod end bearings were the manufacturers of commercial aircraft. Companies such as Boeing, Douglas, Hughes, and Sikorsky that made military aircraft also made commercial aircraft and after the war made the switch to using rod end bearings. Around the same time, manufacturers of smaller commercial aircraft, including Mooney, Cessna, and Piper, installed Heim Joints, and as the 1950s progressed, aviation became a huge buyer of rod end bearings made by the Heim Company.

Spherical Bearings

With his rod end bearings under patent protection, Heim continued to develop and patent novel bearings and the methods for making them. Heim expanded his product line from rod ends to making standard spherical bearings as well as his original 1930s-era rolling bearings made from stamped sheet metal. By the early 1950s, Heim was making large spherical bearings along with select types of ball and roller bearings. He

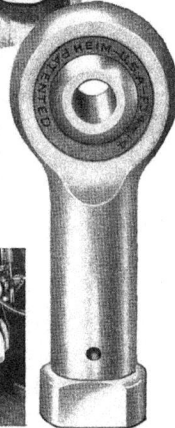

Fig. 13-1: Advertisement for Heim Unibal Rod End Bearings Installed in AMF Company Automatic Pinspotters, ca. 1957.

published catalogs for his various products, such as No. 14, which listed the company's offerings of rod end and spherical bearings using the trademark name "Unibal." In the foreword, Heim made the following statement about the benefits of his bearings:

Heim Unibal Spherical Bearings and Spherical Rod Ends have a greater ca-
pacity than the conventional type since they have a greater surface support-
ing area. Because of this area, they are not subject to false brinelling.* By
their very design, maximum correction of misalignment is obtained. Because
of their revolutionary construction, they will take a greater radial and axial
thrust load with resulting longer life.

(*Brinelling is the permanent indentation of a hard surface caused by ex-
cessive loading. False brinelling is damage caused by wear induced under
load by moving surfaces, which looks similar to but is wholly different from
brinelling.)

For his line of rolling bearings made from stamped sheet metal, Heim published separate catalogs such as No. 23 for his pillow block and flange unit bearings fabricated with balls and rollers. In Catalog No. 23 Heim promotes his roller bearings made from stamped and flanged sheet metal formed in dies and self-aligning ball bearings fabricated with unground balls. An abridged reprint of both catalogs can be viewed in Appendix 4 *(A4-2 and 3)*.

Double-Row Self-Aligning Ball Bearings

In 1948, Heim developed the first of two self-aligning ball bearings incorporating two rows of balls. The bearing was designed for both radial and axial thrust loads while being low cost and simple to assemble and install. The bearing incorporated a single inner race with a convex curvature that merged into grooves for the two rows of balls. The outer race was formed from a pair of flanged metal shells, each with grooves for one row of balls. To absorb both radial and axial forces, Heim designed the inner and outer races to channel the forces to a common point at the center of the bearing bore, as illustrated in Figure 1 of Figure 13-2. Heim designed three variants of this bearing, with one designed with ports for lubrication and the other two with fewer parts constructing the shell. All three designs used outer shells formed from stamped sheet metal and were assembled using Heim's interlocked construction method, with several bearings using this design illustrated in Catalog No. 23. The patent application was filed on August 26, 1948, and granted six years later as patent 2,675,281.

In 1951, Heim developed his second double-row ball bearing adapted to accommodate misalignment of the relatively rotating parts. The bearing *(Figure 13-3)* was designed primarily for vehicles that were driven over rough terrain in which the wheels and axles were subjected to higher levels of strain and to provide a longer useful life over similar bearings available at that time. The bearing consisted of a single inner race with two rows of deep annular grooves. The outer race was formed from a pair of stamped sheet metal shells, each with flanges on both ends. The flange adjacent to the inner race was designed to leave a small gap to accommodate the excessive movement from misalignment forces. To seal the bearing internals from ingestion of dirt and to prevent leakage of lubricant, washers were installed just inside both flanges.

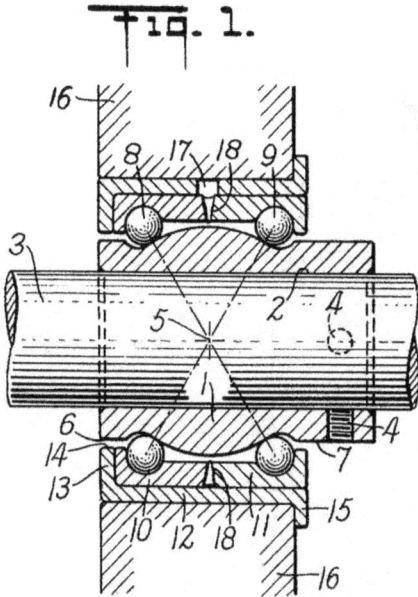

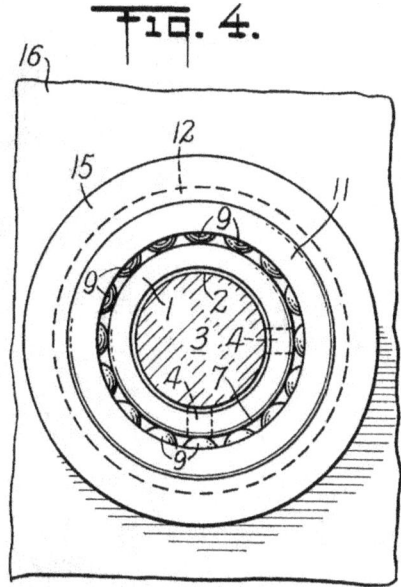

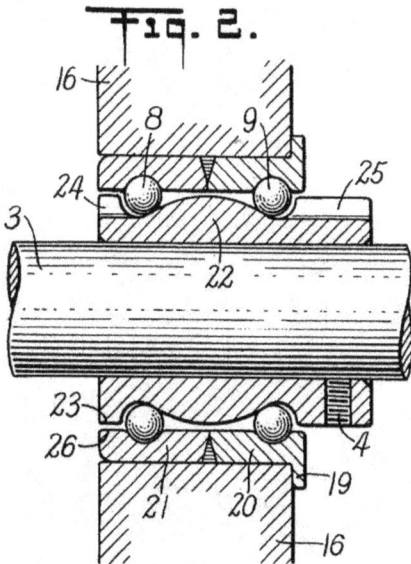

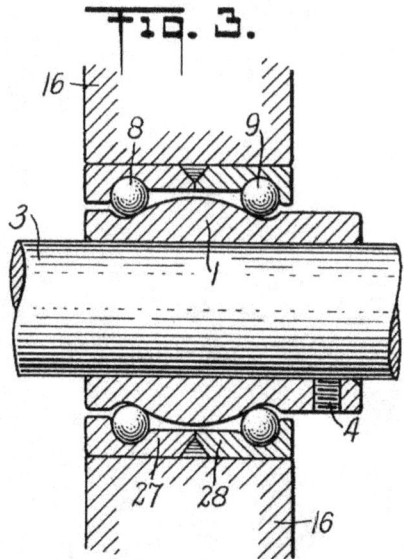

Fig. 13-2: Double-Row Self Aligning Ball Bearing for Radial and Axial Thrust Loads. Patent 2,675,281, filed Aug. 26, 1948.

This invention was filed for patent protection on December 28, 1951, and granted in September 1953 as patent 2,653,064.

As with any company, not all products developed prove successful in the market. At times, Heim would see an opportunity for a special application and design a

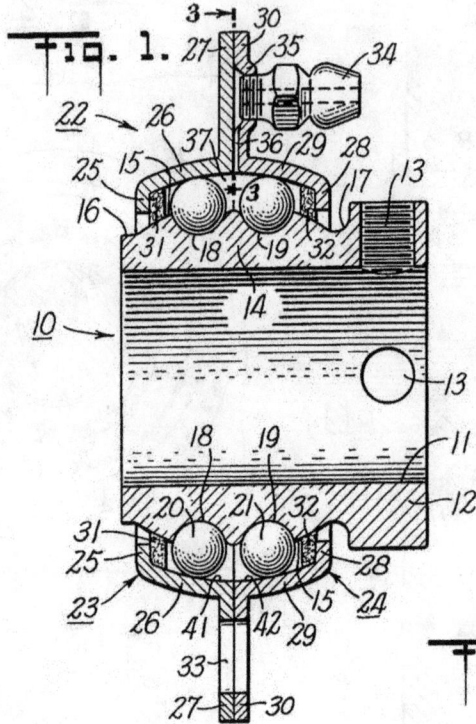

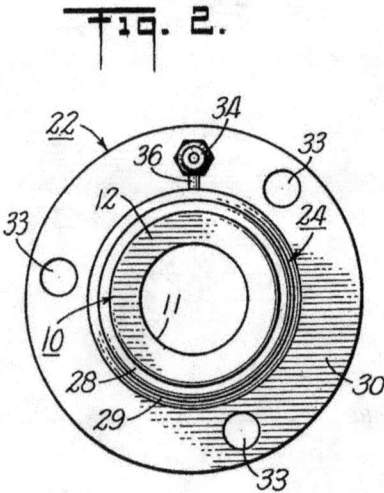

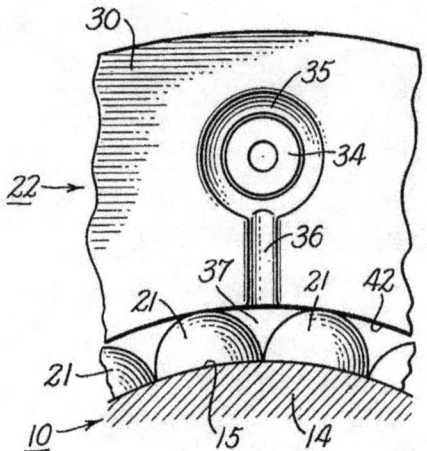

INVENTOR
Lewis R. Heim
BY
Watson, Johnson, Leavenworth & Blair
ATTORNEYS

Fig. 13-3: Double-Row Ball Bearing Designed for Wheel Misalignment Patent 2,653,064, filed Dec. 28, 1951.

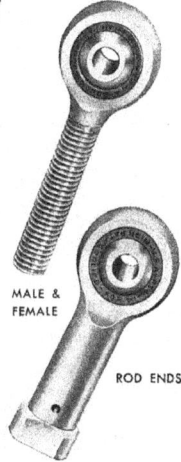

Fig. 13-4: Heim Company Product Advertisement in American Machinist, *Nov. 15, 1952.*

231

bearing for it, as he did with his two self-aligning ball bearing inventions. Whereas the first bearing with widely spaced rows of balls was sold in pillow blocks and flange units *(see Catalog 23)*, the self-aligning ball bearing for wheels was dropped from Heim's product line. An advertisement *(Figure 13-4)* published in *American Machinist* in November 1952 illustrates the many bearing types sold by the Heim Company at that time.

Improved Sweepstick for Power Looms

Sometime in late 1948 or early 1949 the Draper Corporation, a manufacturer of power looms, approached Heim to help solve a design problem involving bearings. Power looms are automated looms used to weave cloth and tapestry, and were one of the key developments in the industrialization of weaving during the First Industrial Revolution. In 1785, Edmund Cartwright, an English inventor, patented the first power loom. Over the years and into the late 1800s, power loom technology slowly advanced. In the late 1940s, there were three major manufacturers of textile machinery in New England, all located in Massachusetts. These were Crompton and Knowles of Worcester that made looms for wool weaving, Whitin Machine Works of Whitinsville that made spinning machinery, and the Draper Corporation of Hopedale.

Draper Corporation has roots dating back to the early 1800s with businesses making threshing machines and weaving cloth. The current company was established in Hopedale around 1841, with George Draper running the company since 1868. In 1889, James Northrop, who worked as a mechanic for George Draper and Sons, developed what became known as the Northrop loom and was a huge success. The Northrop loom improved automation to the extent that it relieved the operator, known as the "weaver," of much of the manual tasks required to keep the loom operating.

One of Northrop's inventions involved the shuttle, a spindle-shaped device that was loaded with threads known as weft. The shuttle carries the crosswise weft threads through the temporary separation *(shed)* between the upper and lower *(warp)* threads. Northrop developed a mechanism that upon exhausting its inventory automatically loaded new thread into the shuttle without stopping. The improved automation allowed the weaver to run up to 16 looms at once.

However, the loom was still a very loud device and created severe vibrations that affected the life of mechanical components. The mechanism that forces the shuttle back and forth consists of a picker stick, a sweepstick, and the picker cam, which is the mechanism for driving the sweepstick. The picker stick is mounted vertically and supported at the bottom by a large main bearing on which it pivots to drive the shuttle back and forth through the shed at speeds in the range of 120 to 250 picks per minute. Connected to and driving the picker stick is the sweepstick that is mounted horizontally just below the midpoint of the picker stick and contacts the picker stick in paced but sharp, high-impact movements in order to drive the shuttle rapidly through the shed. Both the picker stick and sweepstick can be seen in the Draper power loom in Figure 13-5 and the Heim sweepstick advertisement shown in Figure 13-6.

The shuttle mechanism involves considerable shock, vibration, and misalignment of moving parts resulting in periodic downtime of the loom until repairs can be made. Under these conditions, the sweepsticks developed fractures in and around the bear-

Fig. 13-5: Draper Company Dobby Power Loom with Northrop Automatic Bobbin Changer. Early 1900s. Picker stick and horizontal sweepstick located on left side of loom. (Courtesy of the American Textile History Museum).

ing, causing splitting and eventual failure. It was these operating conditions and Heim's claims that spherical bearings were specifically suited to accommodate shock and misalignment that led Draper to Heim hoping that such a bearing would provide the durability and longevity they were seeking.

Heim loved challenges and committed a significant amount of time to help Draper develop an improved sweepstick. Within a short period Heim submitted concepts for modified versions of his spherical bearings and subsequently filed three patent applications, one each in June, August, and September 1949. The patents were unique in that they disclosed new concepts for the sweepstick as well as mounting and supporting of the spherical bearing. Taken from his first patent for an improved sweepstick *(2,662,557)*, Heim described his invention as follows:

> *It is desirable to provide this mechanism (sweepstick) with sufficient flexibility, yielding and resiliency which will absorb much of the shock incident to the operation of the shuttle and operating mechanism, to thus eliminate as far as possible breakage of the parts and reduce wear. It is also desirable to have a bearing in the sweepstick capable of compensating for misalignment*

without cramping or binding. In this invention I have secured these desirable results by making either the whole body of the sweepstick or at least a portion of it of a flexible, resilient, yielding, shock-absorbing material such, for example as vulcanized rubber, a suitable plastic or other suitable material. A sweepstick of this material not only has a certain amount of yield or resiliency longitudinally to absorb longitudinal strains or shocks, but is also capable of a certain amount of lateral bending or yielding, and by mounting in the sweepstick body comprising this material an improved bearing which is capable of lateral rocking movement the construction effectively absorbs both longitudinal and lateral shocks and also compensates for misalignment without binding or cramping effect. This greatly increases the life not only of the sweepstick itself but also of the connected mechanism with which it is used, and improving operation of the device as well as greatly increasing its operative life and efficiency.

All three patents used Heim's four-piece construction consisting of the non-metallic housing *(sweepstick)*, a spherical metal ball with a hole drilled through, and two metal bushings to securely hold the ball in the housing. Whereas in an all-metal four-piece bearing the bushings are made of a softer metal that molds around the ball when inserted under high pressure, Heim used pre-shaped metal bushings and a modified procedure and machine for assembly into the sweepstick.

To accommodate the material used to make the sweepstick, the bushings had to be formed or "molded" to the shape of the ball before assembly since the softer, non-metallic sweepstick opening would not survive the high pressures induced during assembly of an all-metal bearing. To assemble the bearing into the transverse opening in the sweepstick, Heim developed a machine similar to the one used for his all-metal spherical bearing consisting of die sets and a power press. In his machine for sweepsticks, the pre-molded bushings were inserted over the ball from above and below simultaneously. As the bushings were inserted over the ball, the bushing walls expanded slightly against the sweepstick to tightly interlock both the bushing and ball into the sweepstick.

Heim proposed three bearing designs to Draper for testing in their looms. The first design involved spherically shaped bushings with smooth outer walls that tightly interlocked within the walls of the sweepstick. This invention was filed for a patent on June 15, 1949, and included three versions of an improved sweepstick, with one using a single sweepstick body and the other two using two sections that were bolted together. Figure 13-7 illustrates the single sweepstick showing Figures 1 through 4 of the patent. This invention was divided into two patents, one for the device *(2,662,557)* and one for the method of assembly *(2,701,907)*.

The second design proposed by Heim used bushings designed with ridges and grooves to better grip the sweepstick under high loads and misalignment. Heim developed two designs of this improvement, with one using a single molded metal bushing shaped with ridges that insert into grooves in the sweepstick walls. The second design consisted of two molded metal bushings, one inside the other, with grooves and ridges that fit together into the sweepstick walls to provide for greater retention under shock

Fig. 13.6: Advertisement for Heim Sweepstick with Unibal Spherical Bearing, ca. 1957.

forces and misalignment. This invention was filed two months later on August 5 and was also divided into the design *(2,592,566)* and the method for assembly *(2,701,409)*.

For the third invention *(2,665,956)* filed on September 7, Heim designed the bearing to incorporate lubrication. The invention *(Figure 13-8)* used a pair of metal bushings with smooth surfaces that when pressed together deformed outward to create a channel to hold a lubricant as well as securely interlock within the sweepstick

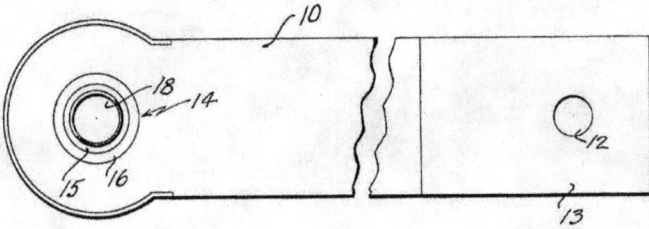

Fig. 1.

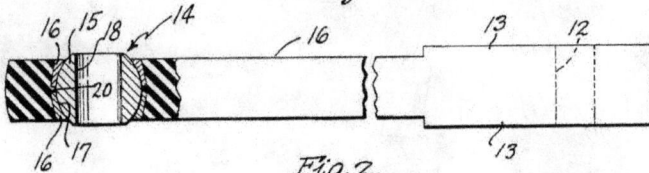

Fig. 2.

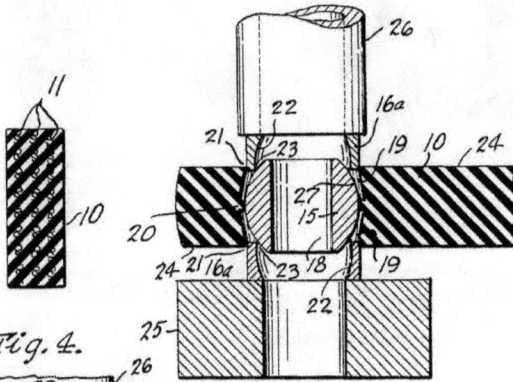

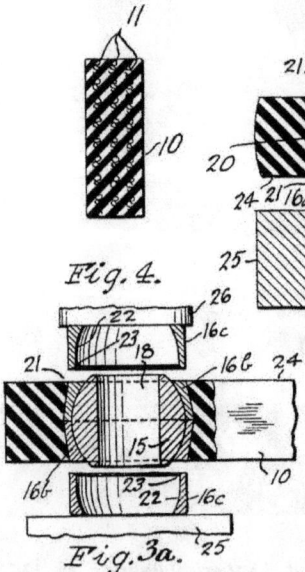

Fig. 4.

Fig. 3.

Fig. 3a.

Inventor

By Lewis R. Heim

Wooster & Davis Attorneys

Fig. 13-7: Sweepstick and Spherical Bearing with Smooth Bushing Surfaces for Power Looms. Patent 2,662,557, filed June 15, 1949.

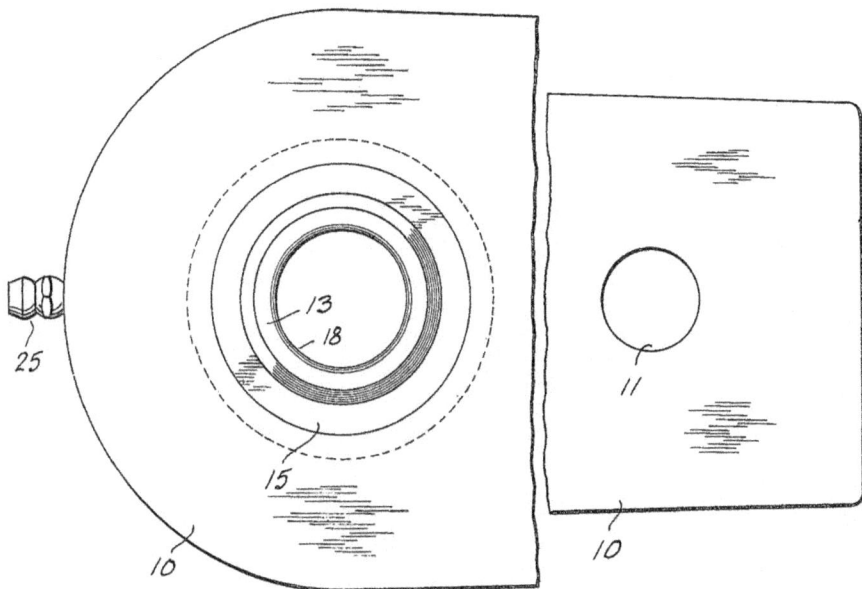

Fig. 1.

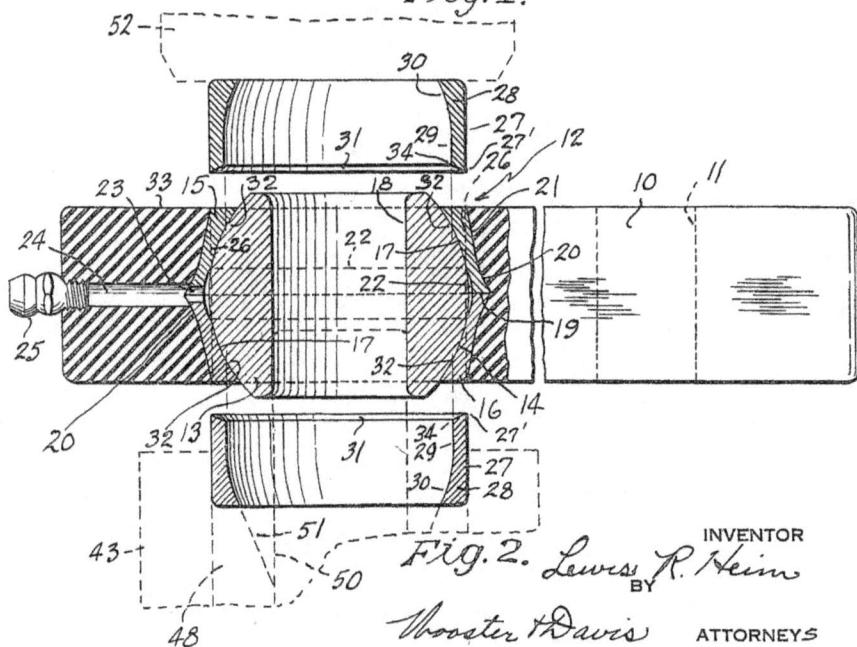

INVENTOR
Fig. 2. *Lewis R. Heim*
BY
Wooster Davis ATTORNEYS

Fig. 13-8: Spherical Bearing with Lubrication Channel in a Sweepstick Used in Power Looms. Patent 2,665,956, filed Sept. 7, 1949.

walls. As a result of deformation, the bushing tightly gripped the ball to prevent free movement and required a second step to loosen the ball, as was done in Heim's rod end bearings.

To achieve this two-step assembly method, Heim developed a machine that performed both processes *(assembly and loosening)* using a modified version of the die set and power press machine developed for rod end bearings. The upper portion of the machine supporting the downward actuator contained two pressing mechanisms, one inside the other, with the outer press *(#52 in Figure 13-9)* for the bushing and the inner press (#54) for the ball. During assembly of the bearing into the sweepstick, the outer press first inserted the two bushings until they made contact with each other. At that point the machine continued to press on the bushings, causing the bushings to grip the ball and to expand outward into the sweepstick walls and at the same time create a channel around the ball for insertion of a liquid lubricant *(#23 in Figure 13-8)*. To loosen the ball against the grip of the bushings, the inner press was actuated downward on the ball with sufficient force that caused the lower bushing to yield slightly, thus loosening the ball for movement.

Heim was not the only inventor of sweepsticks that were patented and assigned to the Heim Company. Edward Dardani, an engineer and foreman at Heim, was granted two patents *(2,601,875 and 2,717,006)* for designs involving a spherical bearing locked into place using a system of brackets, bolts, and rivets. Also, Heim's son, Charles, who was at the time general manager, was granted a patent *(2,738,570)* for a five-piece spherical bearing that was inserted into the sweepstick opening fabricated with a V-groove to improve the locking action of the bushings. According to longtime employees of the Heim Company interviewed in 2016, the supply to the textile industry of sweepsticks containing Heim spherical bearings was a significant source of sales that continued well into the 1970s.[3, 4]

Two-Piece Spherical Bearing

By the early 1950s, the aircraft industry had many applications for small, lightweight, self-aligning bearings capable of withstanding very high loads for their size. Specifications for such bearings required that the bearing be able to withstand loads at least two times the shear value of the supporting bolt or pin. Most bearings capable of meeting this requirement at the time were bulky and/or too heavy for use in aircraft. It was this requirement that led Heim to develop a two-piece version of his four-piece spherical bearing. A two-piece design not only eliminated the extra bulk and weight of soft metal bushings, but also increased their ability to survive under greater relative loadings.

In 1953, Heim filed an application to patent a two-piece spherical bearing that met the greater load-carrying requirements of the aircraft industry. The invention patented the bearing design plus the method and the machine for making the bearing. The bearing consisted of a ball and outer race made of the same high-grade material such as SAE 52100 chrome steel. In the method, Heim used a drilled-through ball that is heat-treated to the required hardness and placed inside an "unhardened" outer race that was pre-machined with a "cylindrical" bore. The method involved two procedures for

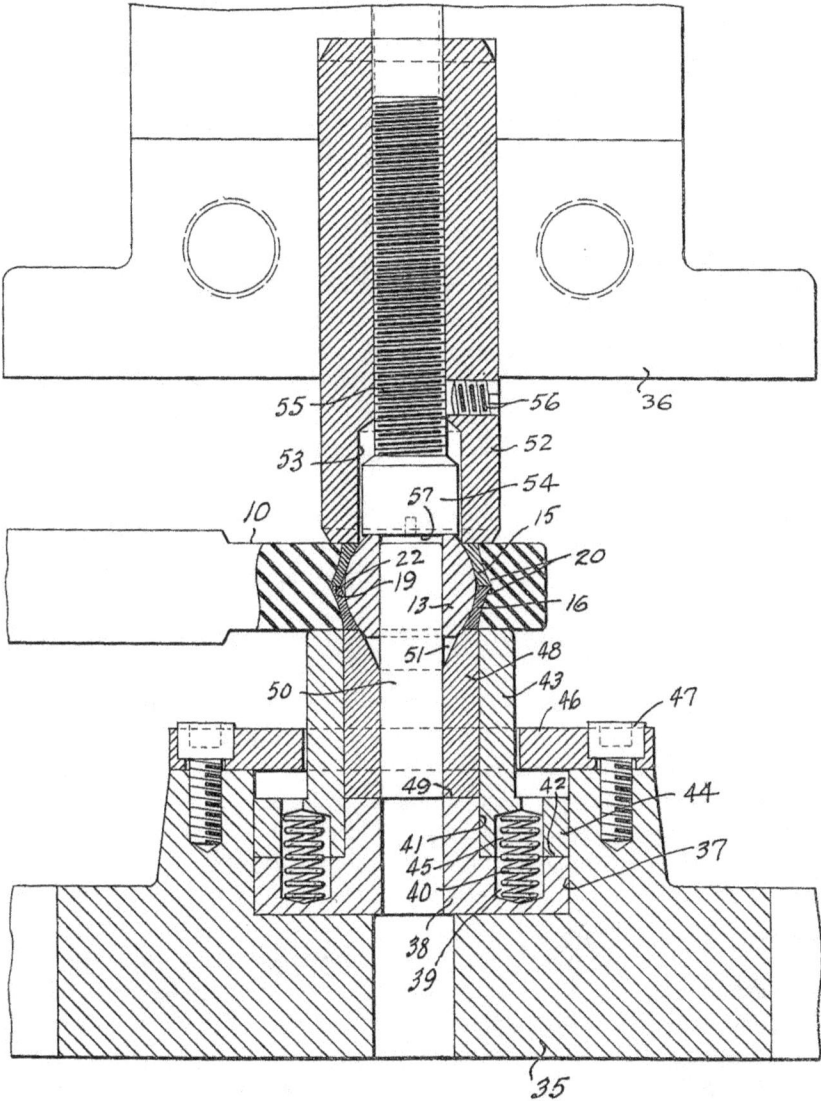

Fig. 13-9: Machine to Insert Metal Bushings and Loosen Ball in Sweepstick. Patent 2,665,956, filed Sept. 7, 1949.

producing an assembled bearing meeting the greater strength requirements. First, the unhardened and ductile *(softer)* outer race was molded or "coined" around the hardened ball using a pair of dies inserted into a machine called a power press. Second, after assembly, the entire bearing was heat-treated to raise the hardness of the outer race to equal that of the ball.

In the first part of Heim's assembly method, the ball and race are placed over a vertical rod and dropped down to rest on a spring-loaded support passing through the lower die. To mold the outer race over the ball, the upper die is actuated down, forcing the race into both upper and lower dies to "coin" the race around the ball. Being unhardened, the softer, outer race deforms from its original cylindrical shape to the spherical shape of the hardened ball without affecting the ball sphericity. The bearing and the machine *(power press with die sets)* are shown in Figures 13-10 and 13-11 obtained from patent 2,787,048.

The standard design for spherical bearings was for the ball diameter to be greater than the depth of the race *(housing)* so that the ball swivel or rod end was capable of movement when the other was fixed in place. In order to achieve this, Heim designed the dies such that once the softer outer race was molded to the spherical shape of the ball, the dies continued to further deform the race to its required shorter depth as shown in Figure 13 of Figure 13-11. Once assembled, the entire bearing was heat-treated to relieve internal stresses and raise the hardness value of the race to match that of the ball. At the same time, heat-treating also increased the race diameter just enough to free the ball to swivel. Heim's invention of the two-piece design along with the methods and machine for assembly was a significant advancement in making spherical bearings supplied to the aircraft industry and made the Heim Company a primary supplier of high-grade bearings used in commercial and military aircraft.

Spherical Bearings for Levers

In 1955 Heim filed an additional application to patent a method for mounting a spherical bearing in a machine element such as a lever. Unlike the sweepstick, the machine element in this invention is made of a relatively hard material such as metal and makes no mention of the need to design for excessive shock forces or misalignment. The invention relates to a two-piece spherical bearing consisting of a drilled-through metal ball that is mounted in the housing that also acts as the outer race. In this invention Heim calls the housing a "bushing" that is designed specifically for mounting inside the opening in the lever.

The novel aspect of this invention, which became patent 2,898,671, was that Heim developed both a method and a machine for mounting the ball into the bushing and the bushing into the lever in a single operation. To accomplish this Heim used a preformed cylindrical bushing that was closed at the lower end to hold the ball and open at the top end into which the ball was placed. The bushing also contains on its outside surface an annular shoulder or ledge onto which the lever was placed before assembly.

To perform assembly, Heim modified the machine described in 2,787,048 for spherical bearings, with dies designed for this application. In operation, the bushing was placed into the lower die of the machine, the ball was inserted into the bushing, and then the lever was inserted over the bushing to rest on the annular shoulder. Once set up, the upper die was lowered and pressed down on the upper part of the bushing, causing the inner bushing wall to deform over the ball, and at the same time, causing the outer bushing wall to deform over the lever, thus locking in both the ball and the lever. Illustrations of the machine and bearing before and after assembly are shown in Figure 13-12.

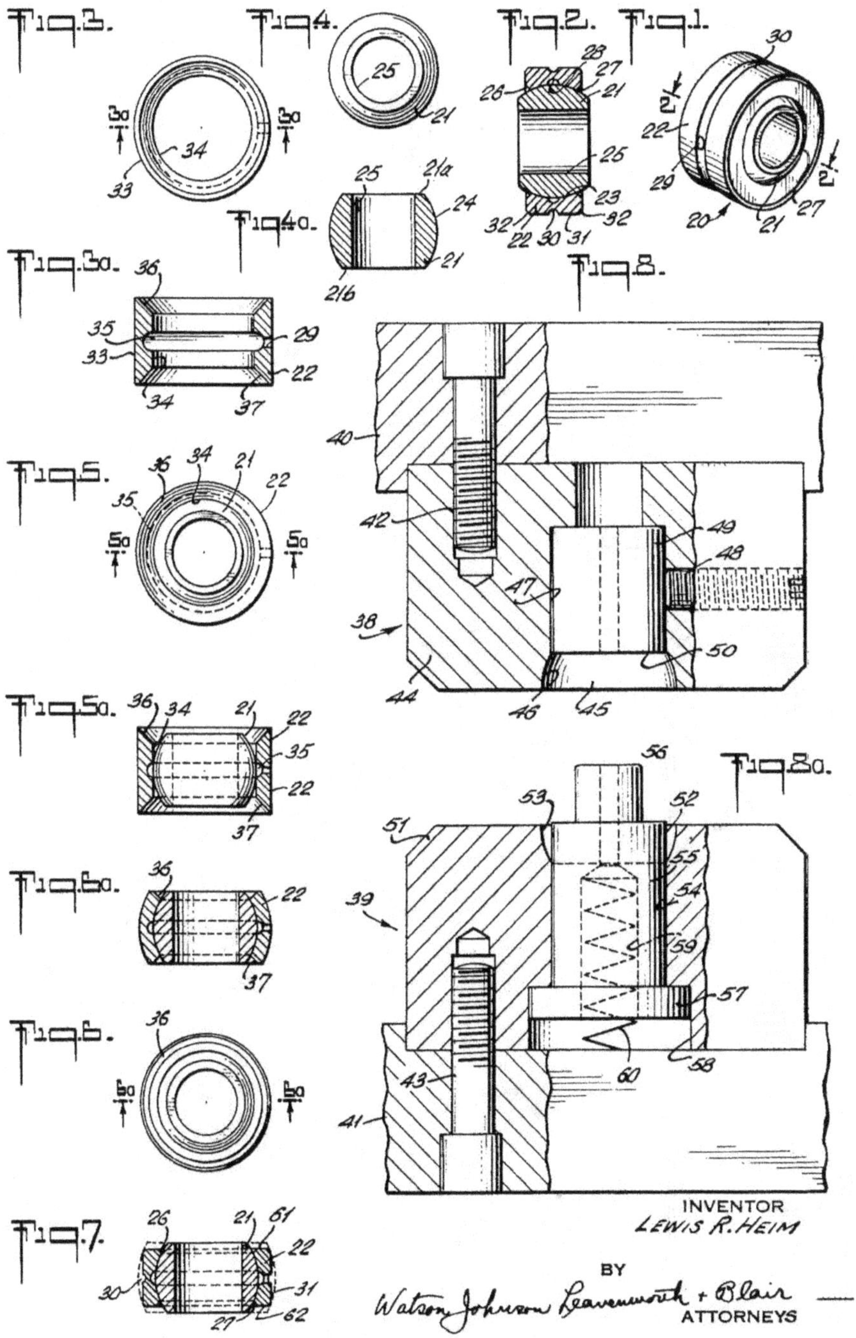

Fig. 13-10: Two-Piece Spherical Bearing for Aircraft Industry with Die Set Pressing Machine to Coin Outer Race Around Ball. Patent 2,787,048, filed Aug. 20, 1953.

Sizing and Forming Drilled-Through Balls

Following the invention of his two-piece spherical bearing, Heim built an improved machine and method to manufacture hardened drilled-through balls. In patent 2,941,290 filed on January 11, 1955, Heim developed a modified version of his power press machine fitted with dies to make balls from standard steel bar stock such as commercially available rods. Heim's invention is a "sizing and forming" operation in

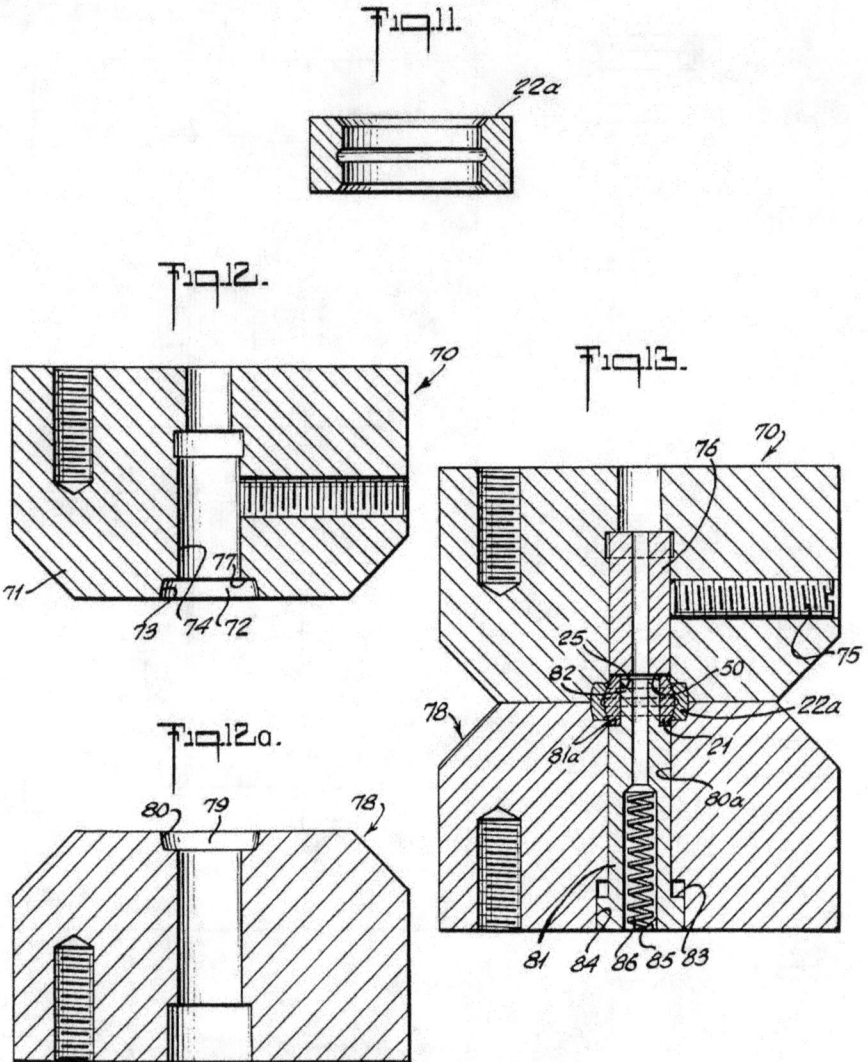

Fig. 13-11: Die Set and Pressing Machine for Two-Piece Spherical Bearing Showing Final Coining of Bearing Housing Around Ball. Patent 2,787,048, filed Aug. 20, 1953.

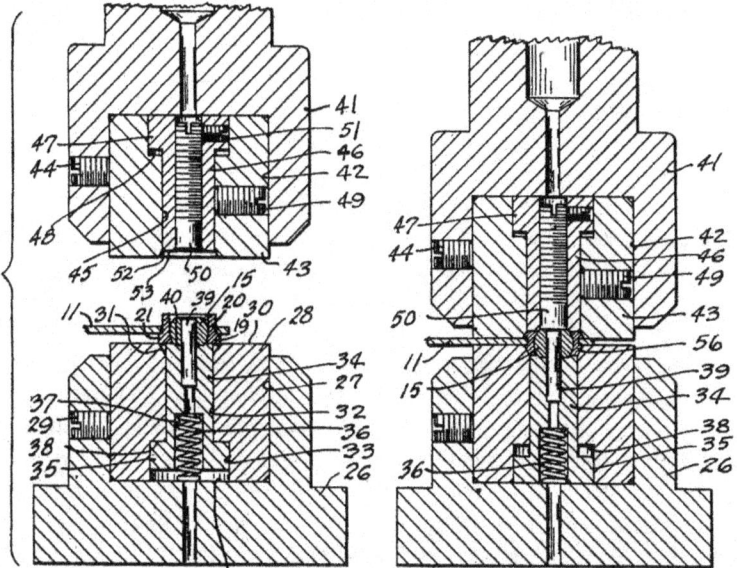

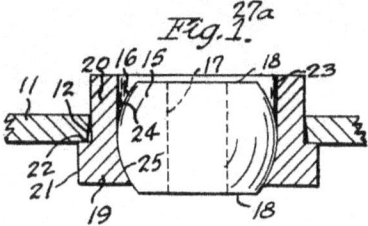

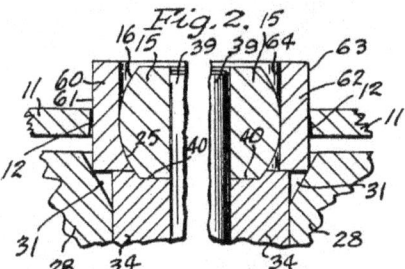

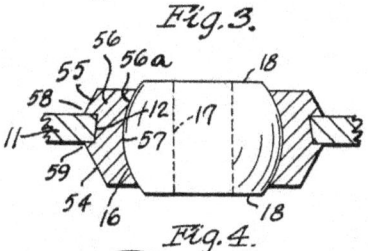

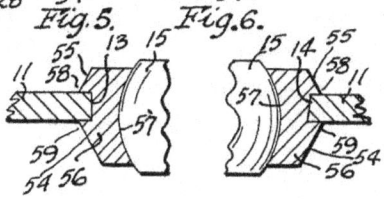

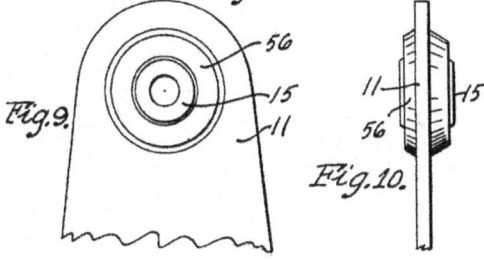

Fig. 13-12: Machine Mounting Spherical Bearing to a Metal Lever. Illustrations Showing Cylindrical Bushing Molded Around Drilled-Through Ball and Lever. Patent 2,898,671, filed Feb. 16, 1955.

which a ductile *(unhardened)* and undersized convex *(ellipsoidal)* roll with an under-sized drilled-through hole is expanded into spherically shaped dies to form a perfectly round drilled-through ball for use in both rod end and standard spherical bearings. The process for fabricating the bearing balls is described below, with the machine and the stages of ball formation illustrated in Figure 13-13.

To initiate the ball-formation process, standard round bar stock with a diameter less than the finished ball is cut to size, then machined to form a convex roll in which its axial length is greater than the finished ball diameter. Next, a hole is drilled through the axial length with a diameter less than that of the finished ball. The convex roll is then placed into the lower die being held in place by a pilot *(rod)* from below. In form-ing the ball, the upper die is actuated downward until it makes contact with the lower die. During this operation, the upper die compresses the roll such that its walls expand into the dies to form a spherical shape. Once the dies make contact and the compres-sion process stops, a tapered punch is actuated downward through the upper die into the undersized hole to expand the diameter to the required size, while further expand-ing the outer diameter into the spherically shaped dies to complete the ball-forming process. After formation in the machine, the ball is suitable as is or heat-treated to harden the metal for use in higher load rated bearings. Heim's invention was based on the same strategy employed to fabricate rolling bearings made from commercially available sheet metal. By using round bar stock, a low-cost commodity item, and his new method, Heim was able to not only eliminate costly machining operations, but also to minimize the total cost of production.

Unbroken Races for Ball and Roller Bearings

In the summer of 1958, Heim's grandson, Carl H. Van Etten, the son of his daugh-ter, Florence, started working at the Heim Company. At that time no rolling bearings were manufactured, which means, given the advertisement shown in Figure 13-4, Heim stopped production of his pillow block and flange unit ball and roller bearings sometime after 1952. Competitive pressure was the primary reason for exiting that market as well as the catalyst for his next invention.[5, 6]

As discussed earlier, ball and roller bearings were made in many configurations, from single and double-row ball bearings to roller bearings made with uniform, ta-pered, convex, and concave rolls. Some bearings were made with fixed axis inner races while others were the self-aligning type. In order to retain the balls or rollers, one or both of the races *(inner and outer)* were formed either with flanged sides, grooves, or a combination of both. Some bearings were made with cages to space the rolling elements evenly around the circumference and others contained a full complement for applications requiring high relative loads.

One of the challenges faced by manufacturers was to assemble a rolling bearing with smooth and unbroken surfaces for the inner and outer races that would allow as-sembly of a full complement of rolling elements. Bearings with unbroken races were highly valued as they provided smooth tracking surfaces and reduced friction of the balls or rollers. Into the 1950s manufacturers of ball and roller bearings used various methods for assembling rolling bearings, with many bearings designed with notches or

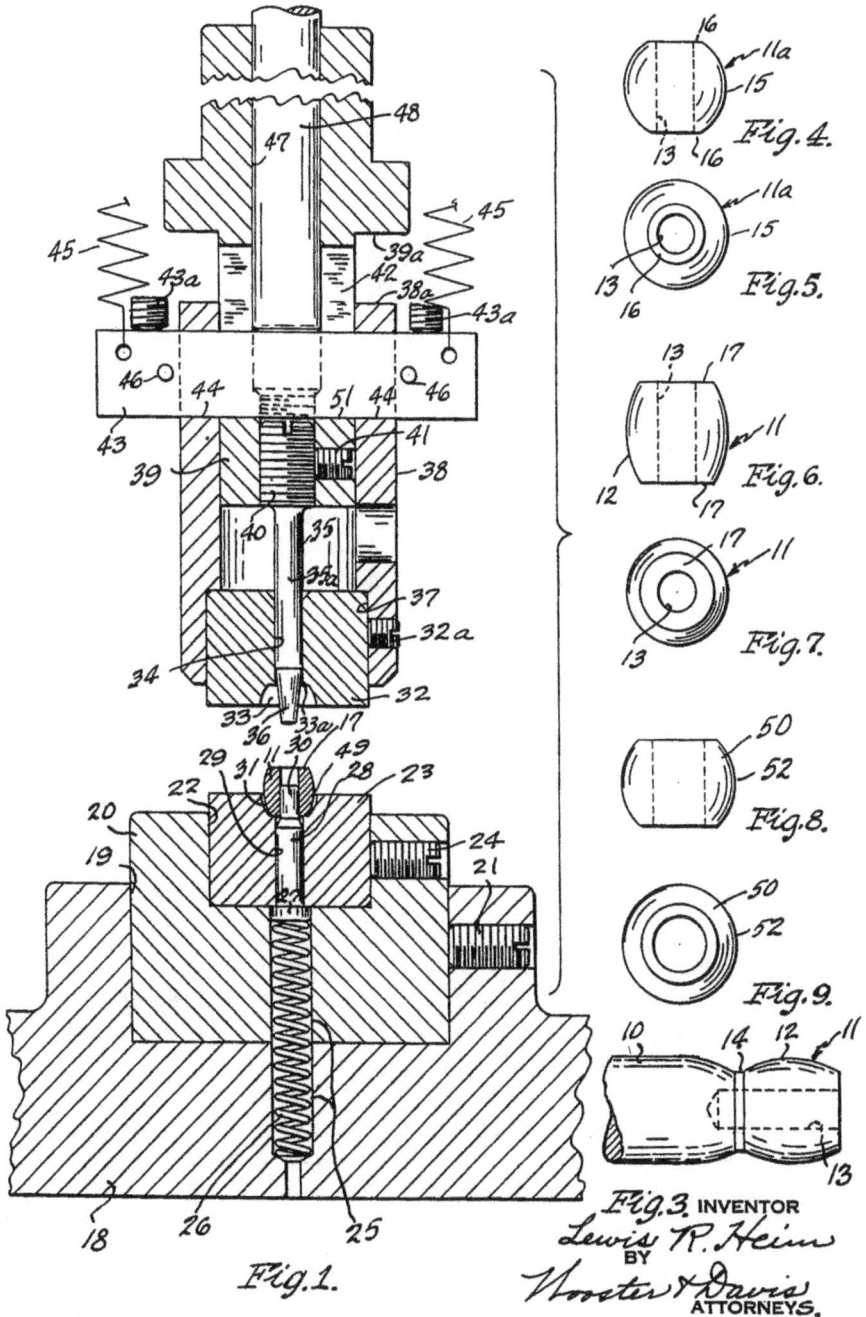

Fig. 13-13: *Machine for Sizing and Forming Ball for Spherical Bearings from Round Bar Stock. Ball Shown in Various Stages of Manufacture: Round Bar Stock, Convex Roll, and Finished Ball. Patent 2,941,290, filed Jan. 11, 1955.*

recesses in the sidewalls or periphery of the races for insertion of the balls or rollers. This process was not only more expensive but the use of insertion notches or recesses would create disturbances in the rolling action and increase the potential for premature failure of the bearing. Insertion notches were nothing new and neither were one-piece grooved races, with many designs invented in the late 1800s and early 1900s. One of the first to invent a rolling bearing designed with insertion notches, unbroken races, and a full complement of balls was Max Mossig of Berlin, Germany. In 1916, Mossig filed an application that became U.S. Patent 1,301,295 in which his bearing *(Figure 13-14)* retained the balls in unbroken races machined with grooves.

After developing the two-piece spherical bearing, Heim turned his attention to making rolling bearings using similar assembly methods. In 1955 Heim filed his first application to patent a bearing and a method for assembly of rolling *(antifriction)* bearings with unbroken races. Taken from patent 2,910,765, Heim made the following statement on his invention.

> *This invention relates to antifriction bearings, and more particularly to a novel method of making and assembling antifriction bearings, either of the ball or roller type, and either of the fixed axis or self-aligning type; and has for an object to provide an improved and effective means and method of making and assembling antifriction bearings, including a series of antifriction rolling elements either in the form of balls or rollers, and using them either with or without a cage or separator.*

> *A further object is to provide a method of assembling antifriction bearings in which a bearing with unbroken raceways may be assembled with its full complement or quota of the rolling elements either balls or rollers, that is, all the rolling elements the raceways will hold with slight permissible clearance between them.*

To assemble rolling bearings with unbroken races, Heim used his concept of controlled expansion, developed for making the drilled-through balls used in spherical bearings. Heim's method started with an undersized and ductile *(unhardened)* inner race then expanded its radius to interlock with the rolling elements and outer race. To accomplish this, Heim modified his previously developed power press and die set processes to accommodate ball and roller bearings. When assembling the bearing, the outer race is placed in the lower die, into which the balls or rollers are inserted and held by some temporary means. The undersized inner race is telescopically inserted inside the rolling elements and over a cylindrical pilot that holds the inner race in position. After positioning the components, the upper die is lowered onto the lower die to lock in and prevent movement and expansion of the outer race. Once locked in place, the tapered punch is actuated downward to displace the pilot and expand the inner race to the correct diameter while interlocking the rolling elements with the outer race. After assembly, and if required, the entire bearing is heat-treated to both relieve internal stresses and harden the metal to raise its load-carrying capability, as done with spherical bearings. Heim's assembly method could be used for a wide range of bearings in-

Fig. 1

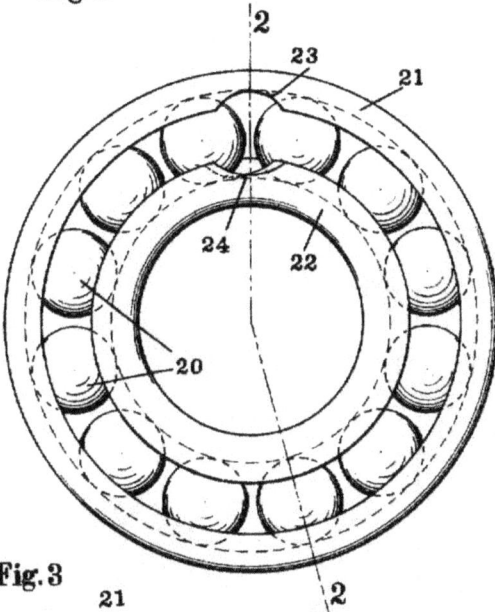

Fig. 2

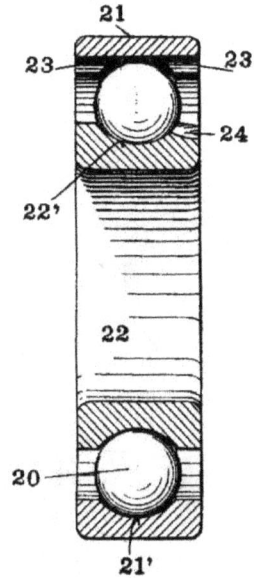

Fig. 3

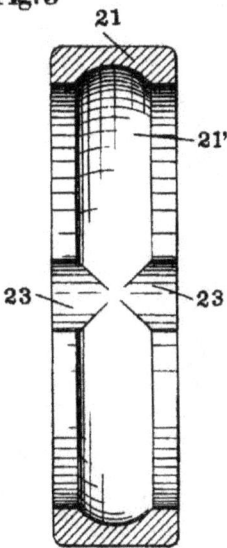

Fig. 4

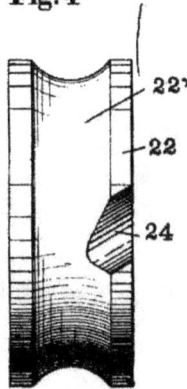

Fig. 5

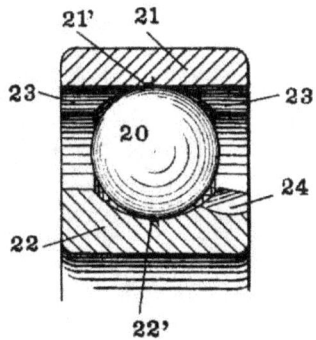

Fig. 13-14: Ball Bearing with Unbroken Races and Insertion Notches. Patent 1,301,295 by M. Mossig, filed Aug. 24, 1916.

cluding ball, roller, and rod end rolling bearings as shown in Figures 13-15 and 13-16.

The deformation of metal through expansion, contraction, and molding *(coining)* are all forms of "swaging" in which metal is deformed and shaped using a tool or die and includes both cold working and hot working *(forging)* processes. Heim was not the original inventor of swaging the inner or outer race to assemble rolling *(antifriction)* bearings.

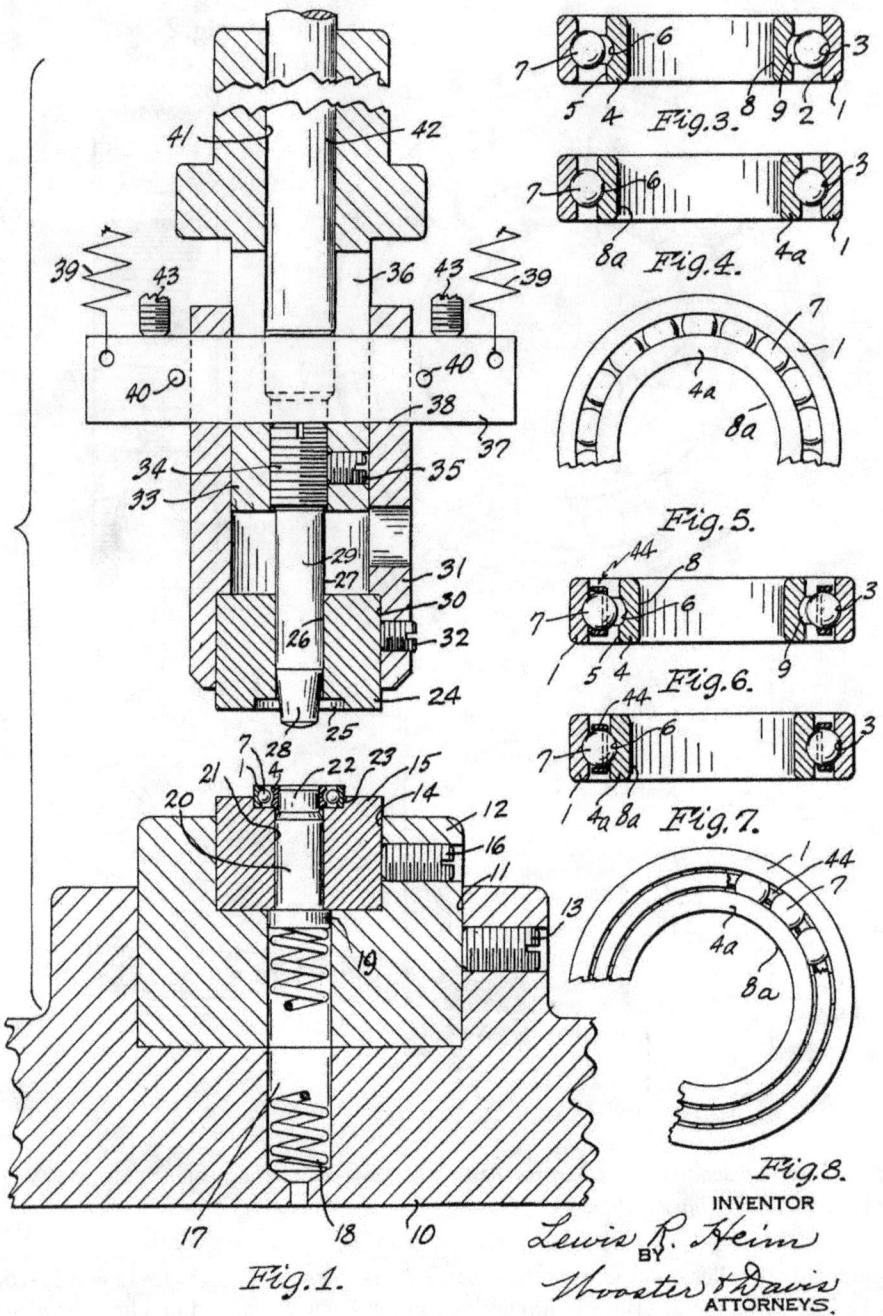

Fig. 13-15: Machine for Expanding Inner Race in Assembly of Ball and Roller Bearings with Unbroken Races. Patent 2,910,765 , filed April 6, 1955, and refiled Aug. 19, 1958.

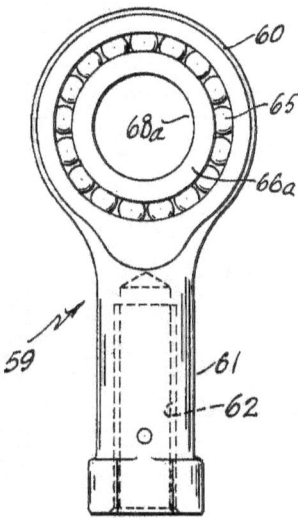

Fig. 15.

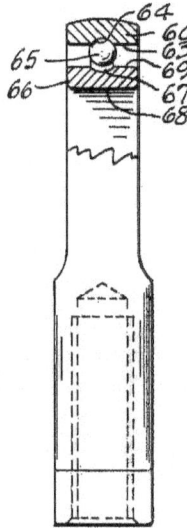

Fig. 16.

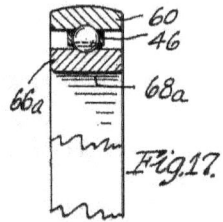

Fig. 17.

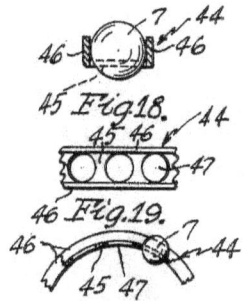

Fig. 18.

Fig. 19.

Fig. 20.

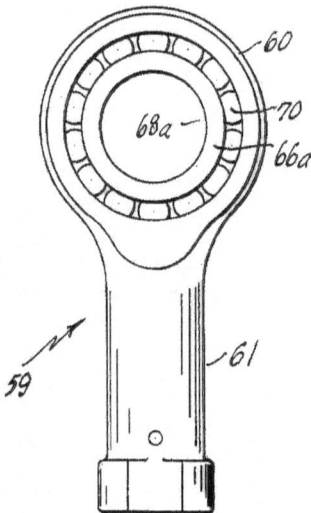

Fig. 21.

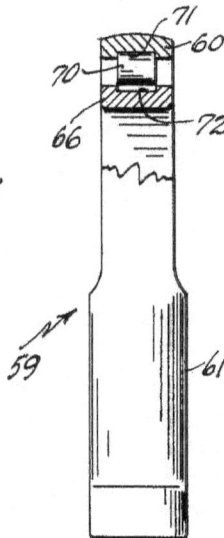

Fig. 22.

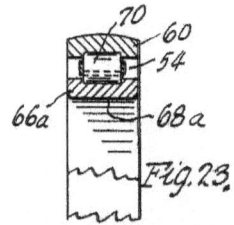

Fig. 23.

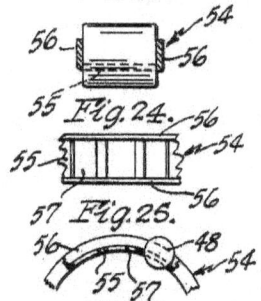

Fig. 24.

Fig. 25.

Fig. 26.

INVENTOR

Lewis R. Heim

BY

Wooster & Davis

ATTORNEYS.

Fig. 13-16: Heim Rod End Bearings with Unbroken and Expanded Inner Races. Patent 2,910,765, filed April 6, 1955, and Aug. 19, 1958.

The concept was first disclosed by Albert S. Reed in 1913 in patent 1,080,169, which described a method for assembling a ball bearing with grooved inner and outer races using either contraction of the outer race or expansion of the inner race.

The Reed method employs a pair of dies in which the lower die is set on a plunger and supported the unassembled bearing components as shown in Figure 2 of Figure 13-17. The upper die was fixed in place and consisted of a cylinder with inwardly tapered walls. In operation, the plunger was actuated upward into the bore of the fixed upper die, which caused the tapered walls to contract the outer race enough to lock the balls into the grooves of the respective races for an assembled bearing.

Heim submitted the application to patent his invention on April 6, 1955, but it was subsequently abandoned and replaced by two applications, one for the "Method" on August 19,1958 *(2,910,765)*, and one for the "Device" *(bearing design)* on September 28, 1959 *(3,123,413)*. The reason for the abandonment and resubmittal was that the examiner from the US Patent and Trademark Office cited the Reed patent as prior art and challenged Heim to prove novelty. Reed states in his patent that he found it preferable to compress the outer race, rather than expand the inner race, but further specifies that his invention is not confined to either method since the essential feature consisted in constructing the races of such difference in diameters as to permit the insertion of balls followed by reducing the space between races to lock the bearing together.

It is obvious that the basic method used by Heim for expansion of the inner race is the same disclosed by Reed. However, the expansion portion of Heim's method is only part of the invention. As Heim states in his patent, the invention is an "improved method for making and assembling antifriction bearings" that involves other features that meet the criteria for novelty. One feature was the machine Heim used to make his bearing. Reed, as stated in his patent, described only the concept of reduction or expansion of one of the races and did not describe in detail the "machine" for executing his method. Heim describes the machine in detail, plus the additional features that improve the method. One improvement that is not included in Reed's method is the ability to control race expansion accurately to achieve a very close, running fit of the bearing. Another feature included in Heim's method, and not at all mentioned in Reed's method, was the procedure for post-assembly heat-treating to remove internal stresses created in the metal during swaging. Still another was to produce a bearing with hardened components that are essential to making bearings durable with relatively high load ratings.

Clearly, the USPTO agreed with Heim and the arguments made for meeting the criteria of novelty since it granted Heim the method patent as well as a second patent for the bearing designs. In support of the USPTO decision, Heim's method and machine were novel enough to attract the attention of the Channing Company of New York City, which, prior to the grant date, bought all rights to Heim's two patents *(2,910,765 and 3,123,413)*, plus a third related invention described in patent 2,913,810. In this patent, Heim expanded the assembly method to include a second machine designed to coin, size, and burnish the raceways to a high degree of precision.

The Channing Corporation was a multifaceted conglomerate set up in 1952 by Kenneth Van Strum and Herbert Towne to service the fast-growing automobile and home-building

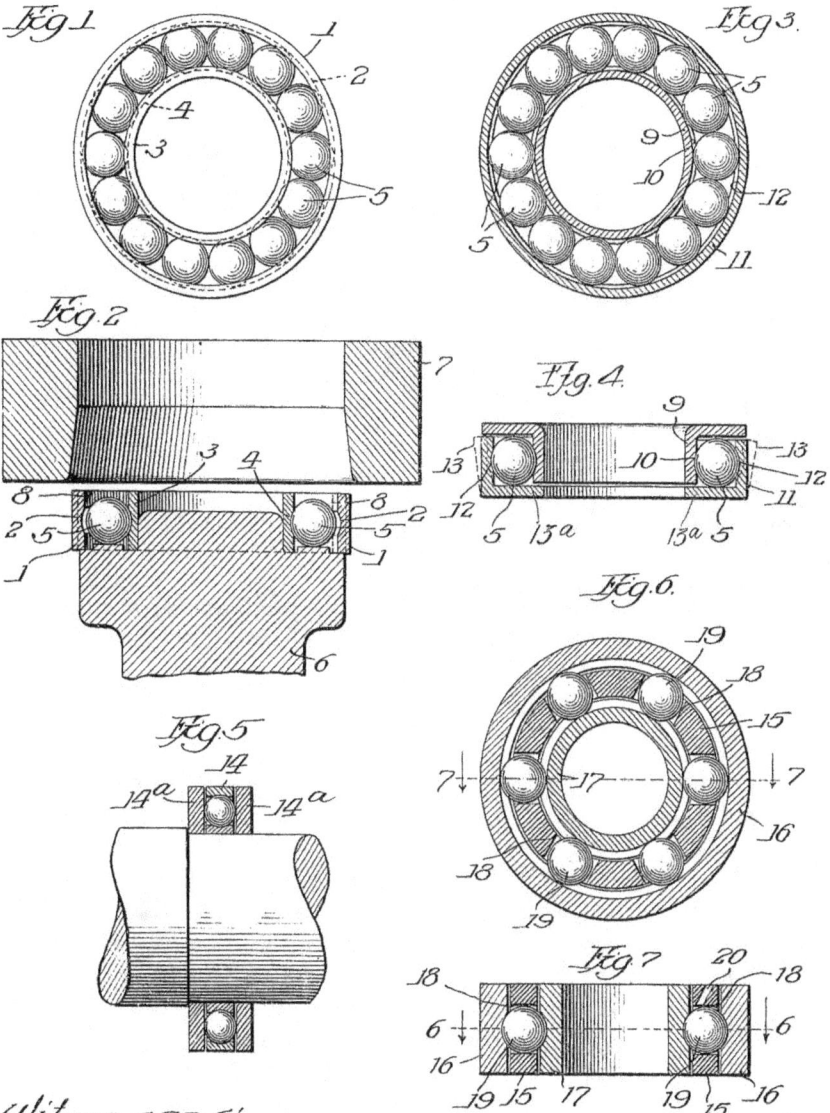

Fig. 13-17: Reed Method for Making Ball and Roller Bearings with Unbroken Races. Figure 2 Illustrates Contraction Die and Plunger. Patent 1,080,169, filed April 8, 1912.

industries.[7] Channing manufactured car parts and other hardware and purchased Heim's patents to support expansion of the new business. The purchase by Channing of the three patents is again another example of the value others found in Heim's inventions.

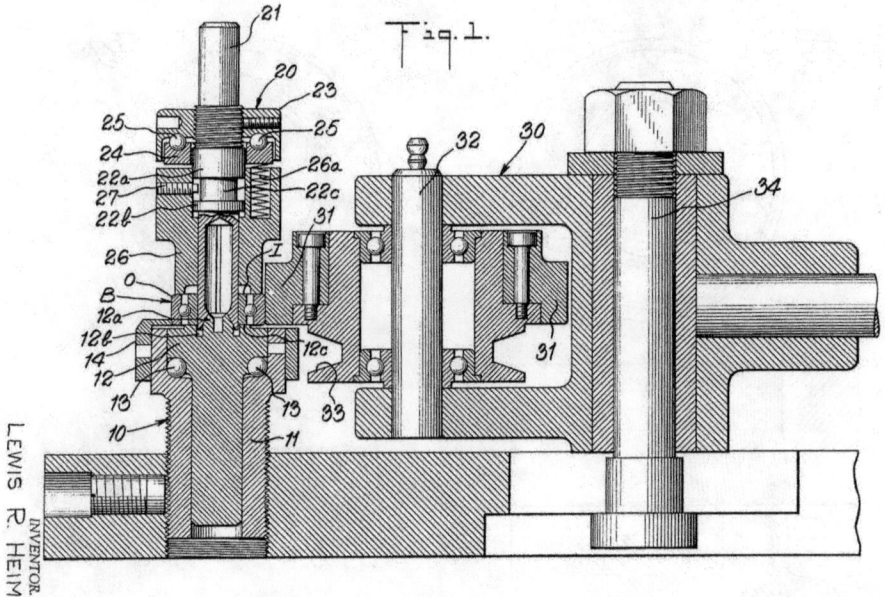

Fig. 13-18: Machine for Burnishing Raceways in Rolling Bearings. Bearing (B) Being Burnished Radially and Axially. Patent 3,093,884, filed June 16, 1959.

Final Patents

From the years 1956 through 1960 and well into his 80s, Heim continued to develop new methods and machines to improve quality and lower the cost to manufacture
spherical and rolling bearings, with six new inventions filed for patent protection.
One patent *(3,088,190)* was an additional method to assemble rolling bearings with
unbroken races by contracting *(swaging)* the outer race over the inner components,
after which one of the races is rotated to free up the rolling elements. This patent was
a continuation of the original expansion method *(2,910,765)* shown in Figure 13-15
for bearings with unbroken races. Next was a method *(3,049,789)* for assembling ball
and roller bearings using a die expander machine that was equipped with a reservoir of
oil used to cool and lubricate both the tapered punch and the bearing during inner race
expansion, resulting in reduced distortion of the inner race diameter.

To support assembly of his bearings, Heim developed two other machines that
performed dedicated operations. One invention *(3,027,590)* was an air-operated device for automatically removing dirt and metal particles after assembly. The other
invention *(3,093,884)* was a machine for burnishing *(polishing)* the entire area of the
raceways as well as the rolling elements of antifriction bearings. This machine *(Figure 13-18)* was unique in that it was able to "rotate both races at the same time while
exerting both radial and axial loading required for burnishing." The machine was a
simple and speedy method to ensure quality in assembled bearings while reducing
total production costs.

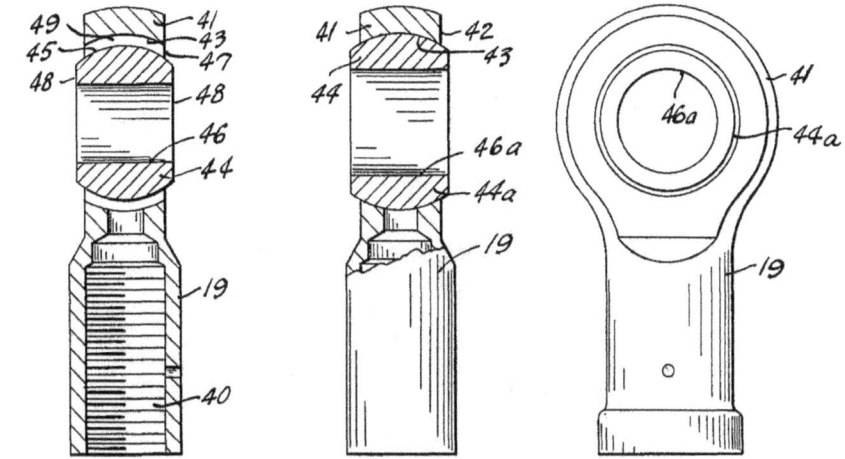

Fig.4.   Fig.5.   Fig.6.

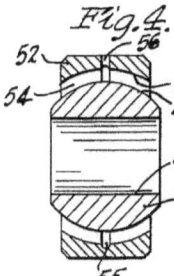 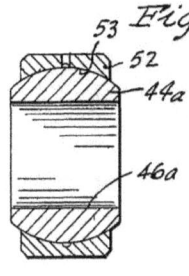 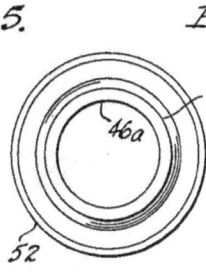 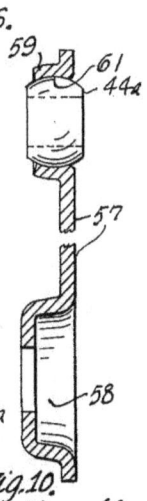

Fig.7.   Fig.8.   Fig.9.

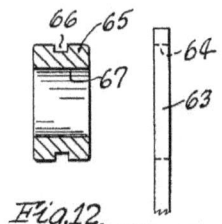 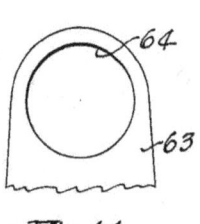 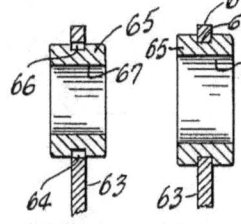 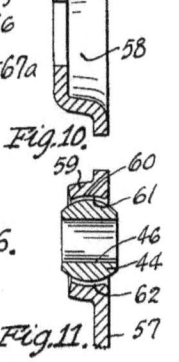

Fig.12.   Fig.13.   Fig.14.   Fig.15.   Fig.16.

 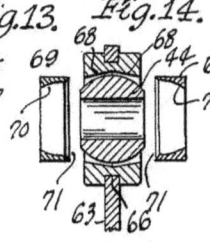 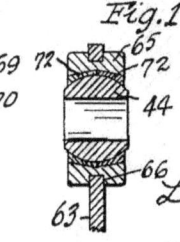  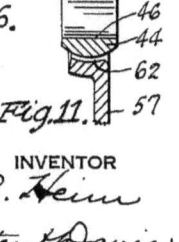

Fig.17.   Fig.18.   Fig.19.   Fig.10.   Fig.11.

INVENTOR
Lewis R. Heim
BY
Wooster & Davis
ATTORNEYS

Fig. 13-19: Variety of Spherical Bearings Made Using Improved Power Press Machine Employing Heim Expansion Methods. Patents 3,221,391 and 3,369,285, filed Nov. 18, 1960.

253

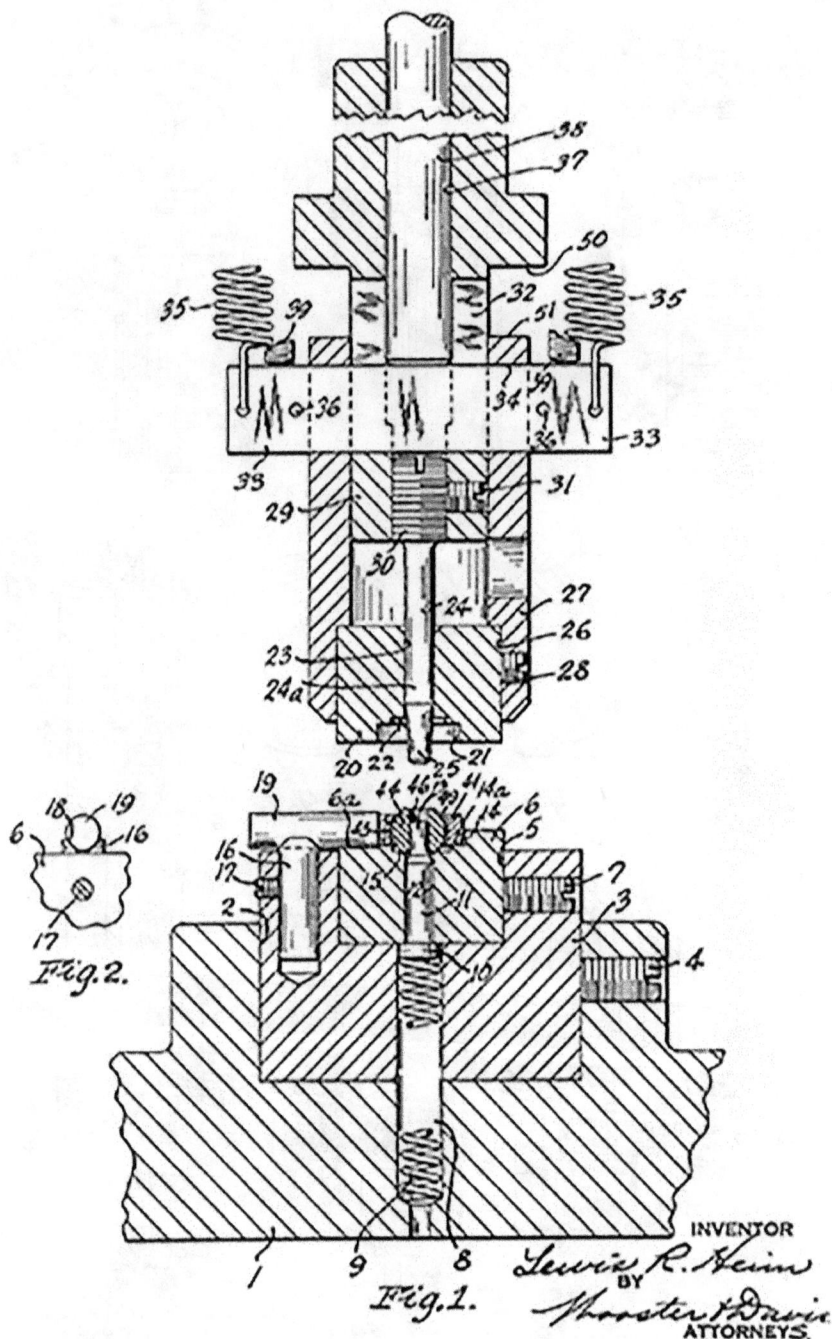

Fig. 13-20: Machine to Insert Spherical Bearings into Machine Elements (Levers) Using Expansion and Contraction Methods. Patents 3,221,391 and 3,369,285, filed Nov. 18, 1960.

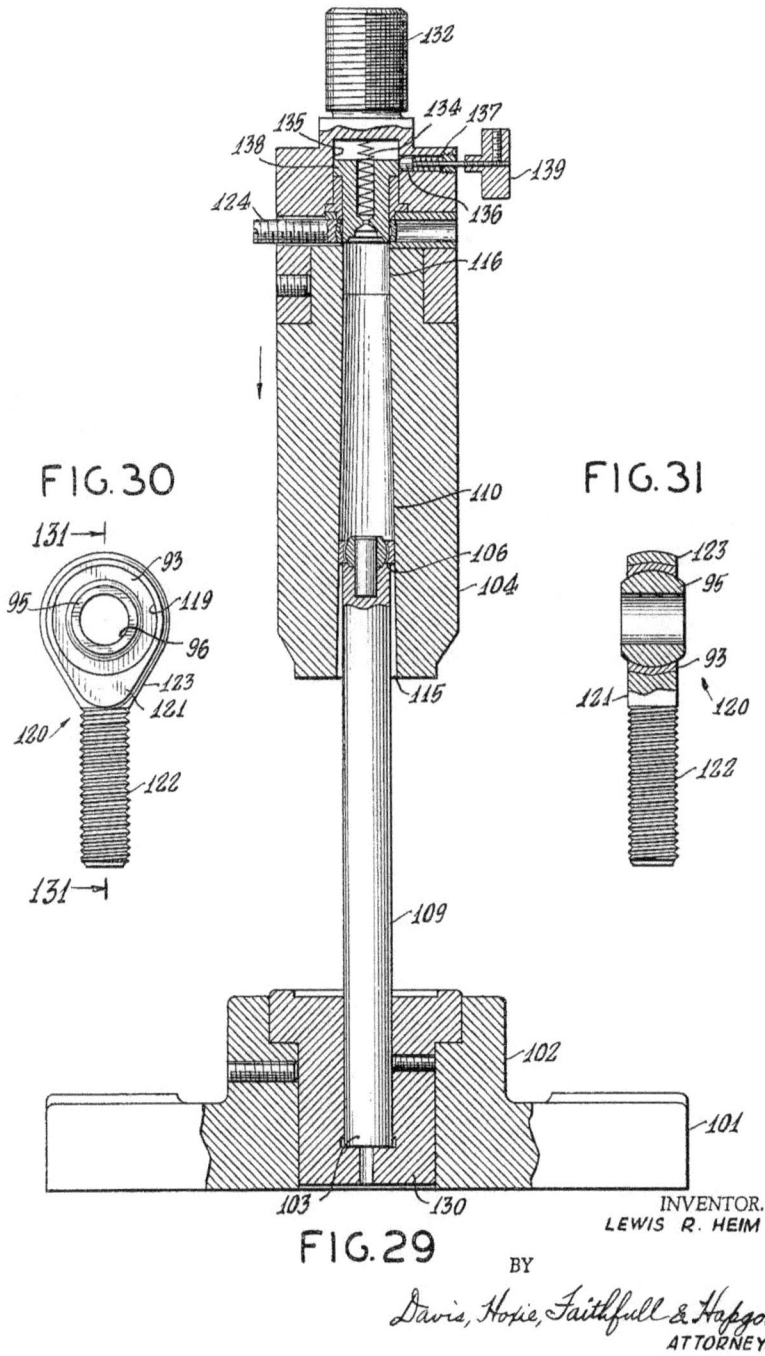

FIG. 30

FIG. 31

FIG. 29

INVENTOR.
LEWIS R. HEIM

BY

Davis, Hoxie, Faithfull & Hapgood
ATTORNEYS.

Fig. 13-21: Machine to Radially Contract Housing Over Ball of Rod End Bearings. Contraction Device Shown in Raised Position Patents 3,221,391 and 3,369,285, filed Nov. 18, 1960.

The last two patent applications filed by Lewis Heim occurred on November 18, 1960, when he was 86 years old. The applications described improvements to his previous inventions for fabricating and assembling spherical bearings, with one disclosure for the methods *(3,221,391)* and the other for the machines *(3,369,285)*. The patent referred to spherical bearings as the "ball and socket" type and covered assembly of the three different types: fixed spherical, rod end, and spherical mounted in levers *(Figure 13-19)*.

There were three machines *(devices)* described in the patents, with the first designed to assemble spherical bearings into a lever or other machine element. The other two devices were designed to assemble rod end and fixed spherical bearings using a contraction *(swaging)* device. The first machine, shown in Figure 13-20, is designed to assemble spherical bearings into a lever and combines the expansion and contraction methods used to manufacture the drilled-through balls of patent 2,941,290 *(Figure 13-13)*. The other two machines were designed for contraction of the housing *(outer race)* over the hardened ball. One machine *(Figure 13-21)* was designed for rod end bearings and the other, which was very similar, for standard fixed spherical bearings.

The machines and methods were invented to address the need for a high level of dimensional control during assembly to ensure contact over the entire ball and socket surfaces. A second objective was to provide a versatile, rapid, and inexpensive method of assembling bearing combinations constructed of unbroken members. These two inventions essentially were a culmination of Heim's work for manufacturing spherical bearings, and the patents were granted posthumously with the methods *(3,221,391)* in December 1965 and the machines *(3,369,285)* in February 1968.

Chapter 14

Accomplishments

Final Years

When Heim left Mount Dora, Florida, in 1932 and returned to Danbury, Connecticut, and the business of making bearings, the decline in the U.S. economy due to the Great Depression was bottoming out. Despite starting in the midst of the worst business conditions of the twentieth century, Heim's new company was able to survive by selling his newly developed sheet metal roller bearings and by performing grinding services on contract from other companies. Success during this period can be credited to Heim's skills as a machinist and native inventiveness that produced not only novel lower-cost bearings, but also the machines and methods required to make them. These innate qualities are seen in the machines he invented for hatting, ironing detachable collars, precision grinding, and fabricating mechanical bearings.

After moving to Fairfield, Connecticut, in 1935, Heim lived at his beach house bought when he operated the Ball and Roller Bearing Company. When not at the office, Heim worked in his studio overlooking Long Island Sound, where he developed drawings for new inventions as well as illustrations for his patents. Heim also spent a lot of his free time in the company tool shop fabricating and testing his ideas. There were even times when, during the night, Heim conceived of a new bearing, machine, or other device and instead of waiting until morning, would drive to the factory to work on it.

With the invention of the spherical rod end bearing in 1942, the Heim Company was no longer just another bearings company in a crowded industry. Being the owner of the patents gave Heim exclusive rights to manufacture and sell his rod end bearings in the United States without competition until the early 1960s. However, as discussed in Chapter 12, due to the needs of British aircraft manufacturers during World War II, Heim sold a manufacturing license to the Rose Bearings Company of Saxilby, England.[1] With the Heim license, the Rose Bearings Company became the first supplier of rod end bearings in Europe. Sometime after the war ended, Heim licensed other European companies including Schaublin of Switzerland and Hirschmann of *(West)* Germany, which produced additional earnings for the Heim Company[2, 3]

After the invention of the rod end bearing and with the company firmly established and successful, Heim, who turned 70 in 1944, began setting the stage for retirement. During the war, responsibility for managing commercial operations was increasingly delegated to his son, Charles, and son-in-law, Carl C. Van Etten. Charles advanced to general manager and treasurer. Carl, who left a teaching job at the Scarsdale public high school, primarily managed the purchasing department, but he also helped with correspondence, accounting, and other office functions. Together they were respon-

Fig. 14-1: Carl Heim Van Etten: Grandson of Lewis Heim, ca. 1970.

sible for managing most of the daily business operations, and by the mid-1950s Lewis had ceded all executive decision making to Charles.

After the war ended, Heim continued to spend part of the winter in Florida and in 1948, with business picking up, hired a young sales engineer by the name of Albert McCloskey. McCloskey, who graduated from Newark College of Engineering with a degree in Mechanical Engineering, initially went to work in New York City at Combustion Engineering, a manufacturer of steam generation equipment for power generation and industrial processes. After working at a second company in New York City making heavy machinery, McCloskey took a job in Bridgeport, Connecticut, with TEK Bearing Company, from which he was hired at Heim.[4]

Under the guidance of Lewis Heim, McCloskey learned the art of making bearings, leading to development of his own inventions, with the first *(2,767,034)* filed in 1953 for a self-aligning rod end bearing. The invention related to a four-piece spherical bearing in which the bushings were made from segmented hardened steel and concurrently press fit over a drilled-through ball and into the outer housing. The design allowed for a simple method of assembly while providing a bearing able to withstand greater loadings than one made with soft metal bushings, such as bronze. In 1957 Heim promoted McCloskey to chief engineer responsible for development and manufacture of rod end and spherical bearings.

In 1958, grandson Carl Heim Van Etten *(Figure 14-1)*, the son of Carl C. Van Etten and Heim's daughter, Florence, graduated from Wesleyan College in Middletown, Connecticut, with a bachelor of arts degree in history and subsequently asked for a job at the Heim Company. The family agreed to hire Carl, but set stringent conditions for employment, with the first being that he learn the manufacturing side of the business, including setup and operation of all machines used to fabricate and assemble bearings. Over the next three years, Carl learned to use screw-making, pressing, grinding, honing, thread rolling, lapping, and milling machines. After that, Carl spent time in the tool room acquiring the skills for making tooling used to fabricate and assemble bearings, including cutting tools and metal-shaping dies. After his apprenticeship in manufacturing,

Carl took a position in the only department in which he had not worked; the shipping department, which was responsible for final inspection, greasing and oiling, wrapping, packaging, and shipping to customers. While in that department, Carl also worked with the shop foreman, who was responsible for developing and issuing daily work orders to all departments. One day around 1962, when Carl reported for work, he was informed that the shop foreman had passed away suddenly and that he was being reassigned to that position. The daily duties of foreman were Carl's entry into managing plant operations, which led to his career in manufacturing management.[5]

Deep Groove Ball Bearing

Following the invention of the widely popular rod end bearing in the early 1940s, Heim continued to develop new and improved bearings and the machines and methods to make them. Standard spherical bearings and sweepsticks followed, as well as improved designs for double-row self-aligning ball bearings that were also patented. Later, rolling bearings with shells fabricated from stamped sheet metal were replaced with shells formed with unbroken races, using Heim's expansion and contraction *(swaging)* methods developed in the mid-1950s.

Around this time, Heim conceived of his final antifriction bearing that also became a huge success. Known as the Unibal Ball Bearing *(Figure 14-2)*, it contained a full complement of "unground" balls inserted between races manufactured with "deep grooves" in which the race dimensions are similar to the ball dimensions. Whereas conventional flat or slightly grooved races make point contact on the ball with very high, localized loads, deep groove races contact a much greater portion of the ball surface, thus spreading the loads over a wider area. This design made the bearing capable of sustaining greater loads both radially and axially than conventional designs and offered performance advantages equal to higher-cost bearings using precision-ground balls. Because of its advantages, Heim's deep groove ball bearings could be used in a wide variety of high-load, low-speed applications such as wheels, casters, power tools, conveyors, pulleys, and many other devices. Even though the concept of ball bearings made with grooved races was patented in the early 1900s, Heim's methods and machines for making such bearings were protected in three separate patents. Many companies now offer similar ball bearings made using Heim's fabrication and assembly methods.

With the development of the Unibal Ball Bearing, Charles Heim decided in 1957 to organize a separate company, the Universal Bearing Company, dedicated to its production. Universal Bearing was established not far from the Heim Company in an existing facility used by the former Connecticut Tool & Die Company at 60 Round Hill Road in Fairfield, Connecticut. Charles was named president of Universal Bearing while Lewis retained the title of president of the Heim Company. Demand for this bearing was slow at first but eventually exceeded expectations and became another success story for Lewis Heim.[6]

In 1962, with the deep groove ball bearing selling well, Heim hired Alfreds Rozentals and Don Miller as manufacturing engineers. Both had experience in making ball and roller bearings. Rozentals and Miller worked at a bearing manufacturer in Poughkeepsie, New York, and were hired to design the tooling *(machines, cutting tools, and dies)* nec-

Heim's UNIBAL® ball bearings offer many of the performance advantages of precision ground bearings at unground bearing prices. Heim's unique, full complement, deep groove design makes this bearing an ideal replacement for costly precision ground bearings in high load, low speed applications. Burnished races provide smooth operation. Heim UNIBAL® ball bearings may be used as an economical upgrade from conventional unground ball bearings where increased precision and smoothness are required.

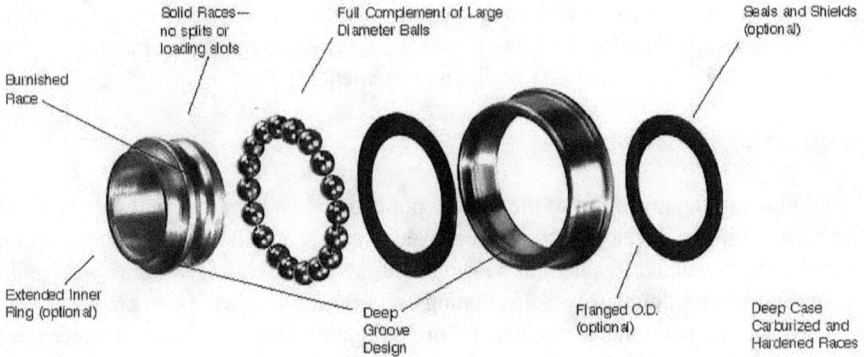

Fig. 14-2: Heim's "Unibal" Deep Groove Ball Bearing from the 2017 Heim Catalog issued by RBC Inc.

essary to make the Unibal Ball Bearing. Rozentals immigrated to the United States from western Poland after World War II and, along with McCloskey, became an inventor of many bearings and bearing-making machines.[7, 8] Miller held several positions and was eventually promoted to vice president and assistant general manager, reporting to Albert McCloskey.[9]

Retirement in Florida

Back in 1933, one year after Lewis returned to Danbury to start the Heim Company, his brother Alfred, along with wife Clara and son Richard, moved into a house on Yale Street in Orlando. After Sylvan Shores was sold off, Alfred also returned to work, and in 1934, bought the Goodwin Laundry Dry Cleaners in Orlando and subsequently changed the name to White Star Laundry. Alfred, as did Lewis, preferred the warmer weather of Florida over the long, cold winters of Connecticut. Whereas Lewis wanted to return to work at his former company in Danbury, Alfred could buy a laundry business locally and avoid the return north. Alfred operated the laundry for nine years and in 1943, at the age of 66, sold the White Star to Acme Laundry and Cleaners and retired.[10]

For many years after moving back to Connecticut in 1932, Lewis and Anna spent some part of the winter at their house in Mount Dora. Heim continued to own the house he built on East Third Avenue but eventually sold it, then rented it back when he visited Florida. A picture of Heim, Anna, son-in-law Carl and granddaughter Elizabeth at the house in 1934 can be seen in Appendix 1 *(A1-2-5)*. After World War II, Lewis and Anna increased their time living in Florida to at least six months, generally leaving Connecticut in October and returning in April. Sometime after Alfred and Clara moved to Orlando, Heim decided to give up renting his old house in Mount Dora and instead rented two rooms at an Orlando hotel. Since the hotel rooms did not have a kitchen, Heim and his wife ate most of their meals at the nearby Morrison's Cafeteria.[11]

Around 1954, Alfred was diagnosed with cancer that was progressing at a slow rate. Soon after, their sister, Minna, relocated from Danbury to Orlando and moved in with Alfred and Clara on Yale Street.[12] Minna was a nurse by profession and moved in to help care for Alfred and ease the caregiving burden on Clara. With Charles running the company, Lewis was free to spend more time in Florida and by November 1960, with the filing of his last two patent applications, had ceased most activities related to new inventions. Then came 1961, which by all accounts had to be the worst year of Lewis Heim's life. First, Heim's sister, Minna, suddenly became ill on April 15 and died two days later. Three months later on July 14, Alfred died from complications of his cancer. Then on August 31, Anna Heim, Lewis's wife of 63 years, died. The deaths devastated Heim, who was very close to his family. Four months later at the end of December, Mr. Heim married a longtime friend, Effie Murray, a minister's daughter and widow who also lived in Orlando.

Frictionless Fishing Reel

One characteristic of inventors is that they never stop inventing. After 1962, even though Heim was not active in his company and was spending most of his time in Florida, enjoying the weather and lawn bowling, he was still engaged in his favorite activity: inventing something new. Heim was an avid saltwater fisherman when he was in Connecticut, which is one of the reasons he bought the beach house in Fairfield. The beach house provided ready access to Long Island Sound and ocean game fish including bluefish, striped bass, and weakfish. Even though he could afford to, he never owned a big boat. Heim preferred smaller boats and owned a 14-foot-long wooden V-hull with overlapping planks, called clinker-built or lapstrake construction, and powered with a 25 horsepower Johnson outboard engine. Heim would take his son-in-law Carl and his grandson Carl out surf fishing using rods fitted with baitcasting reels.

There are many types of fishing reels such as spin cast, bait cast, center spin *(fly reel)*, and others that are used in both freshwater and saltwater fishing. In fishing for blues, stripers, and generally smaller ocean gamefish, many fishermen prefer baitcasting reels with thumb-controlled spools for their ability to cast long distances accurately. A baitcasting reel is a revolving-spool reel fitted with a hand crank. When casting, the spool spins freely to deploy fishing line attached to the bait or lure. Because the spools spin freely, the fisherman must use his thumb to stop the spool once the bait hits the water, otherwise overruns occur, resulting in line tangles known as a bird's nest. Within the reel, the spool is supported on journal or ball bearings and connected to a side-mounted hand-crank that allows the fisherman to quickly reel in the line with minimal turns. When casting and reeling, it is desirable to have the least amount of resistance of the free-spinning spool, with the best reels built with precision-made, low-friction bearings for smooth, quiet, and responsive action.

In a story told by his grandson Carl, Heim conceived of an improved baitcasting reel in which the bearing system was constructed to allow the spool to spin so smoothly and freely that it made the spinning action virtually frictionless. Achieving frictionless spinning would reduce the drag on the spool, thus allowing for longer casts and a more responsive feel when reeling in the fish. However, Heim's idea was met with skepticism by those who knew what he was attempting, with many saying such a reel was not possible. But as always, Heim was undeterred, and with the help of engineers at the company, developed drawings of prototypes that Heim planned to build and test. Heim also wanted to keep his improved reel a secret and asked those involved to destroy the drawings should something happen before he could patent his invention. And that is exactly what occurred. On March 29, 1964, at the age of 89 and at his home on North Shine Avenue in Orlando, Lewis R. Heim died suddenly in his sleep. Soon after, Charles found out about Lewis's instructions and attempted to save the drawings, but the engineers had already destroyed them. Some in the family felt that after seeing the drawings, the frictionless baitcasting reel had the potential to be another great invention, which is why the destruction of the drawings left Charles frustrated and angry.

Sale of the Company

Unlike the Ball and Roller Bearing Company, which had investors when the company was founded, Lewis Heim solely owned the Heim Company. Over the years, he gave a majority of shares to his family, including the grandchildren. In late 1962, for various reasons, the decision was made by Charles to sell the companies. Before the sale, Heim gifted most of his remaining shares to relatives. About the same time, Heim also gifted shares to ten employees, including Albert McCloskey, Alfreds Rozentals, and Don Miller. Even though Rozentals and Miller were relatively new employees, they were considered by Mr. Heim to be key to the continued success of the deep groove ball bearing and he wanted them to participate in its success.[13, 14]

On January 21, 1963, the Heim Company and Universal Bearing Corporation merged to form the Heim Universal Corporation. At the same time, the Draper Corporation of Hopedale, Massachusetts, bought the Heim Universal Corporation in an all-stock transaction. Draper retained the existing management, including Charles Heim as president, Carl C. Van Etten as vice president, and Albert McCloskey as chief engineer.

Draper bought Heim Universal primarily for two reasons. The first was that it wanted to ensure a steady supply of sweepsticks for its power looms.[15, 16] The second was that Draper wanted to diversify away from making textile machines since it was suffering from significant competitive pressure from foreign-made looms.[17, 18] At the time of the purchase, the combined sales of Heim Universal and Draper were in excess of $3,000,000 per year.[19]

In 1964 Draper decided to consolidate the two Heim operations at a single facility and bought land adjacent to the Universal Bearing property at 60 Round Hill Road in Fairfield. In 1965 the plant was expanded to 850,000 square feet and modernized to accommodate the machines and operations of both companies. The original facility on Sanford Street, that was the home of the Heim Company for 30 years, was subsequently closed and the property sold.

On June 30, 1967, Draper was acquired by Rockwell Standard Corporation in an all-stock purchase. After the purchase, Rockwell retained Charles Heim as president and Albert McCloskey as vice president and general manager, but changed the name back to the Heim Company. Three months later on September 22, Rockwell Standard merged with North American Aviation Corporation to form the North American Rockwell Corporation. In the years after 1967, North American Rockwell continued to acquire other companies, including Miehle-Goss-Dexter *(printing presses)* and Collins Radio *(avionics)*. Then in 1973, the company merged with Rockwell Manufacturing to form Rockwell International.

In 1970, Charles Heim retired from the Heim Company, which had $9 million in annual sales and employed 325 people. Albert McCloskey was subsequently promoted to chief executive officer while retaining the titles of vice president and general manager. On March 5, 1976, Charles Heim, who helped his father found, operate, and expand the Heim Company for 37 years, died at his home in Fairfield, Connecticut, at the age of 67.

In January 1978, Rockwell International sold the Heim Company and ten other industrial companies to Incom International Corporation of Pittsburg, Pennsylvania. Incom was organized as a separate company for the purpose of making acquisitions from Rockwell. Other New England companies acquired by the new Incom International included Boston Gear, Acme Chain *(Holyoke, MA)*, and Murray-Carver *(East Bridgewater, MA)*. The combined annual sales of the eleven companies forming Incom were $150 million.[20, 21, 22]

In 1993, Incom sold the Heim Company to RBC Bearings Inc. of Oxford, Connecticut. RBC is an international manufacturer of highly engineered precision bearings and products for a wide range of industrial markets, especially the aerospace and related industries that are highly regulated. RBC also owns Schaublin SA, a Swiss manufacturer of metric rod end and spherical bearings that obtained its original bearing designs through a technology license from the Heim Company.

As of 2017, the Heim Company, now called Heim Bearings by RBC Corporation, continues to produce rod end, spherical, and ball bearings developed by Lewis Heim, with annual sales around $60 million. Many of the methods and machines invented by Heim for the manufacture of bearings also continue to be employed. These include the manufacture of ball and roller bearings with unbroken races using Heim's swaging methods for assembly. In the manufacture of rod end and spherical bearings, Heim's swaging processes for contraction continue to be used, showing their timeless value.

Achievements

Mr. Heim had many successful inventions, starting with hat-making machines, for which the Turner Machine Company bought most of his patents. Heim's invention of the collar roller ironing and forming machine that softened the Gay Nineties' razor-edge stiff collars was sold successfully in the United States, Canada, Europe, and Asia. After exiting that business, Heim's patents for the collar roller and other ironing machines were still valuable when bought by the John E. Fidler Company in 1915. Heim's patent for the invention of the centerless grinder for flat bar stock was purchased by the Cincinnati Milling Machine Company and later sold to the Heald Machine Company, a well-known and successful manufacturer of grinding machines.

Heim's inventions of swaging machines and methods for manufacturing ball and roller bearings with unbroken races were sufficiently revolutionary to capture the attention and purchase of the three patents by the Channing Corporation.

In light of these accomplishments, the two inventions that had the greatest beneficial impact on industry and society are the centerless cylindrical grinder and the spherical rod end bearing known as the "Heim Joint" and "Heim Rod End." The modern rod end bearing and its variants are indispensably used in a wide range of machinery, especially aircraft and other critical applications, and are now widely produced by many bearing manufacturers worldwide.

But the one invention that Heim said was his greatest achievement was the centerless cylindrical grinder that at the time, revolutionized cylindrical grinding with its ability to automatically precision grind many types of cylindrical metal parts in rapid sequence to tolerances of 0.0001 inch *(0.00254 mm)*. It had a tremendous impact on lowering the cost of producing interchangeable parts for a wide range of machines and goods that in turn benefited society by expanding the purchasing power and living standards of consumers. Modern centerless grinders now grind to tolerances better than 0.00002 inch *(0.0005 mm)*, allowing for improved long-term reliability and durability of machines into which the parts are installed.

At the time of its introduction just before 1920, the centerless grinder was disruptive in its impact and quickly copied and manufactured illegally by many competing machine tool manufacturers. The grinder was recognized by the captains of the automobile industry, including Henry Ford and Alfred Sloane, as a must-have machine tool that led to the purchase of Heim's patents by Frederick Geier of the Cincinnati Milling Machine Company. Geier formed the Cincinnati Grinder Company to manufacture and sell the centerless and other grinders and continues in business today as Cincinnati Machines. Centerless grinders are used throughout the world and are considered to be an essential machine tool for the mass production of precision-sized cylindrical metal parts, and the bearings and machine tool industries will always remember the name of Lewis Heim for his achievements and contributions to the advancement of grinding technology. One assurance of that recognition is the historical retention of one of the original Heim centerless grinders by the Smithsonian Institute in Washington, DC, an impressive honor for an inventor.

Fig. 14-3: Lewis R. Heim: September 19, 1874–March 29, 1964.

Appendix 1

Family Information, Photographs, and Patents List

Contents

Middle Row:
Anna Elizabeth Roth Heim *(71)*
Lewis R. Heim *(74)*
Front Row:
Grandson: Carl Heim Van Etten *(12)*
Grandson: John Gary Heim *(9)*
Grandson: Charles Wilkinson Heim *(11)*

A1-3: List of Patents Granted to Lewis R. Heim

A1-1
Family Information

1. Lewis Rasmus Heim
b. September 19, 1874, East Fishkill, New York
d. March 29, 1964 *(Age: 89)*, Orlando, Florida
Height: 5'-11"
Education: Finished Eighth Grade

2. Parents of Lewis Heim
Father: Charles George Heim
b. October 31, 1840, Hesse, Germany: Immigrated to the U.S. in 1870
d. February 8, 1908 *(Age: 67)*, Danbury, Connecticut

Mother: Frederika *(Frances)* Kleifoth Heim
b. January 30, 1844, Hamburg, Germany
d. January 8, 1929 *(Age: 84)*

3. Wilhelmina *(Minna)* Heim *(Sister)*
Occupation: Nurse
b. December 21, 1873, East Fishkill, New York
d. April 17, 1961 *(Age: 87)*, Orlando, Florida

m. June 6, 1914: John F. Heinrich *(1863–1924)*

4. Alfred Henry Heim *(Brother)*
Occupation: Machinist, Businessman
b. May 6, 1877, East Fishkill, New York
d. July 14, 1961 *(Age: 84)*, Orlando, Florida

m. June 12, 1907 to Clara D. Barnum
Alfred *(Age: 30)*, Machinist, from Fishkill, New York
Clara *(Age: 25)*, Stenographer, from Danbury, Connecticut
Spouse to Alfred: Clara D. Barnum:
b. April 26, 1882, Danbury, Connecticut

d. February 21, 1955 *(Age: 72),* Orlando, Florida

Alfred Served in the Spanish-American War: Corporal, First Regiment
Date Enlisted in Connecticut National Guard: March 23, 1897
Date Mustered into Service: July 14, 1898

Children of Alfred and Clara Heim:
Richard Barnum Heim
Occupation: Businessman: Owned Dick Heim's Camera Shop, 1825 Edgewater Drive, Orlando, Florida
b. June 23, 1910, Danbury, Connecticut
d. December 9, 1985 *(Age: 75)*, Pensacola, Florida

Children of Richard B. Heim
Daughter: Peggy Lou Heim *(m. Lowe)*: b. January 1944
Daughter: Bonnie Jean Heim *(m. Perkins)*: b. June 1946

5. William Charles Heim *(Brother)*
Occupation: Machinist, Music Shop Proprietor
b. December 3, 1878, East Fishkill, New York
d. January 4, 1919 *(Age: 40)*, Danbury, Connecticut

m. September 18, 1912 to Ethel N. Sperry
William *(Age: 33)* of Shenandoah, New York
Ethel *(Age: 27)* of Lawyersville, New York

Spouse: Ethel N. Sperry
b. June 30, 1885, West Richmondville, New York
d. December 16, 1967 *(Age: 82)*, Danbury, Connecticut

Children of William and Ethel Heim:
Jeanette Adelaide Heim *(m: Joyce)*
b. August 13, 1914, Danbury, Connecticut

Willa Sperry Heim *(m: Baggett)*
b. June 21, 1919, Danbury, Connecticut

6. Laura Louise Heim *(Sister)*
Occupation: Music Teacher
b. March 28, 1885, East Fishkill, New York
d. November 16, 1987 *(Age: 102)*, Worcester, Massachusetts

m. ca. 1917 to Michael R. Goesch
Occupation: Hatter in Danbury
Children: William Goesch

7. Charles Waters *(Cousin)*
Adopted by Frederika Heim
Occupation: Various Positions *(Secretary, Treasurer)* at the Ball and Roller Bearing Company.

8. Anna Elizabeth Roth *(Wife of Lewis Heim)*
b. May 31, 1877, Brooklyn, New York
d. August 31, 1961 *(Age: 84)*, Fairfield, Connecticut

m. September 21, 1898, Danbury, Connecticut

Father: John Simon Roth: b. 1855 *(Germany)*, d. 1937
Occupation: Hatter
Mother: Anna Elizabeth *(Wicker)* Roth: b. 1856 *(Germany)*, d. 1942
Brother: John Henry Roth: b. 1882, d. 1954
Occupation: Hatter, Treasurer of the Ball and Roller Bearing Company
Brother: Adam A. Roth

9. Effie Murray: Second Wife of Lewis Heim
Married December 29, 1961, in Florida

10. Florence Elizabeth Heim *(Daughter of Lewis Heim)*
Married name: Florence Heim Van Etten
b. October 13, 1901, Danbury, Connecticut
d. December 24, 2003 *(Age: 102)*, Bridgeport, Connecticut
Education: Bennett College, Millbrook, New York
Graduated: 1922, Liberal and Applied Arts

m. September 8, 1928, in Fairfield, Connecticut to Carl Conklin Van Etten of Montclair, New Jersey

11. Carl Conklin Van Etten: Husband to Florence E. Heim
b. September18, 1901, Chester, New York
d. August 23, 1965 *(Age: 63)*, Fairfield, Connecticut

Education:
Lafayette College, Easton, PA, Graduated: May 1923: Bachelor of Arts in History-
Columbia University, New York City: Graduated December 16, 1931: Master of Arts

12. Children of Florence E. H. Van Etten
Elizabeth Ann Van Etten
b. May 13, 1930, Fairfield, Connecticut

d. September 20, 2007 *(Age: 77)*, Davidsonville, Maryland
Education: University of Connecticut, Storrs, Connecticut
Graduated 1952: Bachelor of Arts in Math

Carl Heim Van Etten
b. September 8, 1936, Fairfield, Connecticut
Education: Wesleyan College, Middletown, Connecticut
Graduated 1958: Bachelor of Arts in History

13. Charles Roth Heim *(Son of Lewis Heim)*
Occupation: Machinist, Businessman
b. March 12, 1908, Danbury, Connecticut
d. March 5, 1976 *(Age: 67)*, Fairfield, Connecticut

m. March 20, 1936 to Muriel A. Wilkinson of Carmel, New York
Education: Yale: Sheffield Scientific School
Entered 1929. Graduated 1933, Bachelor of Science in Engineering

14. Muriel Alice Wilkinson *(Wife of Charles R. Heim)*
b. May 4, 1914, Lynn, Massachusetts
d. August 11, 2004 *(Age: 90)*, Maplewood, New Jersey

15. Children of Charles Roth Heim
Charles Wilkinson Heim
b. November 25, 1936

John Gary Heim
b. July 7, 1939

Roth Family Portrait. Danbury, Ct. ca. 1895
Front Row: Parents John Simon Roth and Anna Elizabeth Wicker Roth
Back Row: John Henry, Anna Elizabeth and Adam A. Roth

A1-2-2

Heim Family Portrait. Danbury, Ct. ca. 1909
Left to Right: Florence E., Lewis R. (standing), Anna R. and Charles R.

Heim Family Reunion. Danbury, Connecticut. ca. November, 1916
L to R: Lewis, Laura, William, Wilhemina, Charles Waters (cousin), Frederika and Alfred

A1-2-4

Heim Family Reunion. Danbury, Connecticut. ca. November, 1916
L to R: Lewis, Anna, Charles, Florence, Ethel, William, Jeannette, Minna, John Heinrich and his two children, Francis (Frederika), Laura, Charles Waters' Wife and Children and Charles Waters, Alfred, Richard, Clara.

a. Heim House at 347 East Third Avenue in Mount Dora, Florida. ca. 1928

b. Carl Van Etten, Lewis Heim, Elizabeth Van Etten and Anna Heim in Mount Dora, Florida. ca. 1934

A1-2-6

Heim 50th Wedding Anniversary Family Portrait. September, 1948
Front Row: Carl H. Van Etten, John G. Heim, Charles W. Heim
Middle Row: Anna and Lewis Heim
Back Row: Charles R. Heim, Muriel W. Heim, Florence H. Van Etten,
Elizabeth A. Van Etten and Carl C. Van Etten

A1-3: The Patents of Lewis R. Heim

No.	Patent Title	Patent No.	File Date	Grant Date	Invention Type
1	Hat Stiffening Machine	573,876	1/8/1896	12/29/1896	Hatting Machine
2	Machine For Stiffening Brims Of Hats	580,396	8/14/1896	4/13/1897	Hatting Machine
3	Machine For Stiffening Crowns Of Hat Bodies	581,554	8/18/1896	4/27/1897	Hatting Machine
4	Machine For Stiffening Brims Of Hats	590,370	1/15/1897	9/21/1897	Hatting Machine
5	Hat-Clearing Machine	645,592	6/29/1899	3/20/1900	Hatting Machine
6	Hat Ironing Machine	645,593	9/16/1899	3/20/1900	Hatting Machine
7	Hat Curling Machine	695,393	8/26/1901	3/11/1902	Hatting Machine
8	Picker Roll For Hatting And Fur Refining	761,170	10/7/1903	30-May-1900	Hatting Machine
9	Machine For Shaping Fold-Collars	780,750	17-Jan-1903	24-Jan-1905	Ironing Machine
10	Machine For Ironing Fold-Collars	783,433	7-Apr-1904	28-Feb-1905	Ironing Machine
11	Machine For Ironing Fold-Collars	818,247	20-Mar-1905	17-Apr-1906	Ironing Machine
12	Machine For Ironing Fold-Collars	865,977	8-Jun-1906	10-Sep-1907	Ironing Machine
13	Machine For Dampening The Seams Of Fold Collars	983,051	1-Dec-1905	31-Jan-1911	Ironing Machine
14	Ironing Machine	1,036,111	21-Oct-1908	20-Aug-1912	Ironing Machine
15	Stapling Machine	1,036,841	14-Jan-1910	27-Aug-1912	Wire Bending Machine
16	Automatic Edge Ironer For Collars And Cuffs	1,064,644	28-Mar-1907	10-Jun-1913	Ironing Machine
17	Combined Moistener, Ironer, And Shaper	1,069,548	21-Nov-1911	5-Aug-1913	Ironing Machine
18	Wire Feeding Mechanism	1,119,510	3-Feb-1910	1-Dec-1914	Wire Bending Machine
19	Roller Bearing	1,169,150	6-Mar-1915	25-Jan-1916	Roller Bearing
20	Feeding Device For Grinding Machines	1,210,936	6-Mar-1915	2-Jan-1917	Cenertless Grinding Machine
21	Roll Grinding Machine (Reissue: 15,035)	1,210,937	6-Mar-1915	2-Jan-1917	Cenertless Grinding Machine
22	Ball Cage For Bearings	1,222,760	3-Jul-1916	17-Apr-1917	Ball Bearing
23	Device For Feeding Rolls For End Grinding	1,264,928	13-Jul-1917	7-May-1918	Machine
24	Grinding Machine	1,264,929	13-Jul-1917	7-May-1918	Round Plate Grinding Machine
25	Roll Grinding Machine	1,264,930	13-Jul-1917	7-May-1918	Cenertless Grinding Machine
26	Feeding Mechanism	1,278,463	18-Oct-1917	10-Sep-1918	Cenertless Grinding Machine

A1-3: The Patents of Lewis R. Heim

No.	Patent Title	Patent No.	File Date	Grant Date	Invention Type
27	Method Of Grinding Hardened Rolls	1,281,366	3-Jul-1916	15-Oct-1918	Cenertless Grinding Machine
28	Roll Grinding Apparatus	1,579,932	20-Dec-1920	6-Apr-1926	Cenertless Grinding Machine
29	Apparatus For Grinding Rolls And The Like	1,579,933	18-Feb-1921	6-Apr-1926	Cenertless Grinding Machine
30	Roll Grinding Apparatus	1,579,934	11-Jun-1921	6-Apr-1926	Cenertless Grinding Machine
31	Grinding Apparatus	1,585,982	8-Jan-1923	25-May-1926	Cenertless Grinding Machine
32	Art And Apparatus For Grinding	1,585,983	4-Aug-1922	25-May-1926	Cenertless Grinding Machine
33	Grinding	1,585,984	29-Aug-1922	25-May-1926	Cenertless Grinding Machine
34	Apparatus For Producing Ground Articles	1,586,523	3-Feb-1921	1-Jun-1926	Cenertless Grinding Machine
35	Grinding	1,590,190	23-Oct-1923	29-Jun-1926	Cenertless Grinding Machine
36	Grinding Machine	1,644,057	8-Oct-1921	4-Oct-1927	Cenertless Grinding Machine
37	Grinding	1,647,129	17-Oct-1921	1-Nov-1927	Cenertless Grinding Machine
38	Art Of Grinding	1,647,130	26-May-1922	1-Nov-1927	Cenertless Grinding Machine
39	Grinding	1,647,131	29-Jan-1925	1-Nov-1927	Cenertless Grinding Machine
40	Grinding Apparatus	1,683,974	8-Feb-1921	11-Sep-1928	Cenertless Grinding Machine
41	Grinding Device	1,691,061	17-Oct-1921	13-Nov-1928	Cenertless Grinding Machine
42	Roll Grinding Apparatus	1,692,833	8-Feb-1921	27-Nov-1928	Cenertless Grinding Machine
43	Apparatus For Producing Ground Articles	1,692,834	17-Feb-1923	27-Nov-1928	Cenertless Grinding Machine
44	Apparatus For Producing Ground Articles	1,709,348	23-Feb-1923	16-Apr-1929	Cenertless Grinding Machine
45	Grinding Apparatus	1,733,087	14-Apr-1922	22-Oct-1929	Cenertless Grinding Machine
46	Grinding	1,733,088	20-Oct-1923	22-Oct-1929	Cenertless Grinding Machine
47	Grinding	1,733,089	22-Oct-1923	22-Oct-1929	Cenertless Grinding Machine
48	Grinding	1,741,236	26-Apr-1922	31-Dec-1929	Cenertless Grinding Machine
49	Apparatus For Producing Ground Articles	1,772,042	30-Mar-1922	5-Aug-1930	Cenertless Grinding Machine
50	Truing Device For Abrasive Wheels	1,772,544	16-Feb-1923	12-Aug-1930	Cenertless Grinding Machine
51	Bearing Construction	1,885,914	2-Nov-1928	1-Nov-1932	Roller Bearing & Method
52	Bearing Construction	1,922,805	4-Oct-1930	15-Aug-1933	Roller Bearing

A1-3: The Patents of Lewis R. Heim

No.	Patent Title	Patent No.	File Date	Grant Date	Invention Type
53	Bearing Construction	1,943,864	11-Oct-1928	16-Jan-1934	Roller Bearing & Method
54	Grinding Machinery	1,958,001	9-Aug-1927	8-May-1934	Grinding Machine For Flat Stock
55	Art Of Making Roller Bearings	1,976,019	28-Dec-1928	9-Oct-1934	Roller Bearing
56	Bearing Construction	1,984,213	2-Aug-1930	11-Dec-1934	Roller Bearing
57	Roller Bearing Construction	2,044,168	24-Jan-1931	16-Jun-1936	Roller Bearing
58	Bearing Construction	2,074,182	3-Dec-1930	16-Mar-1937	Ball & Roller Bearings
59	Bearing Construction	2,080,609	2-Jan-1931	18-May-1937	Roller Bearing
60	Method Of Making Roller Bearings	2,102,460	18-Jul-1934	14-Dec-1937	Method & Machine For Making Roller Bearing
61	Method Of Making Roller Bearings	2,160,382	1-Aug-1936	30-May-1939	Roller Bearing and Method
62	Roller Bearing	2,301,399	18-Jul-1934	10-Nov-1942	Roller Bearing Construction
63	Method of Making Bearings	2,366,668	29-Sep-1942	2-Jan-1945	Method and Machine For Making Rod End Bearing
64	Appartus For Polishing	2,372,722	28-Jul-1942	3-Apr-1945	Grinding Machine
65	Rod End Bearing And Method Of Making Same	2,400,506	5-Nov-1942	21-May-1946	Rod End Bearing
66	Internal Grinding Machine	2,430,423	8-Jan-1945	4-Nov-1947	Machine And Method For Internal Grinding
67	Lubricator Head Construction	2,454,252	24-Feb-1945	16-Nov-1948	Rod End Bearings
68	Method For Freeing Mechanical Joints	2,476,728	14-Dec-1942	19-Jul-1949	Machine For Making Spherical Bearings
69	Bearing	2,488,775	24-Jul-1943	22-Nov-1949	Rod End Bearings
70	Method Of Making And Assembling Bearings And The Like	2,541,160	30-Dec-1944	13-Feb-1951	Method For Making Rod End Bearings
71	Sweepstick	2,592,566	5-Aug-1949	15-Apr-1952	Spherical Bearings
72	Ball Bearing	2,653,064	28-Dec-1951	22-Sep-1953	Ball Bearing
73	Sweepstick	2,662,557	15-Jun-1949	15-Dec-1953	Spherical Bearings
74	Bearing	2,665,956	7-Sep-1949	12-Jan-1954	Spherical Bearings
75	Bearing	2,675,279	30-Dec-1944	13-Apr-1954	Method For Making Rod End Bearings
76	Self-Aligning Bearing	2,675,281	26-Aug-1948	13-Apr-1954	Ball Bearing and Method
77	Sweepstick With Resilient Bearing and Means To Prevent Spliting Of Outer Member	2,696,841	11-Sep-1950	14-Dec-1954	Cylindrical Plain Bearing
78	Method Of Making Sweepstick	2,701,409	5-Aug-1949	8-Feb-1955	Method For Making Spherical Bearings

A1-3: The Patents of Lewis R. Heim

No.	Patent Title	Patent No.	File Date	Grant Date	Invention Type
79	Method Of Making Sweepstick	2,701,907	15-Jun-1949	15-Feb-1955	Method For Making Spherical Bearings
80	Method Of Making Bearing	2,759,244	7-Sep-1949	21-Aug-1956	Machine And Method For Making Spherical Bearings
81	Method Of Making Self-Aligning Roller Bearings	2,787,048	20-Aug-1953	2-Apr-1957	Spherical Bearing, Machine And Method Making Bearing
82	Method Of Mounting A Self-Aligned Bearing In A Lever Or Other Machine Element	2,898,671	16-Feb-1955	11-Aug-1959	Machine And Method For Making Spherical Bearings
83	Method Of Making Antifriction Bearings	2,910,765	19-Aug-1958	3-Nov-1959	Machine And Method For Making Ball And Roller Bearings
84	Method Of Sizing And Burnishing Races In A Ball Or Roller Bearing	2,913,810	16-Nov-1955	24-Nov-1959	Method For Making Ball And Roller Bearings
85	Method Of Producing A Sphere With A Hole Therethrough	2,941,290	11-Jan-1955	21-Jun-1960	Method For Making Spherical Bearings
86	Bearing Air-Cleaning Apparatus	3,027,590	6-Oct-1959	3-Apr-1962	Machine For Making Bearings
87	Method Of Making Bearings	3,049,789	6-Oct-1959	21-Aug-1962	Machine And Method For Making Ball And Roller Bearings
88	Method For Making Antifriction Bearings	3,088,190	3-Jul-1959	7-May-1963	Method For Making Ball And Roller Bearings
89	Apparatus For Burnishing Telescoped Bearings	3,093,884	16-Jun-1959	18-Jun-1963	Machine And Method For Making Ball And Roller Bearings
90	Antifriction Bearing	3,123,413	28-Sep-1959	3-Mar-1964	Ball & Roller Bearings
91	Method For Manufacturing Spherical Bearings	3,221,391	18-Nov-1960	7-Dec-1965	Method For Making Spherical Bearings
92	Apparatus For Manufacturing Bearings	3,369,285	18-Nov-1960	20-Feb-1968	Machine For Making Spherical Bearings

Appendix 2

The Heim Machine Company, the Ball and Roller Bearing Company, and the Centerless Grinder

Contents

A2-1: Heim Machine Company Business Envelopes
Correspondence Mailers with Image of Fold Collar Roller
Top: Louis Ritz & Company: Hamburg, Germany: 1909
Middle: Excelsior Carbon Mfg. Co., Lynn, Massachusetts: 1912
Bottom: Norton Company, Worcester, Massachusetts: 1915

A2-2: Advertisements for the Ball and Roller Bearing Company:
This section contains typical advertisements placed by the Ball and Roller Bearing Company in trade magazines between 1914 and 1922. Most advertisements stress the advantages of rolling *(antifriction)* bearings, including reduced friction, lower power consumption, lower operating costs, and longer machine life. The advertisement placed in 1918 mentions that the Ball and Roller Bearing Company has "unusual facilities for the manufacture of cylindrical rollers." This statement is one of the first indications Heim was using his twowheel centerless grinders to make precision-ground cylindrical parts.

A2-2-1: 1914 and 1917
    a. "Friction Is the Enemy," *American Machinist*, July 2, 1914
    b. "Is Grindipuss Eating Away Your Profits in Friction?" *American Machinist,* July 1917

A2-2-2: 1917 and 1918
    a. "Keeping Machinery Running Smoothly," *Machinery*, August 1917
    b. "Quality Bearing for Every Bearing Need," Machinery, August 1918

A2-2-3: 1919 and 1920
    a. "It's a Foregone Conclusion," *Machinery*, August 1919
    b. "Accuracy," Machinery, August 1920

A2-2-4: 1922
    a. "Plug the Leaks," *Machinery*, August 1922

A2-3: Ball and Roller Bearing Company Catalog No. 11, ca. 1920
This catalog contains technical information for ordering ball thrust, roller thrust, and journal roller bearings, as well as steel balls and locating washers. Selected pages from this catalog are provided to illustrate the types of bearings manufactured, grinding accuracies, guarantees, and other information required for ordering.

Pages include pages 1–6, 10, 14, 17, 33, 49 – 52, 54 – 56, 59 – 60, and 63 plus the company logo.

Company Logo: The company logo placed inside the catalog cover illustrates the buildings comprising the Ball and Roller Bearing Company. The original building, located on Maple Street, is shown on the right, behind the railroad tracks. This is the same building in which Heim founded the Heim Machine Company in 1904. In 1910, Heim founded the Ball and Roller Building, which shared the same facilities with the Heim Machine Company. The office is shown on the corner of Maple and Crosby Streets. The building behind the office on Crosby Street, indicated by the smokestack, houses the furnaces and other equipment for hardening and heat-treating bearing components. The building on the left was built by Heim around 1919 to increase manufacturing capacity.

Radial Roller Bearing: On pages 54 and 55, the catalog describes a new bearing designed by Heim for use on heavy motor vehicles, heavy machinery, and journals subject to heavy loads. The bearing is designed with thin races for installation where space is so limited that there is no room for a radial ball bearing strong enough for the load. Heim published an article on this bearing in the September 1916 issue of *Machinery* magazine.

Unusual Facilities: On page 59, the catalog makes a statement that the B&RB have unusual facilities for the production of small hardened and ground cylindrical rolls, piston pins, valve lifters, and other cylindrical metal parts. The use of "unusual facilities" in this catalog and in bearing advertisements indicated that Heim was using his two-wheel centerless grinder to mass-produce precisionground cylindrical parts for the automotive and roller bearings industries.

A2-4: Court Opinion: Ball and Roller Bearing Company v. Commissioner of Internal Revenue, 1929. This opinion summarizes the appeal made by Heim against the Internal Revenue Service related to the transfer of Heim's original patents for the centerless grinder to the Ball and Roller Bearing Company. This document contains information related to the history of the Ball and Roller Bearing Company, its partners, and the invention of the centerless grinder.

A2-5: Articles
Contains six articles related to centerless grinding. Three articles were published by Heim to promote the capabilities of centerless grinding and his machines.

One article, was published by an editor of *American Machinist* magazine that reports the use of centerless grinders sold by various manufacturers to the automotive industry in the early 1920s. The last article documents the sale of Heim Grinder to the Cincinnati Grinder Company owned by Frederick Geier.

1. Heim Centerless Roll Grinder, *Machinery,* Feb. 1921
2. "Production of Small Parts by Centerless Grinding," *American Machinist,* March 8, 1923
3. "Heim Centerless Grinding Methods," *Machinery*, May 1923
4. "Heim Improved Centerless Grinding Machine," *American Machinist*, Dec. 11, 1924
5. "Heim Centerless Grinding Machine," *Machinery,* Sept. 1926
6. "Cincinnati and Heim in Consolidation," *American Machinist*, Sept. 16, 1926

A2-6: Centerless Grinder Product Information
Contains bulletins, specifications, and other product information for Heim centerless grinders.

1. Ball and Roller Bearing Co. Bulletin 110 for Heim CG: Dec. 1920
2. Ball and Roller Bearing Co. Bulletin 140 for Heim CG: ca. 1922
3. Heim Grinder Price List: May 1924
4. Heim Grinder Product Sheet: ca. June 1924
5. Heim Grinder Circular HG1: Sept. 1924
6. Sales Letter by L. G. Henes: May 22, 1926
   The specification referred to by Henes is Heim Grinder circular HG1 *(Item #5)*

A2-7: Advertisements
Contains advertisements for centerless grinders placed by Heim and competitors in *American Machinist* and *Machinery* between 1920 and 1927. The advertisements illustrate the capabilities of centerless grinders in the 1920s, including production rates, grinding tolerances, surface quality, and productivity gains.

1. Sanford: "Precision Centerless Grinders" by Russell Holbrook & Henderson, *Machinery*, 1920
2. Detroit Single-Wheel: "Tireless Workers," *American Machinist*, 1921
3. Sanford: "Grinding File Blanks," *American Machinist*, May 1921 and "Famous 3-Point Contact," *American Machinist*, June 9, 1921
4. Sanford & Detroit Machine Tool: *American Machinist*, June 17, 1921
5. Heim: "A Demonstrative Test," *Machinery,* Aug. 1921
6. Heim: "Don't Miss the Opportunity," *Machinery,* Oct. 1921
7. Sanford Model B: "An Unusual Operation," *Machinery*, Aug. 1922
8. Heim: "Enormous Quantities of Piston Pins," *Machinery*, Sept. 1922
9. Reeves: "Grinding True Tapers without Centering," *American Machinist*, Oct. 1922
10. Heim: "Finishing Difficult Work at Low Cost," *Machinery*, March 1923

11. Heim: "Why Did They Keep It?" *Machinery,* April 1923
12. Heim: "Greater Profits Lower Costs," *Machinery,* Aug. 1923
13. Reeves Grinder and Variable Speed Transmission, *Machinery,* Aug. 1923
14. Detroit 4C: "High Production Machinery," *American Machinist,* April 2, 1925, p. 1.
15. Detroit 4C: "High Production Machinery," *American Machinist,* April 2, 1925, p.2.
16. Heim: "Shoulder Work," *American Machinist,* May 7, 1925
17. Heim: "Twist Drills," *American Machinist,* June 4, 1925
18. Cincinnati: "Can You Beat This?" *Machinery,* Aug. 1925
19. Heim: "The Automobile Designer," *Machinery,* Oct. 1925
20. Cincinnati: "The Best Piston," *Machinery,* Aug. 1927

A2-8: Photographs of Original Heim Centerless Grinders
Photographs 1 and 2 as well as Figure 6-3 in Chapter 6, were obtained from court records of the Ball and Roller Bearing Company v. F. C. Sanford Manufacturing Company. They show the original Heim experimental two-wheel peripheral centerless grinder built by Heim in 1915. The machine shown in the pictures was reconstructed in May 1921 at the Ball and Roller Bearing Company for the trial as Plaintiff Exhibit B-1.

Photographs 3, 4, and 5 were obtained from court records of the Ball and Roller Bearing Company v. F. C. Sanford Manufacturing Company. The pictures show the Heim double-ring wheel centerless grinder installed at the Ball and Roller Bearing Company in 1921. These machines *(Plaintiff Exhibit D-1)* were used to grind rolls used in roller bearings sold by Heim.

Photographs 6 through 11 were professionally made around 1922 for Heim's advertising media and were given to the author by the Ball and Roller Bearing Company of New Milford, Connecticut, in March 2013.

Photographs 12 and 13 were taken in March 2013 at a storage facility of the American Precision Museum in Windsor, Vermont.

Photographs 14 and 15 were taken in June 2016 at the exhibits storage facility of the Smithsonian Institute in Hyattsville, Maryland.

1. Reconstruction of the Original *(Experimental)* Heim Peripheral Wheel Centerless Grinder: Two pictures shown at various angles.
2. Reconstruction of the Original *(Experimental)* Heim Peripheral Wheel Centerless Grinder: Top view showing path of rolls and date of reconstruction.
3. Heim Double-Ring Wheel Centerless Grinder: Front view and side view.
4. Batteries of Heim Double-Ring Wheel Centerless Grinders: Two views.
5. Heim Centerless Grinder: Front-side view.
6. Heim Centerless Grinder: Rear-side view showing tilted regulating wheel

shaft.

7. Heim Centerless Grinder: Upper: Rear view showing helical gear system for automated loading and unloading of work. Lower: Carrier for spot grinding of headed work.

8. Heim Centerless Grinder Carrier #3 Regular and Special and Carrier #8.

9. Heim Centerless Grinder #158 and Nameplate at the American Precision Museum in Windsor, Vermont, 2014: Shown with Heim's great-grandson Robert Jacobs *(author)*.

10. Restored Heim Centerless Grinder #209 and Nameplate at the Smithsonian Institute in Maryland, 2017.

A2-9: Ekholm Letter: May 11, 1966

Letter written in 1966 by Carl Gustav Ekholm, who was the individual primarily responsible for bringing centerless grinding to Europe in the mid-1920s. Ekholm, who is from Sweden and worked for machine toolmaker Lidkoping Mekaniska Verkstad *(LMV)*, tells the story of how he learned of the design for building a centerless grinder and how the machine was introduced to Europe.

A2-1

Heim Machine Business Envelopes Showing the Collar Roller

a. American Machinist: 2 July 1914

b. American Machinist: July 1917

A2-2-2

a. Machinery: August 1917

b. Machinery: August 1918

a. Machinery: August 1919

b. Machinery: August 1920

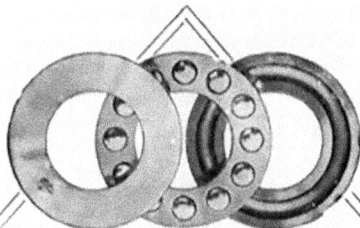

Machinery: August 1922

291

CATALOGUE No. 11

Superior
Ball and Roller Bearings

The Ball and Roller Bearing Co.
Danbury, Conn., U. S. A.

Manufacturers of

BALL THRUST BEARINGS
With Flat Ball Race

BALL THRUST BEARINGS
With Grooved Ball Race

BALL THRUST BEARINGS
With Grooved Ball Race and Spherical Seat

ROLLER THRUST BEARINGS   **JOURNAL ROLLER BEARINGS**

ANNULAR ROLLER BEARINGS

ANTI-FRICTION BEARINGS TO SPECIFICATIONS

CYLINDRICAL ROLLERS TO SPECIFICATIONS

CARBONIZING, HEAT TREATING AND GRINDING
OF STEEL PARTS

A2-3-2

The Ball and Roller Bearing Company's Plant
Danbury, Connecticut
Established 1904

THE BALL AND ROLLER BEARING CO.

GUARANTEE

We guarantee our bearings against imperfection in material and workmanship, and undertake to replace any that may show defects within a reasonable length of time from date of supplying them, provided they are returned to us carriage prepaid for our inspection and decision. This guarantee covers bearings only which are used under loads, speeds, and in mountings sanctioned by us, but does not cover bearings which have failed by reason of misapplication, misuse or neglect.

SHIPPING DIRECTIONS

Unless shipping instructions are given we will use our best judgment as to route and conveyance.

SAMPLES

We will be pleased to furnish samples of any standard size bearing listed in this catalogue.

We will also be pleased to design and furnish blue print of any special size or design, providing quantity required is reasonably large, which must be specified.

QUOTATIONS

All prices quoted are F. O. B. Danbury, Conn.

The prices on thrust bearings and certain sizes of journal roller bearings listed in this catalogue are subject to a discount; which may be changed without notice. Send for latest discount sheet.

All agreements are contingent upon strikes, accidents or other reasonable delays beyond our control.

TERMS

Terms are 1% 10 days, net 30 days from date of invoice, unless otherwise specified.

Orders for bearings from customers not having a satisfactory commercial rating must be accompanied by cash or money order.

THE BALL AND ROLLER BEARING CO.

A2-3-4

INTRODUCTION

We believe in the merits of our line of manufacture, and for that reason can approach the public with a clear conscience; and, without reserve, recommend our product. Not only are we sure that Ball Bearings and Roller Bearings possess merit and for that reason should receive consideration by manufacturers, but the severe tests to which Ball and Roller Bearings have been subjected have proven beyond a doubt their practicability and durability over the Plain Bearings, not only because of their Anti-Friction quality and great saving of power, but also because of the elimination of wear which is present in the Plain Bearings, as well as overcoming the troublesome feature of over-heating of high speed journals.

There is much that can be said in favor of Anti-Friction Bearings without exaggeration, and it only remains for the purchaser to see that he gets the best that the market affords. We manufacture a superior line of Ball and Roller Bearings and use high grade steels best adapted for the different types, and our supreme efforts during the years which we have been manufacturing them has been to the end of furnishing bearings of superior quality. Not only do we use high grade steels, but we have provided modern facilities for heat treating of the steels entering our line of Ball and Roller Bearings; this being of vital importance, as the life of the bearing is entirely dependent upon its being properly heat treated.

Precision in Ball and Roller Bearings is also of primary importance. If the raceways and surfaces are not properly ground you cannot expect the maximum amount of work out of the bearings. This we have also given very careful attention and have designed special machinery for machining and grinding them which enables us to attain a very fine degree of accuracy and superior finish.

The balls and rollers used in our bearings are made from a special alloy steel. They are strictly high grade and guaranteed to be extremely accurate to size.

THE BALL AND ROLLER BEARING CO.

Index
BALL AND ROLLER BEARINGS

FLAT SEAT
Ball Thrust Bearings, Flat Race, English Dimensions

FLAT SEAT
Ball Thrust Bearings, Grooved Race, English Dimensions

FLAT SEAT
Ball Thrust Bearings, Grooved Race, Metric Dimensions

SPHERICAL SEAT
Ball Thrust Bearings, Grooved Race, Metric Dimensions

DOUBLE DIRECTION, FLAT SEAT
Ball Thrust Bearings, Grooved Race, Metric Dimensions
Middle Washer Locked to Shaft

DOUBLE DIRECTION, SPHERICAL SEAT
Ball Thrust Bearings, Grooved Race, Metric Dimensions
Middle Washer Locked to Shaft

DOUBLE DIRECTION, FLAT SEAT
Ball Thrust Bearings, Grooved Race, Metric Dimensions
Middle Washer Locked in Housing

A2-3-6

Index—Continued

BALL AND ROLLER BEARINGS

DOUBLE DIRECTION, SPHERICAL SEAT
Ball Thrust Bearings, Grooved Race, Metric Dimensions
Middle Washer Locked in Housing

THE BALL AND ROLLER BEARING COMPANY

SUPERIOR
BALL THRUST BEARINGS
With Bronze Ball Cage

Light Series

FLAT SEAT   FLAT BALL RACE

Number of Bearing	Inside Diameter	Outside Diameter	Width	Number of Balls	Ball Diameter	Price
A 01	3/16"	7/16"	3/16"	9	1/16"	$.75
A 02	3/16"	17/32"	3/16"	9	1/16"	.75
A 03	3/16"	5/8"	1/4"	9	3/32"	.75
A 04	3/16"	15/16"	3/8"	12	1/8"	.75
A 05	1/4"	9/16"	7/32"	8	3/32"	.75
A 06	1/4"	13/16"	3/8"	11	1/8"	.75
A 07	1/4"	27/32"	3/8"	9	1/8"	.75
A 08	5/16"	5/8"	1/4"	9	3/32"	.75
A 09	5/16"	27/32"	3/8"	9	1/8"	.75
A 010	3/8"	11/16"	9/32"	11	3/32"	.75
A 011	3/8"	1 1/32"	1/2"	10	3/16"	.75
A 012	5/8"	1 3/32"	1/2"	9	3/16"	.75
A 013	7/16"	13/16"	5/16"	9	1/8"	.75
A 014	7/16"	1 3/32"	1/2"	9	3/16"	.75
A 015	1/2"	7/8"	1/4"	10	1/8"	.75
A 016	1/2"	7/8"	3/8"	10	1/4"	.75
A 1	1/2"	1 7/32"	9/16"	11	3/16"	.80
A 2	9/16"	1 7/32"	9/16"	11	3/16"	.80
A 3	5/8"	1 11/32"	9/16"	13	3/16"	.80
A 4	11/16"	1 11/32"	9/16"	13	3/16"	.80
A 5	3/4"	1 15/32"	9/16"	14	3/16"	.90
A 6	13/16"	1 15/32"	9/16"	14	3/16"	.90
A 7	7/8"	1 27/32"	5/8"	14	1/4"	.90
A 8	15/16"	1 27/32"	5/8"	15	1/4"	.90
A 9	1"	1 31/32"	5/8"	15	1/4"	.90
A 10	1 1/16"	1 31/32"	5/8"	15	1/4"	.90
A 11	1 1/8"	2 3/32"	5/8"	16	1/4"	1.00
A 12	1 3/16"	2 3/32"	5/8"	16	1/4"	1.00
A 13	1 1/4"	2 11/32"	5/8"	20	1/4"	1.00
A 14	1 5/16"	2 11/32"	5/8"	20	1/4"	1.00
A 15	1 3/8"	2 15/32"	5/8"	21	1/4"	1.30
A 16	1 7/16"	2 15/32"	5/8"	20	1/4"	1.30
A 17	1 1/2"	2 19/32"	5/8"	24	1/4"	1.30
A 18	1 9/16"	2 19/32"	5/8"	24	1/4"	1.30
A 19	1 5/8"	2 31/32"	13/16"	20	5/16"	1.75
A 20	1 11/16"	2 31/32"	13/16"	20	5/16"	1.75
A 21	1 3/4"	3 3/32"	13/16"	23	5/16"	1.75

Carried in Stock

A2-3-8

SUPERIOR

BALL THRUST BEARINGS

With Retaining Band

Light Series

FLAT SEAT   GROOVED BALL RACE

Number of Bearing	Inside Diam.	Outside Diam.	Width	Number of Balls	Ball Diam.	Price
D 1	$\frac{1}{2}''$	$1\frac{7}{32}''$	$\frac{9}{16}''$	15	$\frac{3}{16}''$	$1.60
D 2	$\frac{9}{16}''$	$1\frac{7}{32}''$	$\frac{9}{16}''$	15	$\frac{3}{16}''$	1.60
D 3	$\frac{5}{8}''$	$1\frac{11}{32}''$	$\frac{9}{16}''$	17	$\frac{3}{16}''$	1.60
D 4	$\frac{11}{16}''$	$1\frac{11}{32}''$	$\frac{9}{16}''$	17	$\frac{3}{16}''$	1.60
D 5	$\frac{3}{4}''$	$1\frac{15}{32}''$	$\frac{9}{16}''$	19	$\frac{3}{16}''$	1.80
D 6	$\frac{13}{16}''$	$1\frac{15}{32}''$	$\frac{9}{16}''$	19	$\frac{3}{16}''$	1.80
D 7	$\frac{7}{8}''$	$1\frac{27}{32}''$	$\frac{5}{8}''$	18	$\frac{1}{4}''$	1.80
D 8	$\frac{15}{16}''$	$1\frac{27}{32}''$	$\frac{5}{8}''$	18	$\frac{1}{4}''$	1.80
D 9	$1''$	$1\frac{31}{32}''$	$\frac{5}{8}''$	20	$\frac{1}{4}''$	1.80
D 10	$1\frac{1}{16}''$	$1\frac{31}{32}''$	$\frac{5}{8}''$	20	$\frac{1}{4}''$	1.80
D 11	$1\frac{1}{8}''$	$2\frac{3}{32}''$	$\frac{5}{8}''$	21	$\frac{1}{4}''$	2.00
D 12	$1\frac{3}{16}''$	$2\frac{3}{32}''$	$\frac{5}{8}''$	21	$\frac{1}{4}''$	2.00
D 13	$1\frac{1}{4}''$	$2\frac{11}{32}''$	$\frac{5}{8}''$	23	$\frac{1}{4}''$	2.00
D 14	$1\frac{5}{16}''$	$2\frac{11}{32}''$	$\frac{5}{8}''$	23	$\frac{1}{4}''$	2.00
D 15	$1\frac{3}{8}''$	$2\frac{15}{32}''$	$\frac{5}{8}''$	24	$\frac{1}{4}''$	2.60
D 16	$1\frac{7}{16}''$	$2\frac{15}{32}''$	$\frac{5}{8}''$	24	$\frac{1}{4}''$	2.60
D 17	$1\frac{1}{2}''$	$2\frac{19}{32}''$	$\frac{5}{8}''$	26	$\frac{1}{4}''$	2.60
D 18	$1\frac{9}{16}''$	$2\frac{19}{32}''$	$\frac{5}{8}''$	26	$\frac{1}{4}''$	2.60
D 19	$1\frac{5}{8}''$	$2\frac{31}{32}''$	$1\frac{3}{16}''$	24	$\frac{5}{16}''$	3.50
D 20	$1\frac{11}{16}''$	$2\frac{31}{32}''$	$1\frac{3}{16}''$	24	$\frac{5}{16}''$	3.50
D 21	$1\frac{3}{4}''$	$3\frac{3}{32}''$	$1\frac{3}{16}''$	25	$\frac{5}{16}''$	3.50

Carried in Stock

SUPERIOR

BALL THRUST BEARINGS

With Bronze Ball Cage

FLAT SEAT GROOVED BALL RACE

Number of Bearing	Inside Diam.	Outside Diam.	Width	Number of Balls	Ball Diam.	Price
C 1	½"	1 7/32"	⅝"	7	¼"	$1.60
C 2	9/16"	1 7/32"	⅝"	7	¼"	1.60
C 3	⅝"	1 11/32"	⅝"	8	¼"	1.60
C 4	11/16"	1 11/32"	⅝"	9	¼"	1.60
C 5	¾"	1 15/32"	⅝"	10	¼"	1.80
C 6	13/16"	1 15/32"	⅝"	10	¼"	1.80
C 7	⅞"	1 27/32"	¾"	10	5/16"	1.80
C 8	15/16"	1 27/32"	¾"	11	5/16"	1.80
C 9	1"	1 31/32"	¾"	12	5/16"	1.80
C 10	1 1/16"	1 31/32"	¾"	12	5/16"	1.80
C 11	1⅛"	2 3/32"	¾"	12	5/16"	2.00
C 12	1 3/16"	2 3/32"	¾"	12	5/16"	2.00
C 13	1¼"	2 11/32"	¾"	13	5/16"	2.00
C 14	1 5/16"	2 11/32"	¾"	13	5/16"	2.00
C 15	1⅜"	2 15/32"	¾"	15	5/16"	2.60
C 16	1 7/16"	2 15/32"	¾"	15	5/16"	2.60
C 17	1½"	2 19/32"	¾"	16	5/16"	2.60
C 18	1 9/16"	2 19/32"	¾"	16	5/16"	2.60
C 19	1⅝"	2 31/32"	⅞"	14	⅜"	3.50
C 20	1 11/16"	2 31/32"	⅞"	15	⅜"	3.50
C 21	1¾"	3 3/32"	⅞"	16	⅜"	3.50

Carried in Stock

A2-3-10

SUPERIOR

BALL THRUST BEARINGS

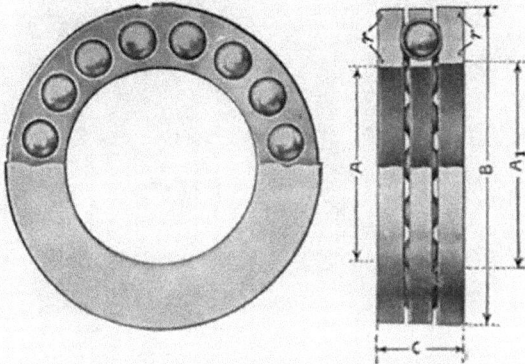

Type AA—Special Light Series

FLAT SEAT

Number of Bearing	Dimensions in Millimeters					Number of Balls	Diameter of Balls Inches	Price
	Inside Diameter A	Inside Diameter A1	Outside Diameter B	Width C	Radius r			
AA 10	10	11	26	12	1	8	7/32	$3.30
AA 12	12	13	28	12	1	8	7/32	3.50
AA 15	15	16	31	12	1	9	7/32	3.60
AA 18	18	19	35	12	1	11	7/32	3.80
AA 20	20	21	37	12	1	12	7/32	4.00
AA 22	22	23	42	14	1	12	1/4	4.40
AA 25	25	26	45	14	1	13	1/4	4.60
AA 30	30	31	50	14	1.5	15	1/4	5.00
AA 35	35	36	55	16	1.5	15	9/32	5.80
AA 40	40	41	60	16	1.5	17	9/32	6.00
AA 45	45	46	68	16	1.5	20	9/32	6.40
AA 50	50	51	74	18	1.5	19	5/16	7.40
AA 55	55	56	78	18	1.5	20	5/16	8.00
AA 60	60	61	82	18	1.5	22	5/16	8.90
AA 65	65	66	90	20	1.5	22	11/32	10.10
AA 70	70	71	95	20	1.5	23	11/32	11.10
AA 75	75	76	100	20	1.5	24	11/32	11.90
AA 80	80	81	110	22	1.5	24	3/8	13.50
AA 85	85	86	115	22	1.5	26	3/8	14.50
AA 90	90	91	120	22	1.5	27	3/8	15.60
AA 95	95	96	130	25	1.5	25	7/16	18.80
AA 100	100	101	135	25	1.5	26	7/16	19.80
AA 105	105	106	140	25	2	27	7/16	21.00
AA 110	110	111	145	25	2	28	7/16	22.00

Tables of equivalents given on pages 72, 73, 74, 75

SUPERIOR

BALL THRUST BEARINGS

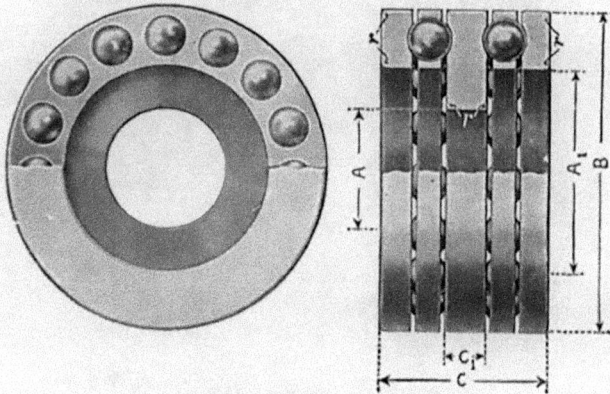

Type CL—Light Series

DOUBLE DIRECTION, FLAT SEAT
MIDDLE WASHER LOCKED TO SHAFT

Number of Bearing	Dimensions in Millimeters						Number of Balls	Diameter of Balls Inches	Price
	Inside Diam. A	Inside Diam. A1	Out-side Diam. B	Width C	Thick-ness C 1	Radius r			
CL 15	15	21	40	26	6.7	1.	11	¼	$7.20
CL 20	20	26	45	26	6.7	1.	13	¼	8.10
CL 25	25	32	53	27	6.7	1.	16	¼	9.40
CL 30	30	42	64	32	7.3	1.	17	5⁄16	12.50
CL 35	35	47	73	36	7.9	1.	16	⅜	15.00
CL 40	40	52	78	36	7.9	1.	18	⅜	16.60
CL 45	45	57	88	42	8.4	1.5	17	7⁄16	19.5C
CL 50	50	62	90	42	8.4	1.5	17	7⁄16	21.7C
CL 55	55	72	103	48	10.	1.5	18	½	26.60
CL 60	60	77	110	48	10.	1.5	19	½	29.30
CL 65	65	82	115	52	10.	1.5	20	½	31.50
CL 70	70	88	125	58	10.1	1.5	18	⅝	40.80
CL 75	75	93	132	62	12.2	1.5	19	⅝	45.70
CL 80	80	98	140	67	13.7	1.5	18	11⁄16	50.10
CL 85	85	103	148	68	14.7	1.5	19	11⁄16	56.90
CL 90	90	110	155	72	15.25	2.	18	¾	61.75
CL 95	95	115	160	72	15.25	2.	19	¾	66.80
CL 100	100	120	165	76	14.75	2.	18	13⁄16	71.20
CL 105	105	125	170	80	15.75	2.	19	13⁄16	75.10
CL 110	110	130	175	82	16.4	2.	19	⅞	81.15
CL 115	115	140	190	86	16.4	2.	19	⅞	94.30
CL 120	120	150	200	92	17.6	2.	19	1	105.30
CL 130	130	160	220	98	17.6	2.	20	1	140.50

Tables of equivalents given on pages 72, 73, 74, 75

A2-3-12

SUPERIOR

STEEL BALLS

PRICE PER 1,000

These balls are made from High Carbon Chrome stock and are extremely accurate both as to size and sphericity, and will not vary to exceed .0001″ from size.

Diameter in Inches	Price	Diameter in Inches	Price	Diameter in Inches	Price
1/8	$ 6.00	11/16	$ 90.00	1 1/2	$ 860.00
5/32	7.50	23/32	101.00	1 9/16	980.00
3/16	9.00	3/4	110.00	1 5/8	1,102.00
7/32	10.50	25/32	120.00	1 11/16	1,220.00
1/4	12.50	13/16	130.00	1 3/4	1,350.00
9/32	14.40	27/32	142.00	1 13/16	1,480.00
5/16	16.70	7/8	160.00	1 7/8	1,620.00
11/32	20.00	29/32	174.00	1 15/16	1,760.00
3/8	24.00	15/16	198.50	2	1,900.00
13/32	28.20	31/32	219.00	2 1/8	2,140.00
7/16	32.20	1	252.00	2 1/4	2,450.00
15/32	38.00	1 1/16	310.00	2 3/8	2,860.00
1/2	44.90	1 1/8	370.00	2 1/2	3,345.00
17/32	51.50	1 3/16	434.00	2 5/8	3,395.00
9/16	60.00	1 1/4	500.00	2 3/4	4,600.00
19/32	66.50	1 5/16	578.00	2 7/8	5,520.00
5/8	75.00	1 3/8	663.00	3	6,600.00
21/32	77.00	1 7/16	756.00		

Subject to Discounts

SUPERIOR
ROLLER THRUST BEARINGS

STRAIGHT ROLLERS

The Plain Roller Thrust Bearing consists of two (2) hardened and ground steel plates with a bronze roller cage; the latter being self-contained so that there are no loose rollers. Its construction is theoretically wrong, but works perfectly in practice and we guarantee satisfactory results. The steel rollers are straight, but are cut into short sections so that they turn readily, and any slippage that takes place does not cause trouble. In actual practice these bearings will sustain loads approx. four times the weight that a Plain Ball Thrust Bearing of the same dimensions will carry. We have many thousands of these bearings in use and nearly all of them in places where Ball Bearings were not sufficiently strong to carry the load. This style of bearing can be made to carry with perfect satisfaction loads of most any weight. They are specially adapted for use in connection with Drives such as worm wheels, turbines, centrifical pumps, etc., and have given universal satisfaction for taking the thrust of propeller shafts. These bearings are very well suited for any place where duty is extremely heavy. The steel plates are carefully hardened, tempered and ground as are also the rollers used in these bearings. On the opposite page we show side sectional view of this bearing with a spherical seat; also the same bearing with spherical seat and locating washer. Also please note: these bearings are provided with a hardened steel ring which is locked in a groove between the two halves of the roller cage. The outward thrust of the rollers is taken up by this ring and prevents any wear taking place at the end of the roller pockets which would be present were it not for this hardened steel ring, and assures the user of this bearing against any possible wear which would take place were it not for this construction.

A2-3-14

The bearings listed on pages 52 and 53 are carried in stock; and, as a rule, can be furnished in most any reasonable quantity promptly on receipt of order, but in ordering we would advise you furnishing us the weight that the bearings are to carry and the R. P. M. at which they will turn. We can then advise you whether the bearings are suitable for the load, and if not, recommend proper sizes. We are prepared to furnish these bearings to carry any load at any speed that may be specified and can also furnish these bearings with a spherical seat, as can we also with spherical seat and locating washer. We can also supply these bearings with phosphor bronze bushings for heavy work, so that the hardened steel washers will not cut the shaft. The results obtained by these bearings have been remarkable and we recommend engineers and machine builders to try them in places where all other types of bearings have failed.

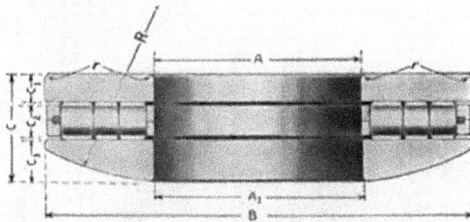

Bearing With Spherical Seat

We would recommend for bearings that are intended for extremely heavy duty to provide them with self-aligning washer. By its use the Bearing is always kept in true alignment with the shaft, insuring an even distribution of the load upon the bearing. We recommend its use in any construction where there is difficulty in installing a bearing and securing perfect alignment with the shaft.

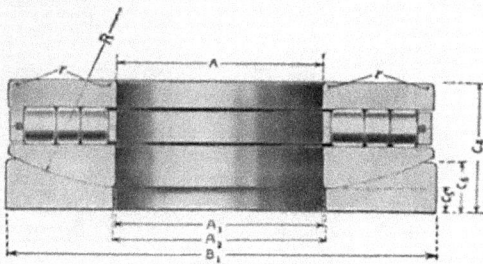

Bearing With Spherical Seat and Aligning Washer

SUPERIOR
ROLLER THRUST BEARINGS

Light Series

PRICE LIST

No. of Bearing	Inside Diam.	Outside Diam.	Width	Diam. of Rollers	Price
E 18	7/8"	1 27/32"	5/8"	1/4"	$3.75
E 19	15/16"	1 27/32"	5/8"	1/4"	3.75
E 20	1"	1 31/32"	5/8"	1/4"	3.75
E 21	1 1/16"	1 31/32"	5/8"	1/4"	4.00
E 22	1 1/8"	2 3/32"	5/8"	1/4"	4.50
E 23	1 3/16"	2 3/32"	5/8"	1/4"	4.50
E 24	1 1/4"	2 11/32"	5/8"	1/4"	5.25
E 25	1 5/16"	2 11/32"	5/8"	1/4"	5.25
E 26	1 3/8"	2 15/32"	5/8"	1/4"	6.00
E 27	1 7/16"	2 15/32"	5/8"	1/4"	6.00
E 28	1 1/2"	2 19/32"	5/8"	1/4"	6.50
E 29	1 9/16"	2 19/32"	5/8"	1/4"	6.50
E 30	1 5/8"	2 31/32"	13/16"	5/16"	7.00
E 31	1 11/16"	2 31/32"	13/16"	5/16"	7.00
E 32	1 3/4"	3 5/32"	13/16"	5/16"	7.50
E 33	1 13/16"	3 5/32"	13/16"	5/16"	7.50
E 34	1 7/8"	3 7/32"	13/16"	5/16"	8.00
E 35	1 15/16"	3 7/32"	13/16"	5/16"	9.00
E 36	2"	3 11/32"	13/16"	5/16"	9.00
E 37	2 1/8"	3 19/32"	13/16"	5/16"	9.50
E 37s	2 3/16"	3 19/32"	13/16"	5/16"	9.50
E 38	2 1/4"	3 23/32"	13/16"	5/16"	10.00
E 39	2 3/8"	3 27/32"	13/16"	5/16"	11.00
E 39s	2 7/16"	3 27/32"	13/16"	5/16"	11.00
E 40	2 1/2"	3 31/32"	13/16"	5/16"	12.00
E 41	2 5/8"	4 11/32"	1"	3/8"	12.50
E 42	2 3/4"	4 15/32"	1"	3/8"	13.00
E 43	3"	4 23/32"	1"	3/8"	13.25
E 44	3 1/4"	4 31/32"	1"	3/8"	13.75
E 44s	3 7/16"	5 7/32"	1"	3/8"	14.00
E 45	3 1/2"	5 7/32"	1"	3/8"	14.00
E 45s	3 9/16"	5 7/32"	1"	3/8"	14.00

Bearings comprise one roller cage and two collars.

Bearings are made large enough to easily fit on the shafts as listed above.

Furnished with Spherical Seat, if desired, also Locating Washer.

Carried in Stock

A2-3-16

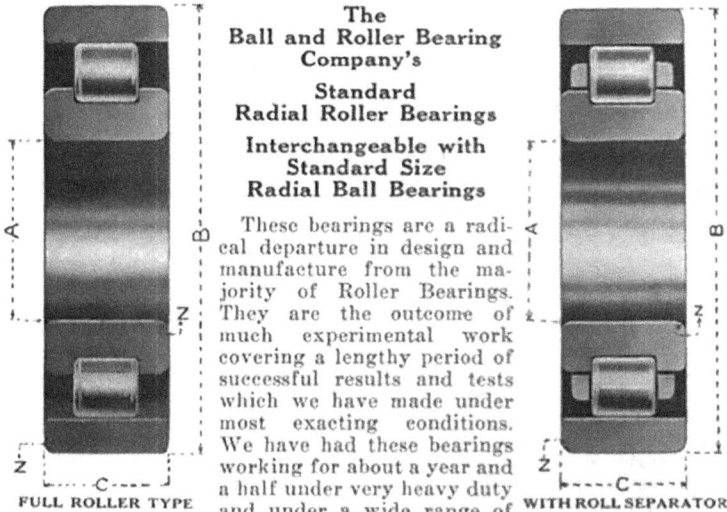

FULL ROLLER TYPE

WITH ROLL SEPARATOR

**The
Ball and Roller Bearing
Company's**

**Standard
Radial Roller Bearings**

**Interchangeable with
Standard Size
Radial Ball Bearings**

These bearings are a radical departure in design and manufacture from the majority of Roller Bearings. They are the outcome of much experimental work covering a lengthy period of successful results and tests which we have made under most exacting conditions. We have had these bearings working for about a year and a half under very heavy duty and under a wide range of speeds, and have therefore confidence in recommending them for heavy motor vehicles, heavy machinery and journals subject to severe loads, they are to be employed where space is so limited that there is no room for a Radial Ball Bearing strong enough for the load and for work beyond the capacity of a Ball Bearing of reasonable size. We are quite certain that the introduction of this Radial Roller Bearing has long been desired and will be appreciated by designers who have been handicapped by the limitations of Radial Ball Bearings.

DESIGN

We can furnish these bearings in two designs; one a full roller type, the other with roll separator. The inner race is channeled to form a groove or track for the rollers. The outer race is a straight cylinder, so that the rolls are free to take up their correct position in this outer race.

The separator or roller cage is made of a high grade bronze cast metal carefully machined over all. It is so designed as to float freely with the rolls and there is thus no rubbing friction between the cage and the rolls.

MOUNTING OF RADIAL ROLLER BEARINGS

We would call the attention of the designers to the fact that the method of mounting these Radial Roller Bearings differs somewhat from that employed with Radial Ball Bearings, and the following points should therefore receive careful attention. It is essential for the satisfactory working of these bearings that the tracks or paths of the rollers in the inner and outer race should be perfectly parallel with each other, and accurate machining is therefore very necessary. The outer race should be a good fit in the housing, free from shake, and must be located exactly opposite the inner race so that the path of the rollers is in the center of the outer race.

SIDE THRUST

It will be obvious that side thrust can, under no condition, be imposed upon these bearings, even to locate the shaft endwise, and therefore both the inner and outer rings may be press fits; shoulders preferably being provided to locate them in their respective positions. Care should be taken, however, when this is done to see that the bearing has not been unduly tightened by the process. Where end thrust exists Ball Thrust Bearings or Roller Thrust Bearings must be provided to take the side thrust load.

CARRYING CAPACITY

Because of the greater carrying capacity of these Radial Roller Bearings they are employed where the conditions are very severe and can take the place of large and expensive Plain Bearings which would, of necessity, have to be very carefully fitted, scraped and bedded down. The use of Roller Bearings dispenses with these operations and the cost is therefore largely reduced. If there is practically no side thrust or only end location is required, Plain Thrust Collars are often sufficient. On no account should the shaft be allowed to move endwise when the load is on the bearings.

The load which may safely be put upon these Roller Bearings varies considerable with the conditions under which they are required to run. We therefore much prefer that customers submit their designs to us, with full particulars of speed and load, previously to actually settling on the size to be used, so that we may recommend the most suitable size and type. It is quite safe to assume, however, that the Radial Roller Bearing will carry 50% more load than a Ball Bearing of the same dimensions.

LIMITS OF ACCURACY

The limits of accuracy to which these bearings are ground both for the bore and outside diameter of the rings is within .0003" minus .0002" plus.

The thickness of the rings are ground to standard dimensions within a limit of .005" in either direction.

LUBRICATION OF THE BEARINGS

As regards to lubrication—Roller Bearings should be treated in precisely the same way as Ball Bearings, and the same lubrication used except for high speeds, when a mineral oil should be used. It will be noted that these bearings have the same advantage in only requiring the renewal of lubrication at long intervals.

Protection from dust and moisture is essential and a suitable method for guarding against this is to bring the housing down to within .004" of the shaft and provide grooves which form oil pockets into which the grease finds its way and makes a more or less perfect seal against moisture or dirt.

NOTICE—For certain exceptional cases a bearing lighter than the standard medium type is sometimes required. To meet this demand we have listed a light series of these Radial Roller Bearings, but we do not recommend or advise their use where this can be avoided. The races are, of necessity, thin in cross section, and are readily distorted by inaccuracies in the shafts and housings, and such distortion subjects the bearings to permanent loads, the extent of which it is almost impossible to estimate. In all cases we recommend the medium or heavy type bearings.

A2-3-18

RADIAL

ROLLER BEARINGS

Interchangeable with Radial Ball Bearings

LIGHT SERIES WIDE TYPE

FULL ROLLER TYPE   **WITH ROLL SEPARATOR**

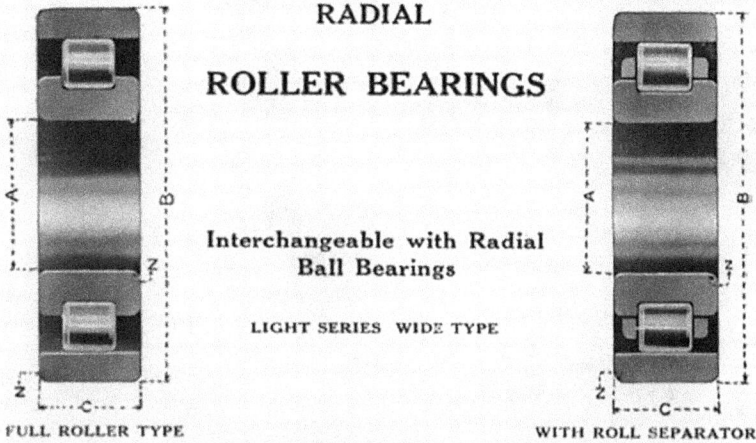

Full Roller Type is furnished unless Roll Separator is specified.

PRICE LIST

No. of Bearing	Bore Inch	Bore M.M.	Diameter Inch	Diameter M.M.	Width Inch	Width M.M.	Dia. Rolls	Price
R 204	0.7874	20	1.8504	47	0.5512	14	¼	$4.75
R 205	0.9843	25	2.0472	52	0.5905	15	¼	5.25
R 206	1.1811	30	2.4409	62	0.6299	16	⁵⁄₁₆	6.25
R 207	1.3780	35	2.8346	72	0.6693	17	⁵⁄₁₆	6.85
R 208	1.5748	40	3.1496	80	0.7087	18	⅜	7.50
R 209	1.7717	45	3.3465	85	0.7480	19	⅜	9.00
R 210	1.9685	50	3.5433	90	0.7874	20	⅜	10.00
R 211	2.1653	55	3.9370	100	0.8268	21	⁷⁄₁₆	12.50
R 212	2.3622	60	4.3307	110	0.8661	22	⁷⁄₁₆	14.50
R 213	2.5591	65	4.7244	120	0.9055	23	½	16.50
R 214	2.7559	70	4.9213	125	0.9449	24	½	18.75
R 215	2.9528	75	5.1181	130	0.9843	25	½	21.50
R 216	3.1496	80	5.5118	140	1.0236	26	⅝	24.10

When ordering use the letter R and number of bearing for full roller type.

When ordering bearings with roll separator use R. S. and number of bearing.

Full roller types have greater carrying capacity than the bearings with roll separator.

These bearings carry at least 50% greater load than Radial Ball Bearings of same dimensions.

The dimensions in inches given are nearest to the mm. sizes.

Pages 54-55 give directions for mounting and care of bearings.

We have unusual facilities for the production of:

      Small Hardened and Ground Cylindrical Rolls

      Piston Pins

      Valve Lifters

      Valve Lifter Rolls

      Valve Lifter Roller Pins

      Steel Bushings, etc.

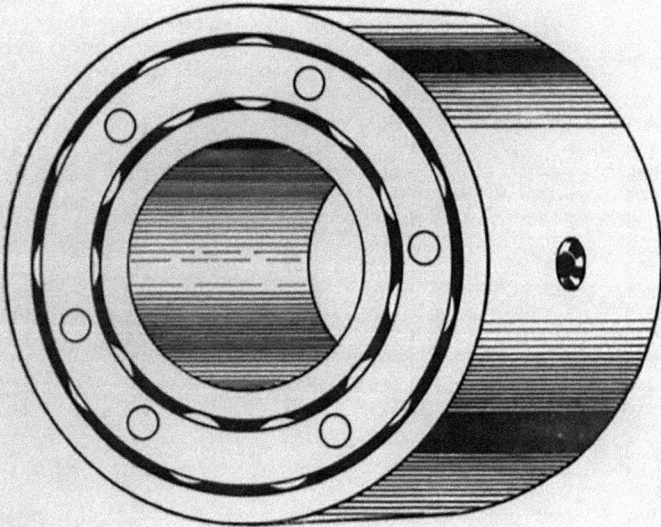

A2-3-20

SUPERIOR
JOURNAL ROLLER BEARINGS

Rollers Not Hardened

For Agricultural, Mining, Industrial, R. R. Cars, Baggage Trucks, etc.

PRICE LIST

Length of Bearings

Code Words			nard	nape	nail	nam	naw	navy	near	nep	ned	nest	news	nice	nigh
Code Words	Dia. Shaft	Dia. Rolls	1″	1¼″	1½″	1¾″	2″	2¼″	2½″	2¾″	3″	3½″	4″	5″	6″
gaff	½″	3/16″	.50	.55	.60	....	....	....	....	....	....	....	....	....	....
gage	⅝″	3/16″	.50	.55	.60	....	....	....	....	....	....	....	....	....	....
gain	¾″	3/16″	.55	.60	.65	....	....	....	....	....	....	....	....	....	....
gale	⅞″	¼″	.60	.65	.70	.80	.90	....	....	....	....	....	....	....	....
game	1″	¼″	.65	.70	.75	.85	.95	1.05	1.15	1.20	1.25	....	....	....	....
gang	1 1/16″	¼″	.65	.70	.75	.85	.95	1.05	1.15	1.20	1.25	....	....	....	....
gard	1⅛″	¼″	.65	.70	.75	.85	.95	1.05	1.15	1.20	1.30	....	....	....	....
gasp	1 3/16″	¼″	.65	.70	.75	.85	.95	1.05	1.15	1.20	1.30	....	....	....	....
gato	1¼″	¼″	.70	.75	.80	.85	.95	1.05	1.15	1.20	1.30	....	....	....	....
gaze	1 5/16″	¼″	.70	.80	.90	1.00	1.10	1.20	1.30	1.40	1.50	....	....	....	....
gear	1⅜″	¼″	.75	.85	.95	1.05	1.15	1.25	1.35	1.45	1.55	....	....	....	....
gent	1 7/16″	¼″	.75	.85	.95	1.05	1.15	1.25	1.35	1.45	1.55	1.75	....	....	....
giddy	1½″	¼″	.80	.90	1.00	1.10	1.20	1.30	1.40	1.50	1.60	1.80	....	....	....
gift	1 9/16″	5/16″	.90	1.00	1.10	1.20	1.35	1.50	1.65	1.80	1.90	2.15	....	....	....
gild	1⅝″	5/16″	.90	1.00	1.10	1.20	1.35	1.50	1.65	1.80	1.90	2.15	....	....	....
gird	1 11/16″	5/16″	.95	1.05	1.15	1.25	1.40	1.55	1.70	1.85	1.95	2.20	....	....	....
give	1¾″	5/16″	.95	1.05	1.15	1.25	1.40	1.55	1.70	1.85	2.00	2.25	....	....	....
glee	1 13/16″	5/16″	....	....	1.30	1.45	1.60	1.75	1.90	2.05	2.20	2.50	....	....	....
glen	1⅞″	5/16″	....	....	1.30	1.45	1.60	1.75	1.90	2.05	2.20	2.50	....	....	....
glib	1 15/16″	5/16″	....	....	1.35	1.50	1.65	1.80	1.95	2.10	2.25	2.55	2.85	3.45	....
glow	2″	5/16″	....	....	1.40	1.55	1.70	1.85	2.00	2.15	2.30	2.60	2.90	3.50	....
glum	2¼″	5/16″	....	....	....	....	1.80	1.95	2.10	2.25	2.40	2.70	3.00	3.60	....
glut	2½″	5/16″	....	....	....	....	1.90	2.05	2.20	2.35	2.50	2.80	3.10	3.70	....
gnaw	3″	⅜″	....	....	....	....	2.25	2.40	2.55	2.70	2.85	3.15	3.45	4.05	....
goal	3½″	⅜″	....	....	....	....	....	....	2.75	2.90	3.05	3.35	3.65	4.25	....
goer	4″	½″	....	....	....	....	....	....	....	....	3.70	4.00	4.30	4.90	5.50
gone	5″	½″	....	....	....	....	....	....	....	....	4.50	4.80	5.10	5.70	6.50

Intermediate sizes and lengths at price of next larger size.
Order by Code Words.

SUPERIOR

JOURNAL ROLLER BEARINGS

WITH CASINGS AND SLEEVES

Cases, Sleeves and Rollers of Steel

Hardened, Tempered and Ground

PRICE LIST

Length of Bearings

Code Words				latin	laugh	laver	leafy	learn	leash	ledge	leech	legal
Code Words	Shaft Diam.	Outside Diam.	Diam. Rollers	1″	1½″	2″	2½″	3″	3½″	4″	5″	6″
each	½″	1¼″	3/16″	$3.00	$3.20	......	......	......	......	......	......	......
earn	⅝″	1⅜″	3/16″	3.30*	3.50	......	......	......	......	......	......	......
ease	¾″	1½″	3/16″	3.60	3.90	......	......	......	......	......	......	......
east	⅞″	1⅞″	¼″	4.00	4.30	$4.50	......	......	......	......	......	......
easy	1″	2″	¼″	4.32	4.86*	5.40*	$6.48*	8.10	......	......	......	......
ebny	1 3/16″	2 3/16″	¼″	4.56	5.04	5.64	6.72	8.34	......	......	......	......
echo	1¼″	2⅜″	¼″	4.56	5.04*	5.64*	6.72*	8.34*	......	......	......	......
edor	1 7/16″	2 9/16″	¼″	4.86	5.40*	5.94*	6.85*	9.02	11.88	......	......	......
eddy	1½″	2⅝″	¼″	4.86	5.40*	5.94*	6.85*	9.02	11.88	......	......	......
edit	1¾″	3⅛″	5/16″	......	5.94*	6.72*	7.44	9.54	12.30	......	......	......
egad	1 15/16″	3 3/16″	5/16″	......	6.72	7.44	8.16	10.08*	12.96	14.50*	16.00*	......
else	2″	3⅜″	5/16″	......	6.72	7.44	8.10*	10.08*	12.96*	14.50*	16.00*	......
envy	2¼″	3⅝″	5/16″	......	......	8.10*	9.54*	10.80*	13.44*	15.90	17.20	......
epha	2 7/16″	3 15/16″	5/16″	......	......	8.70	10.08	11.46	14.82	16.80	18.30	......
espy	2½″	4″	5/16″	......	......	10.26	11.46	12.30	15.66*	17.55*	20.21*	......
etch	3″	4⅝″	⅜″	......	......	8.70	10.08	11.46*	14.82*	16.80	18.30	......
even	3½″	5¼″	⅜″	......	......	......	12.12	13.44	16.20	19.89*	23.40*	......
ever	4″	6″	½″	......	......	......	......	19.50*	21.60	26.32*	29.25	33.15
exit	5″	7¼″	½″	......	......	......	......	22.30	29.52	34.10	37.70	39.00

Intermediate sizes and lengths at price of next larger size.
Order by Code Words.

*These sizes carried in stock and listed on discount sheet. Other sizes will quote.

A2-4
15 B.T.A. 862 (1929)

BALL & ROLLER BEARING CO., PETITIONER,
v.
COMMISSIONER OF INTERNAL REVENUE, RESPONDENT.

Docket No. 15233.

Board of Tax Appeals.

Promulgated March 14, 1929.

Barry Mohun, Esq., Henry F. Parmelee, Esq., and Avery Tompkins, Esq., for the petitioner.

B. M. Coon, Esq., and William J. Carroll, Esq., for the respondent.

This is a proceeding for the redetermination of deficiencies in income and profits tax for the calendar years 1920 and 1921 in the amounts of $2,473.88 and $551.11, respectively. The point in issue is the actual cash value on May 1, 1916, of an application for patent paid in to the petitioner corporation in exchange for shares of stock for the purpose of computing invested capital for the years 1920 and 1921, and also the allowance for exhaustion, if any, deductible from gross income of the taxable years in respect of a patent issued on January 2, 1917, upon the application for patent theretofore filed.

FINDINGS OF FACT.

The petitioner is a Connecticut corporation with principal office at Danbury. It was incorporated in 1914, with an authorized capital stock of $300,000 divided into 3,000 shares of a par value of $100 each.

From 1909 to 1916 Lewis R. Heim, of Danbury, was engaged in the design and manufacture of laundry machinery, conducting his business as a sole proprietor under the trade name of Heim Machine Co. In 1910 Heim began the manufacture of ball bearings and roller bearings and in 1913 discovered the principle governing the socalled two-wheel centerless grinder, since known as the Heim centerless grinder. From 1913 to 1916 he devoted most of his time to the development of this invention and expended large amounts of money for the purpose of perfecting it, which were charged to operating expenses of the Heim Machine Co.

The invention above referred to constitutes a basic improvement in the art of grinding cylindrical rollers for roller bearings. It is generally referred to as centerless grinding" and represents a radical departure from prior grinding practice, taking the place in the art of cylindrical grinding. Generally speaking, in order to grind the outer surface of articles under the prior art it was necessary to mount them on centers or in a chuck, very much as they are mounted in a lathe, and to treat their surface with a grinding wheel individually; then remove the article being ground, replace it with another article to be ground, and so on. The broad idea of Heim's invention of the two-wheel centerless grinder is simply to put such articles between two wheels rotating with their surfaces in opposite directions, so that one wheel would do the grinding and the other would not only rotate the work but feed or propel the work and govern it. The two-wheel centerless grinder did automatically about everything that the old grinding process would do with a certain number of operations and a very large number of elements. By the use of the two-wheel centerless grinder it became possible to take articles to be ground and run them through the machine in a stream, end to end, entirely separate; the machine would automatically take entire charge of them and turn them out at a rapid rate. The effect was to revolutionize the art of grinding cylindrical articles, manufacturing roller bearings and other sundries, where a high degree of accuracy is essential, and increasing production many fold and producing greater accuracy and a much finer finish than under the prior art of center grinding.

In 1914 Heim turned over to the petitioner merchandise of a value of approximately $10,000 in exchange for $10,000 par value of capital stock, less qualifying shares.

On March 6, 1915, Heim applied for United States letters patent upon his invention of the two-wheel centerless grinder and purported to assign his invention to the petitioner in such application. He received no consideration at this time for the assignment of the invention.

The Heim Machine Co. engaged only in the manufacture of laundry machinery. On or prior to May 1, 1916, Heim disposed of his entire laundry machinery business to one Fiddler. On May 1, 1916, he transferred to petitioner all of the assets of the Heim Machine Co. remaining after the sale of the laundry business. In exchange for these assets he received capital stock of the petitioner of the par value of $290,000, which, with the $10,000 capital stock theretofore received by him in 1914, less qualifying shares, aggregated $300,000 in par value. Of this amount $100,000 was at the time of the transfer attributed to the value of the intangible assets transferred, including the application for patent, and $200,000 was attributed to the tangibles, such estimate of $200,000 being based upon an appraisal made for the petitioner by a firm of certified public accountants engaged for that purpose. In the books of account opened for the petitioner in 1916, $100,000 was attributable to "patents and good will." It was the intention of all parties concerned that the $100,000 should be regarded as the value of the application for patent upon the Heim centerless grinder.

The business of the petitioner, commencing May 1, 1916, was that of manufacturing ball bearings and roller bearings and Heim centerless grinding machines made in accordance with the application for letters patent above referred to. On January 2, 1917, United States Letters Patent No. 1,210,937 were issued upon the application thereto-

314

fore made by Lewis R. Heim. Letters patent were issued to the petitioner as assignee of Lewis R. Heim.

The invention of the Heim centerless grinding machine effects very great savings over the prior art. In some cases one centerless grinding machine has displaced from 6 to 20 men who were grinding work on the center grinders. The labor cost of finished articles by this machine is but one-sixth or one eighth of the cost of finishing on the center grinder and, under particularly favorable circumstances, much less than such fraction. The petitioner made large profits from the manufacture of centerless grinding machines during the period from 1916 forward. The actual profits attributable to the manufacture of these machines can not be stated with accuracy, but apparently reliable estimates placed the average annual profits of the petitioner, attributable exclusively to the invention from 1916 to 1923, at approximately $28,000. From 1917 to 1919 the petitioner was manufacturing ball bearings solely on requisitions or orders approved by the United States Government, the petitioner having placed its entire production to the support of the Government. Sales of Heim centerless grinders, commencing in 1921, were as follows:

1921 _____	$ 47,950.04
1922 _____	$137,833.45
1923 _____	$247,258.31

Petitioner's profits were seriously impaired, even during the war period, by the unlawful competition of four principal infringers. Petitioner commenced litigation in the United States District Court for the District of Connecticut against some of these infringers, which litigation was decided against the petitioner, but on appeal the decision of the lower court was reversed by the Circuit Court of Appeals for the Second Circuit in 1924, in the case of Ball & Roller Bearing Co. v. Sanford Manufacturing Co., 297 Fed. 163.

The actual cash value of the application for patent of the Heim centerless grinder paid in to the petitioner in 1916, in exchange for shares of stock, was $100,000 and the actual cash value of Letters Patent No. 1,210,937, issued to the petitioner on January 2, 1917, was at least $100,000.

OPINION.

SMITH:

That an application for letters patent is property subject to inclusion in invested capital has been many times decided by this Board. *Individual Towel & Cabinet Service Co., 5* B. T. A. 158; *Starbuck, Administrator*, 13 B. T. A. 796; Hershey Mfg. Co., 14 B. T. A. 867. The respondent contends that the application for patent on the Heim centerless grinder was paid in to it for the nominal consideration of one dollar and not for shares of stock. The evidence, however, completely refutes this contention. There is no question but that Heim was fully cognizant of the value of his invention and that he intended to and did pay it in to the petitioner corporation in exchange for $100,000 par value of capital stock. Letters patent were not granted on this invention until 1917. It is apparent, however, from the entire record that the invention was something absolutely new in the art of grinding cylindrical bearings. The field for the use of such bearings was practically unlimited. Numerous competent witnesses have testified as to the value of the invention and some of these competent witnesses have testified that in their opinionthe value was much in excess of $100,000. The petitioner has sustained the burden of proving an actual cash value of the application for patent of at least the $100,000 contended for.

The Letters Patent No. 1,210,937, issued on the application for patent on January 2, 1917, had a cash value of $100,000. A reasonable allowance for the exhaustion of the patent for each of the taxable years was one-seventeenth of $100,000.

Judgment will be entered under Rule 50.

February, 1921   MACHINERY   583

moved without disturbing any of the vertical adjustments of the support as set for the job at hand. The support arm is simply unclamped and swung to one side where it is out of the way. After mounting a new blank, the arm is swung back in place and again tightened. Larger gears up to 60 inches diameter are supported by height blocks and jacks.

The machine is provided with automatic trips to the vertical cutter movement and the horizontal work movement. The vertical trip always operates to stop the entire machine. The machine is rated for cutting 2½ diametral pitch in steel or 2 diametral pitch in cast iron at a rapid production rate, and with a minimum grinding of hobs.

Additional specifications are as follows: Face capacity, 18 inches; minimum centers between work- and cutter-arbors, 2½ inches; hob capacity, 7¾ inches diameter by 9 inches long; diameters of cutter-arbors, 1¼ and 1½ inches; work-arbor diameter, 2¼ inches; motor drive, when furnished, requires constant-speed 7½- to 10-horsepower motor with a speed of 1100 to 1200 revolutions per minute. The machine weighs 14,000 pounds.

HEIM CENTERLESS ROLL GRINDER

A grinding machine for automatically grinding cylindrical work without placing such work on centers, is now being manufactured by the Ball & Roller Bearing Co., Crosby St. and Maple Ave., Danbury, Conn. This machine is known as the Heim centerless cylindrical roll grinder, and is the invention of L. R. Heim, general manager of the company. The machine is adapted for grinding work such as rolls for roller bearings, wrist-pins, camshafts, valve lifters, valve-lifter roll-pins, pistons, shackle bolts, spring bolts, roller chain studs, or in fact any cylindrical part that requires grinding on only one diameter. A large battery of these grinders is in use in the Ball & Roller Bearing Co.'s plant in connection with the manufacture of journal roller bearings, radial roller bearings, and various small cylindrical parts manufactured for the trade by this company.

General Features of Construction

Figs. 1 and 2 show front views of the latest design of this machine. In Fig. 1 the machine is shown with the doors or covers removed from the wheel housing so that the position of the regulating wheel and the grinding wheel may be seen. This illustration also shows the roll-supporting fixture, and the chute in which the rolls are placed and from which they pass into the roll support, and thence between the grinding and regulating wheels. It will be noticed that there is a platform or shelf at the front of the machine on which the rolls

Fig. 2. Heim Centerless Roll Grinder made by Ball & Roller Bearing Co.

or supply of work can be placed in a convenient position for the operator. In feeding, the operator places the pieces to be ground in the chute, which is supported by a rod clamped to the shelf, as shown. Underneath the work-holding shelf is a water tank provided with a tray into which the work is automatically dropped after being ground. The tray into which the work falls is hung from the top of the water tank, and can be easily removed and dumped when it is full.

The roll-regulating wheel and the grinding wheel are adjusted in respect to each other and the roll-supporting fixture, by means of the two handwheels shown at the right-hand end of the machine. The grinding wheel and regulating wheel are each mounted on a slide, which, in turn, is mounted on the top of the main bed of the machine. These slides, the housings for the wheels, and the regulating spindles are of heavy construction, as will be readily apparent by referring to Fig. 2. On each side of the machine base, there is a large opening. The openings at the ends of the base and the one at the rear are provided with metal doors. Over the doors of the openings at the ends of the machine are hung trays for holding tools. The opening at the rear of the machine gives easy access to the mechanism inside the frame. The water tank, fastened to the front of the bed, serves as a cover for the front opening in the base. Attached to the water tank is a centrifugal pump, driven from the main drive shaft of the machine. This pump supplies the grinding and regulating wheels with a constant flow of coolant.

Design of Driving Mechanism

The drive shaft for driving the grinding spindle and also for conveying power to the regulating wheel spindle and coolant pump is hung in the main bed of the machine, and is mounted on radial roller bearings, which are enclosed in a substantial housing and rotate in oil. From this drive shaft, power is conveyed to another shaft on which is mounted a speed reduction gear. This gearing and the shaft to which it is attached are enclosed in a housing and run in oil. By the use of this gearing, four speeds can be transmitted to the regulating wheel spindle. The design of the machine is such that the driving belts of the regulating wheel spindle and the grinding wheel spindle pull from underneath, and therefore have a tendency to hold the wheel-slides down in contact with the bed of the machine, thus adding to the rigidity of the slides in which the wheel-spindles are mounted. The slack in these driving belts is taken up by idlers mounted on anti-friction bearings.

The wheel-spindles are made from a special alloy steel, hardened, tempered, and ground, and they run in large phosphor-bronze bearings. These bearings provide automatic

Fig. 1. Heim Centerless Roll Grinder with Wheel Housing Cover removed

Fig. 3. Rear View of Heim Centerless Grinder

lubrication for the spindles, and the design is such that grit and dirt are absolutely excluded. The lateral thrust on the spindles is taken by ball bearings.

Operation of Machine

One operator can take care of a machine; that is, under normal conditions, he can keep the chute full of work and attend to the gaging and adjusting of the wheels. In some instances, it would be possible for one operator to attend to more than one machine, depending upon the nature of the work to be ground. When very long runs of any one size of work are to be ground, automatic feeders can be provided, so that one operator can run several machines, it merely being necessary for him to adjust the wheels, gage the work, and fill the magazines at regular intervals. The machine can be provided with tight and loose pulleys to be driven from a countershaft, or it can be driven by a motor either hung to the ceiling, set on the floor, or mounted on the machine at the end of the bed.

A diamond wheel-dresser shown in the lower right-hand corner of Figs. 1 and 2 is furnished with the machine. This device is placed between the two wheels and fastened to the machine, after first removing the roll-supporting fixture. The truing diamonds are passed back and forth across the faces of the wheels by operating the handwheel at the end of the truing fixture. In this way the faces of both wheels are dressed parallel with each other.

The changing of the speed of the regulating wheel is accomplished by the use of a handwheel shown on the left side of the main bed of the machine near the coolant tank. By turning this handwheel in one direction, the speed of the regulating wheel can be changed from that employed for grinding to that employed for dressing the wheel. By a slight turn of this wheel, the power drive can be disconnected from the regulating wheel.

It is claimed by the manufacturers that exceptionally high production rates can be obtained by the use of this machine, the increase in some instances ranging as high as from 100 to 500 per cent over the ordinary method of grinding, and at the same time a high degree of accuracy can be maintained. The grinding and regulating wheels employed on this machine are of the regular type such as are used on external grinders, and the life of the grinding wheel is approximately the same as that of an external grinding wheel, while the life of the regulating wheel is much greater. This machine occupies a floor space of 4 by 6 feet, and when equipped for belt drive it weighs approximately 4500 pounds.

MONARCH GAP BED LATHE

The gap bed lathe shown in the accompanying illustration is one of five sizes now being manufactured by the Monarch Machine Tool Co., 109 Oak St., Sidney, Ohio. The carriage is of unusually heavy construction, and the rear carriage wings are extended at the tailstock end in order to form a long bearing that will withstand heavy service. The carriage is gibbed, both at the front and the rear, and the cross-bridge and power cross-feed slide are extended to permit taking facing cuts the full swing of the gap without undue stress or chatter. The carriage is also built so that it will overhang the gap, thus permitting the tool to be used close to the faceplate, if necessary.

The gears contained in the apron are of drop-forged steel. The apron is double-walled, permitting bearings at both ends for the studs and shafts. It is provided with a feed reverse, and interlocking device which prevents the feed-rod and lead-screw from being engaged at the same time. The rack and pinion disengages when screw-cutting.

The headstock is of the solid bowl-shaped box housing construction. The quick-change gear-box is of simple design, and the thread and feed range is sufficient to cover all ordinary requirements. The change-gears are made of steel, and are of coarse pitch; they are wide-faced, bushed with bronze, and run on vanadium-steel shafts. The spindle is made of 50-point carbon crucible steel; it is accurately ground and runs in a phosphor-bronze bearing.

The back-gears are locked in and out of position by a spring plunger. The double back-gears are of the positive geared type. The compound rest is gibbed throughout, and has liberal bearing surfaces. The usual swivel and cross-feed graduations are provided. Drilled and tapped holes are provided on the back of the carriage for receiving a taper

Gap Bed Lathe built by the Monarch Machine Tool Co.

American Machinist

Volume 58   NEW YORK, MARCH 8, 1923   Number 10

The Production of Small Parts By Centerless Grinding

By HOWARD CAMPBELL
Western Editor, *American Machinist*

Some examples of centerless grinding, together with figures on production—Methods of handling odd work—Grinding tungsten bars

---

THE ART OF grinding metals, which for many years was considered so standard that there appeared to be only one way of handling a grinding job, has undergone so many changes in the last three or four years, that no one can say with certainty what can be done or what can be expected next. In the old days, grinding constituted a trade in itself, but now the latest types of machines have reduced what was almost a science to an operation. The human element which entered into the grinding of a piece of work by an operator who placed one piece of work at a time into the machine and turned the feed wheel as his conscience dictated, has been eliminated by a machine into which the work is fed through a trough. And there will undoubtedly be other changes, just as radical, in the next few years.

One job, which, because of its simplicity of design is being done almost universally on centerless grinding machines, is the automotive piston pin. Some shops use only the centerless machines while others, such as the Standard Gear Co., Detroit, Mich., use a center machine for the first operation and centerless machines for the subsequent operations. This shop uses a Norton grind-

ing machine for the first operation, as shown in Fig. 1. The diameter of the pin is 1⅛ in. and the amount of stock removed is 0.025 in. The production on this operation is 100 per hr., the wheel (a Norton combination M) being dressed for every 65 pins.

The second operation is shown in Fig. 2, the machine

FIG. 2—SECOND OPERATION—CINCINNATI CENTERLESS GRINDING MACHINE

being a Cincinnati centerless grinding machine. The work is held between the wheels by a rest, as shown and is passed into the machine at the front side and fed through and out of the rear of the machine by the action of the feed wheel. The wheel used is an Aloxite 80 K.C.

The production on this operation is 250 per hour and when the pins leave the machine they are within 0.001 in. of the desired size. Fig. 3 shows the work leaving the machine. The third or finishing operation, shown in Fig. 4, is being performed on a Detroit centerless grinding machine. In the Detroit machine, the grinding wheel is directly over the feed wheel, which controls the rotation of the work as well as the speed at which it passes through the machine. The grinding

FIG. 1—ROUGH GRINDING PISTON PINS

FIG. 3—WORK LEAVING THE MACHINE

and are passed through the machine twice, removing about 0.005 in. the first time, and approximately 0.001 in. the second time, obtaining a fine finish. The total production is 275 pieces per hr. The production is raised to 450 pieces per hour on similar work on which the requirements as to finish are not quite so strict, making it possible to finish the work in one pass. The feed wheel is a 120-C "Vulcanite" (New York Belting and Packing Co.), and the grinding wheel is a 6,600-L Norton.

This machine is also used to grind flange yoke-pins, which are 0.6875-0.6865x3.936 in. These pins are fed through the machine three times at a rate of 650 pieces per hr. for each pass. In the first pass, 0.005 in. of stock is removed, in the second pass, 0.0025 in. and in the third pass 0.0005 in., making 0.008 in. altogether. It is confidently expected that production on this job will be brought up to 3,000 pieces per day.

The job shown in Fig. 6 is that of grinding motor-

wheel rotates at the same peripheral speed as would be used under similar conditions in a standard machine of the center variety while the operator has a choice of three rotating speeds, according to the finish required and the amount of stock to be removed. The speed at which the work travels through the machine is regulated by the angle at which the feed wheel is set. The face of the feed wheel is twice the width of the grinding wheel so that the work will be revolving both when it engages and leaves the grinding wheel, thus precluding the possibility of flat spots on the ends of the work. The wheel is special for this machine and the production is 350 per hr. The pins are held to a tolerance of 0.002 inch in this operation.

The machine shown in Fig. 5 is a Heim centerless grinding machine, at work on gear shifter rods in the plant of the Mechanics Machine Co., Rockford, Ill. The rods are of steel, 0.687-0.687x5⅜ in.

FIG. 5—GRINDING SHIFTER RODS ON A HEIM CENTERLESS GRINDING MACHINE

FIG. 6—GRINDING PISTONS ON A REEVES ROLL GRINDING MACHINE

FIG. 4—FINISHING OPERATION—DETROIT CENTERLESS GRINDING MACHINE

cycle pistons in the plant of the Excelsior Motor & Manufacturing Co., Chicago, Ill. The pistons are of cast iron, 2 11-16 in. in diameter, and are ground in three passes, removing 0.005 in. of stock at each pass. They pass through the machine at the rate of 270 per hr. in each of the first two passes and then the wheels are adjusted so that 70 per hr. are produced in the last pass,

March 8, 1923   *Build Bigger Profits with Better Equipment*   **359**

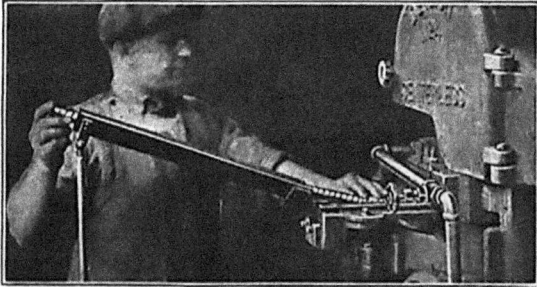

FIG. 7—GRINDING ROLLER BEARINGS

hour. The amount of stock to be removed averages 0.015 in., and is removed in one cut, the finish size being held within 0.001 inch.

The illustration, Fig. 9, shows a Detroit machine set up to grind cam rollers. The rollers are ¼ in. thick and 0.748 in. in diameter, with an allowance of plus 0.002 in. This machine is of a different type and has no feed wheel. A block is bored out to a sliding fit for the rollers, with the front exposed so that the wheel can contact properly with the work which is fed to the machine through the steel tubes, as shown. The tubes are loaded and are then slipped into holes in the end blocks as may be seen in Figs. 9

the extra amount of time required to remove the same amount of stock being sufficient to produce the finish required. Norton wheels are used, a 60-M wheel for grinding and a 120-W wheel for feeding.

The operator shown in Fig. 7 is feeding roller bearings to a Detroit No. 4 centerless grinding machine. The diameter of the stock as it comes to the machine is 0.953-0.960 in., and the machine is set so that the first pass reduces the diameter to 0.950 in. In the second pass, 0.007 in. of stock is removed, 0.005 in. in the third and the fourth or finishing pass reduces the size to 0.9368 in., with a minus allowance of 0.0017 in. The pieces pass through the machine at the rate of ten per minute, which is 600 per hour. The total production is 150 finished pieces per hour. The grinding wheel used on this job is a No. L5-0 Carborundum, running at

FIG. 9—GRINDING CAM ROLLERS

and 10, and a rod is used to push them along. The rod is shown sticking out of the tube in Fig. 9. As the rollers pass the wheel, they slip into another tube and when this is full it is removed.

Among other interesting jobs done on the Detroit machine is that of grinding a number of tungsten bars for the Edison Lamp Works. These bars come to the machine in sections three feet long and approximately 0.016 in. in diameter. The size is reduced by 0.004 in. in the first pass and 0.002 in. is taken off in the second pass, these amounts being just enough to "clean up."

FIG. 8—VIEW SHOWING OPPOSITE SIDE OF MACHINE

5,000 ft. per minute. The peripheral speed of the feed wheel varies from 35 to 45 ft. per min. on a job of this kind, depending on the amount of stock to be removed and the finish required. The illustration Fig. 8 shows the work coming from the machine.

Rollers of the same type as those shown in Fig. 7 but only ⅝x1⅛ in. are ground in the machine shown in Fig. 8, at the rate of 3,600 finished pieces per

FIG. 10—WORK LEAVING THE MACHINE

May, 1923   MACHINERY   721

Centerless Grinding Methods

CENTERLESS grinding, as the name implies, is the grinding of cylindrical work without supporting it on centers in the usual way. In the Heim centerless grinding machine, on which the operations described in this article are performed, this is accomplished by employing two abrasive wheels. One of these wheels rotates in contact with the work at a very slow speed and serves only to rotate the work and advance it through the machine; the other wheel, located directly opposite it as shown in Fig. 2, does the grinding. It is possible to handle any plain cylindrical work without shoulders or tapers, but the best results are obtained on short cylindrical parts, such as rollers for roller bearings, wrist-pins, short shafts, and similar work of comparatively short lengths.

Before describing some of the details regarding the work of the centerless grinder, the design of this class of machine, in so far as it departs from older principles of grinding will be briefly described. In Fig. 1, which is a view from the driving side of a Heim centerless grinder, built by the Ball & Roller Bearing Co., Danbury, Conn., it will be seen that the end of one of the two wheel shafts is considerably lower than the other shaft. This shaft A supports the feed or regulating wheel at the opposite side of the machine, and it is inclined downward at a slight angle. This provides for spinning the work to advance it along between the wheels.

The work is loaded in a suitable support B, extending outward from the wheels, as illustrated in Fig. 2, and this support passes through the machine to the driving side.

Each wheel is provided with a diamond dresser, which may be independently adjusted by means of a handwheel C, Fig. 1, as the diameters of the wheels are decreased through constant wear and truing. In dressing the regulating wheel, which, on account of not doing the cutting, wears slowly and consequently requires little truing, the handwheel shown at the front of the machine in Fig. 5 is operated to release a friction clutch, thereby increasing the speed of the regulating wheel during the dressing operation. A graduated collar facilitates the setting of the wheel-truing device, and a lever arm or handwheel is used to pass the diamond over the face of the wheel by turning the gear segment D, Fig. 1, which engages a rack for this purpose.

The normal operating speed of the grinding wheel is 1100 revolutions per minute, and this speed is constant, while the maximum speed employed for the regulating wheel is only 28 revolutions per minute. The regulating wheel is driven from a cone pulley within the frame of the machine, so that three speeds are available, although for the majority of the work the maximum speed is employed. The wheel speeds are calculated to give standard grinding speeds, and the changing of the regulating wheel speed is all that is necessary to give the correct grinding speed for work of different diameters and for roughing or finishing cuts.

If the work is not too long, it is usually fed into the support through a chute, as shown in the heading illustration, and for this class of work it is desirable to employ two operators, owing to the rapidity with which these parts are passed between the revolving wheels. A compound hand-

Fig. 1. Driving Side of Centerless Grinder, showing Grinding Wheel and Regulating Wheel Shafts and Wheel-truing Device

Fig. 2. Operating Side of Machine, with Grinding and Regulating Wheels exposed—also showing Fixture for grinding Rods

A2-5-3-2

wheel at the end of the machine, which may be seen by referring to Fig. 5, is used to adjust the positions of the regulating and grinding wheels relative to the work, by moving the slides on which the wheel-heads are mounted. During the operation of the machine a stream of coolant is directed down between the two wheels (see valve *E*, Fig. 2) directly on the revolving work. The work revolves clockwise, as indicated by the arrows in Fig. 3, so that the grinding grit passes down and is washed away.

Types of Work-supports

In order to describe the work-supports used, it is necessary first to broadly classify the work. Short cylindrical parts up to and including, say, 5 inches in length and 1½ inches in diameter can be conveniently handled by feeding them through a chute into a support of the type shown in Fig. 4; the longer shafts up to possibly 3 feet in length must be passed through horizontally by hand. This is illustrated in Fig. 2, where the support for handling long work is shown attached to the machine.

In general appearance, the work-supports are very similar. An inspection of Fig. 3 will show the construction and ad-

In the case of short cylindrical work, the fixture has no rolls, and is made similar to the central portion of the fixture for rods, but in this case the steel strip has an angular face inclining toward the regulating wheel and extending above the support about 0.005 inch. The vertical support itself is not made of sections, like the rod fixture, but the strip mentioned is composed of small casehardened steel sections. This construction is a decided advantage, because when wear occurs, the worn sections can be replaced without disturbing the remainder of the support. The same wedge arrangement shown in Fig. 3 is used for raising or lowering the vertical support. Either type of fixture may be quickly attached to the side of the machine and adjusted to the diameter of work to be ground.

Setting up the Machine

In setting up the grinder, after the work-support has been attached, the necessary adjustments of the wedge are made to bring the work to the proper height. To do this, the bolts that clamp the vertical supports in position are loosened, and the wedge moved along. It will be noticed that the wedge *A*, Fig. 4, is graduated to facilitate this set-

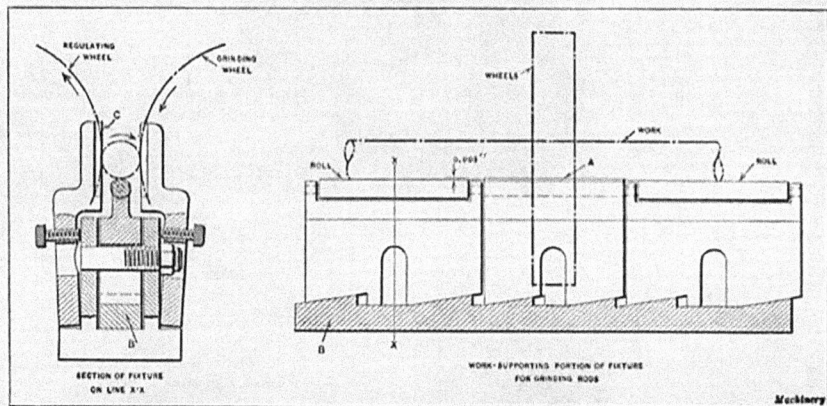

Fig. 3. Diagrams showing Construction of Work-holding Fixtures for Centerless Grinding and Relation of Work to Regulating and Grinding Wheels

justments incorporated in the fixtures for work of both classes. The two side members of these fixtures may be adjusted by means of the knurled screws shown on the sides, to give the proper opening for the diameter of work for which they are to be used. These side members have steel faces at the top on the inside, against which the work bears as it is fed through the machine.

The position of the work is indicated in broken outline as it rests between the steel faces, supported from beneath; also the relative position of the grinding and regulating wheels, as well as the directions of rotation, are shown. The under support may be either a roll or a stationary straight steel piece, depending on the work. If the work is long, the fixture contains at least three supporting sections. The middle section is located directly beneath the wheels and carries a steel strip *A* instead of rolls. These steel strips project slightly above their vertical support, and in the case of small-diameter work, the strip may be quite narrow at the top and slightly concave. The two end sections of the fixture carry rolls, the top surfaces of which are 0.003 inch above the strip *A*. The vertical supports for the rolls and for strip *A* are adjustable for height by means of a wedge member *B*.

ting. Then the side members of the fixture are brought together sufficiently by means of knurled screws on the sides, to permit the work to be passed through without obstruction.

The operator next adjusts the handwheel at the end, as shown in Fig. 5, to bring the grinding wheel against the work, which has been pushed through the fixture. The grinding wheel forces the work against the left-hand face *C*, Fig. 3, of the fixture, leaving a space between the work and the right-hand face of 0.003 inch in width. A feeler gage of this thickness is then passed between the work and the face of the fixture in the manner shown in Fig. 5. This is the allowance recommended so that the work will pass freely through the fixture.

It may be necessary sometimes not only to adjust the sides of the fixture, but also to move the vertical support sidewise slightly to bring it exactly central relative to the two wheels, but this is usually unnecessary after the fixture has once been set and clamped together. The work is next passed through by hand, and the regulating wheel brought into contact with it by turning the smaller of the two handwheels at the end. The machine is started and further adjustments made if necessary, to increase the contact

Fig. 4. Typical Cylindrical Work for the Centerless Grinder and Fixture used to guide Work of this Class

between the work and the regulating wheel enough to feed the work along.

When the regulating wheel has been set correctly, its face will extend in past face C, Fig. 3, about 0.003 inch. This amount, it will be remembered, is the same as that allowed for clearance between the work and the opposite face of the fixture in setting up. When the machine is set in this way, the work is ready to be ground and the operator need not make further adjustments until one cut has been taken on all the parts in a lot, at which time it is necessary to move the grinding wheel in for taking subsequent finishing cuts. Adjustment of the wheels is also necessary, of course, every time they are dressed.

Grinding Type-bed Rollers for Printing Presses

The Heim centerless grinders illustrated in this article are installed in the plant of the American Type Founders Co., Jersey City, manufacturer of Kelly printing presses. The rollers on which the type-bed of this press operates constitute a representative example of centerless grinding work. They are 1 inch in diameter, 1½ inches long, and must be ground uniformly and without taper to within a tolerance of 0.0002 inch on the diameter. There are thirty-six of these bed rollers on a machine, nine in a row, and they must be as near the exact size specified as possible so that the bed will not bear unevenly on them.

It is usual to allow about 0.006 inch on the diameter for grinding, and to employ three or possibly four cuts. In roughing, 0.004 inch is removed, and on the second cut 0.001 inch. The operator then examines his machine adjustments before taking the finishing cut. In cases where some of the parts do not come within the requirements, it is sometimes necessary to take a light fourth cut.

These rollers were formerly ground between centers at an average rate, for roughing and finishing, of 80 grinds per hour. When thus ground, centers had to be machined in the ends and a driving hole drilled, all of which is unnecessary with centerless grinding. Work of this size and type can be fed into a chute and passed through a centerless grinding machine at the rate of about 45 per minute, which is 2700 per hour for each pass, or a production of over 600 complete rollers an hour, allowing four passes per roller.

Finishing Rods by Centerless Grinding

The gripper rods used on Kelly printing presses represent an example of long work which must be fed through the machine horizontally by hand. These rods are about 0.50 per cent carbon steel and carry fingers for gripping the paper as it is passed over the press cylinder. The rods are shown in Fig. 2—one in the fixture and others on the shelf beneath it. They are ⅝ inch in diameter and about 25 inches long, and an absolute measurement of 0.625 inch on the diameter is the standard in grinding them. The same amount of stock, 0.006 inch, is left for finishing as on the rollers, and the same number of cuts is usually required.

A great deal more care must be exercised in performing the grinding operation on these rods than is necessary for the short rollers. In the first place, it has been found that irregularities on the work cannot be corrected by this method of grinding, and as a consequence the rods must be carefully straightened before each grinding operation, if a smaller tolerance than 0.002 inch is required.

There is another limitation that must be considered in grinding long work by the centerless method, and that is its weight. It has been found, when grinding heavy shafts, considerably more than an inch in diameter, that the weight is greater than the regulating wheel can drive without causing the work to lift. In fact, it would seem that this class of work is rather beyond the scope of centerless grinding, at least in its present stage of development, although there has been some heavy work ground on this style of grinder, 2 to 3½ inches in diameter and up to 39 inches long, using a fixture especially designed for this class of work. The production time for the gripper rods on a cylindrical grinder was forty-five minutes, as compared with from fifteen to eighteen minutes apiece by centerless grinding, for rods 25 inches long.

The speed and accuracy which can be obtained on rolls and similar work by centerless grinding constitute its chief advantage. The machines can be arranged for chute feed without difficulty, and a high production obtained. On the other hand, when long cylindrical parts are to be ground, it is necessary to take into consideration the extra work required between grinds in straightening, and the need of a helper, as well as a trained operator, for grinding this class of work. Although the production is high, even on rods and shafts, as compared with straight cylindrical grinding,

Fig. 5. Setting up the Machine, using a Feeler Gage between the Work and One Face of the Fixture

May, 1923

724 MACHINERY

the additional cost of a helper, special fixtures, and straightening, detract somewhat from the advantage of greater production.

While it is customary in the plant referred to in this article to use a helper and an operator on all work, it is possible for one operator to attend to a machine alone, and under a favorable arrangement of machines to attend to two. The work of the helper is largely that of attending to the feeding while the operator, working from the opposite side, is gaging and clearing away the fast accumulating rollers and other parts fed to the machine from the chute, to prevent them from interfering with the flow of work and to safeguard them from injury in falling into a box or other receptacle.

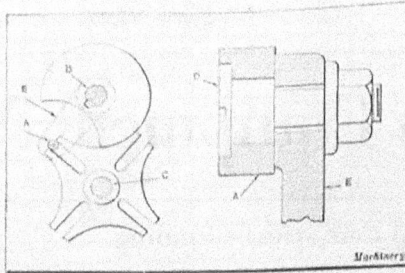

Diagram illustrating Use of Dummy Stud in assembling Geneva Wheels

finished stud, is then inserted in the arm E, an enlarged section of which is shown in the view at the right-hand side of the illustration. The roll A is placed in position on the dummy stud before the arm is assembled on its shaft. By turning the wheel and the arm, and adjusting the stud D, so that the roller is brought into exact alignment with the slot in the star wheel at its point of entrance, the position of the roller can be easily established.

The arm E is next removed from its shaft and set up in a lathe chuck, using an indicator for centering the roller A, which is clamped to the arm by the dummy stud D. The dummy stud is next removed, and the hole in the arm rebored to fit the shank of the finished stud.   P. R. H.

* * *

ASSEMBLING GENEVA STOP-WHEEL MECHANISM

In order to insure the proper operation of a Geneva stop-wheel mechanism, such as shown in the view at the left-hand side of the illustration, it is necessary that the

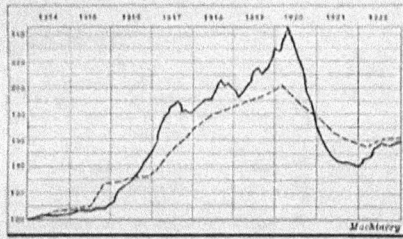

Fig. 1. Comparison between the Price of a Line of Machine Tools and the Price of General Commodities

driving roller A be very accurately located on its arm. It is common practice for the draftsman to specify the distance between the centers of the arm hub B and the roller A in thousandths of an inch, and the shop man takes particular care to work closely to this dimension. When this method is employed, it is essential that an equal degree of accuracy be maintained in the center distance between the hub B of the arm and the shaft C on which the star wheel is mounted, since any error in this dimension would render useless the care taken in locating the roller on the arm. When jigs are not used, time can be saved by working to a nominal dimension for the center distance between hub B and wheel shaft C, and then locating the roller on its arm in the manner described in the following.

The stud hole in the arm is first located by scale measurement and bored 1/16 inch under size. A dummy stud, such as shown at D, having a shank ⅛ inch smaller than the

PRICES OF MACHINE TOOLS AND COMMODITIES COMPARED

In order to show in a graphic manner the relation between machine tool prices and the prices of general commodities, the builder of a well-known line of machine tools

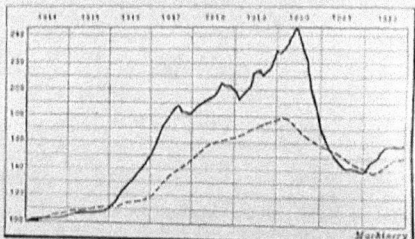

Fig. 2. Chart showing how a Line of Machine Tools increased 80 Per Cent in Price, as compared with 145 Per Cent for Commodities

prepared the three charts shown herewith. In these charts the full lines show the price fluctuations in general commodities, based on the wholesale prices as published by the United States Chamber of Commerce. These prices include the cost of clothing, farm products, house furnishings, etc.—briefly, the necessities of life. The dotted lines in each chart show the price of one specific line of machine tools built by this manufacturer. It will be seen that during the entire period of 1917, 1918, 1919, and 1920, when prices of other products mounted to such high levels, the prices of machine tools were far behind the cost of general merchandise. It will also be seen that at the end of 1922, the average prices of machine tools were slightly below the average price level, as far as the machines produced by this manufacturer were concerned, because one line of his product is ten points below the average, while the other two lines are only two points and four points above.

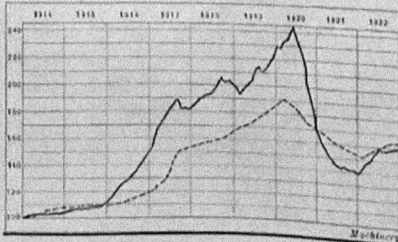

Fig. 3. Diagram showing how the Price of a Line of Machine Tools ran far behind the Average Level for Five Consecutive Years

Shop Equipment News

Heim Improved Centerless Grinding Machine

An improved centerless grinding machine, as shown in the illustration, has recently been brought out by the Heim Grinder Co. of Danbury, Conn.

Besides round work of uniform diameter, the machine will grind tapers, shouldered pieces, irregular contours and parts that do not present a complete circumference; as twist drills that have been fluted and relieved.

The base of the machine is of the cabinet type, housing the main shaft from which the grinding-wheel spindle is driven, and also a differential countershaft that is one of the factors in determining the rate of feed. The machine may be driven by either motor or belt.

An endless belt within the base drives the grinding wheel at a uniform speed. The differential unit is belt-driven from the first shaft through a pair of cones, and is provided with an internal-gear speed-reduction device. From this unit, the spindle of the second, or "regu-lating" wheel is driven at a speed that may be varied to suit the nature of the work. To otain a high speed for dressing the wheel, the speed-reducing device may be thrown out of action by means of a conveniently located handwheel.

Both wheel spindles are carried in slides bearing upon the upper surface of the machine bed and capable of movement horizontally at a right angle to the center line of the wheel spindles. These slides are independent of each other and are actuated by means of screws operated by hand-wheels. The position of the slides governs the diameter of the work, and the smaller handwheel is graduated to assist in maintaining accuracy in this respect.

The spindle of the regulating wheel may be inclined in a vertical plane to any angle from 0 to 6 degrees. The angle of inclination is one factor determining the feed in "through" grinding; the other is the rotative speed of the regulating wheel.

The spindles run in split-sleeve bearings of bronze, tapered upon the outside and fitted with ring adjusting nuts to compensate for wear. The journals of the grinding-wheel spindle are longer than the bearings and provision is made for automatically oscillating the spindle endwise for securing a smooth surface upon the work and to prevent the wheel from wearing unevenly. Ball thrust-bearings, fixed in the case of the regulating wheel spindle and adjustable in the case of the grinding wheel spindle, confine the spindles against endwise movement; the former at all times and the latter when the shape of the work will not permit such movement.

Upon each of the main slides there is mounted a wheel-truing fixture consisting of two members, the lower one resting upon the top of the wheel slide and pivoted so that it may be swiveled in a horizontal plane through an arc of 10 deg. to either side of the central position.

On the inner end of the upper slide is mounted a circular diamond-carrying bar having a movement at a right angle to that of the slide. One end passes inside the wheel housing, and at the other end is a stud that bears upon a formed guide-plate which determines the contour produced by the diamond upon the face of both regulating and grinding wheels.

The diamonds are carried in adjustable holders and are always in position for immediate use, so that a cut may be taken across the face of either wheel without disturbing the set-up. Each holder is connected to the coolant system.

The upper face of the rest for grinding straight work is a rectangular strip of hardened steel and is so held that it can be turned to bring any one of its four faces uppermost. When all faces show signs of wear it may be redressed, or a new strip substituted. The rate of feed may be adjusted from zero to 80 ft. per min., depending upon the speed of the regulating wheel and the angle to which its axis is inclined.

Various fixtures are provided to hold tapered or shouldered work.

Heim Improved Centerless Grinding Machine

A2-5-4-2

940   AMERICAN MACHINIST   Vol. 61, No. 24

Provision is made to handle work automatically, the fixture in this case being actuated by a mechanism on the back of the machine and driven through helical gears from the rear end of the regulating wheel spindle.

In grinding parallel shouldered-work, the faces of both wheels are trued parallel with their axes and a slight oscillation of the grinding wheel spindle is permissible. For tapered work, the face of the regulating wheel is beveled to the included angle of the taper. The spindle of the grinding wheel may be allowed to oscillate, but no endwise movement of the work can be permitted.

To grind work of irregular shape, both wheels are dressed to the corresponding contour, and the upper face of the wearing strip must be machined accordingly. No end movement either of work or wheel spindle may be allowed.

The grinding wheels are inclosed in a housing that effectually confines the cooling medium. The front of the housing is closed by easily removable covers, permitting ready access to the wheels for changing. A coolant tank, provided with straining and settling chambers, is bolted to the base of the machine in such position that the liquid drains from the hood by gravity. A centrifugal pump furnishes the means of circulation.

The machine covers a floor space of 3 ft. 9 in. by 7 ft., and weighs 4,850 lb. net.

Long & Allstatter No. 2 Combination Shear and Punch

A punching and shearing machine for general service is being brought out by the Long & Allstatter Co., of Hamilton, Ohio. It is claimed that this machine performs effectively the following operations: punches holes in plates, angles, structurals, and other shapes; shears angles (square or mitered), rounds, squares and flats; and split plates. It is not necessary to change tools for any of the operations mentioned because the punching tools, as well as the various blades, are always set up and ready for use.

The main frame is an annealed casting. The shear slide is of steel and slides in bronze liners. The punch slide has a lowering device which enables the operator to gage more closely the punch with the work. Blades have four cutting edges, and are made of tool steel of plain shapes for the purpose of lowering the cost of renewal. The changing of blades, when necessary, is easily and quickly done. The gearing is of steel with machine cut teeth. All shaft bearings are bronze bushed.

Sizes with capacity for punching ⅞-in. round holes through ⅞-in. material, up to 1¼-in. holes through 1-in. material, with the shearing in proportion, will be manufactured.

Buhr "Multi-Head" Drilling Machine

The J. F. Buhr Machine Tool Co., Roosevelt Square at 14th and Dalzelle, Detroit, Mich., has recently placed on the market a drill press to

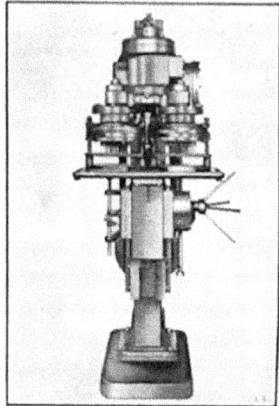

Buhr "Multi-Head" Press

be known as the "Multi-Head." It was designed for the drilling of stove burners but can be used for other kinds of work.

The machine illustrated is equipped with two multiple heads. A single head can be used if desired. Ball bearings are provided throughout, and the machine can be furnished semi or fully automatic, with belt, gear or motor drive.

Long & Allstatter No. 2 Combination Shear and Punch

New Machinery and Tools

The Complete Monthly Record of New Metal-working Machinery

HEIM CENTERLESS GRINDING MACHINE

Hydraulic control of the regulating-wheel slide is one of the important new features of an improved centerless grinding machine recently developed by the Heim Grinder Co., Danbury, Conn. This control is used in spot or in-feed grinding, eliminates all waste time in the movements of the regulating-wheel slide and thus increases the rate of production. There is also a hydraulic control of the truing diamonds by means of which the diamonds may be fed uniformly either at a fast rate across the wheels or with an almost imperceptible movement. Other improvements embodied in the machine include Texrope drives to both the regulating wheel and the grinding wheel, a quick-change gear-box, visible settings and simple adjustments. The machine has a greater capacity than previous models, accommodating work from 1/32 to 6 inches in diameter. In through-pass grinding, bars of any length can be handled, and in in-feed or spot grinding, work up to 8 inches under the head can be ground. The operation of the machine is, of course, based on the basic Heim principle of centerless grinding.

From a 15-horsepower motor mounted on the right-hand end of the base the drive is through a silent chain to a drive shaft the rotation of which is controlled by means of a Carlyle-Johnson friction clutch. This clutch may be conveniently engaged or disengaged by operating the foot-treadle on the front of the machine. The motor is mounted on a hinged bracket which permits of adjusting the tension of the driving chain. On the drive shaft is a 5-groove sheave which is connected to a similar sheave on the grinding-wheel spindle by means of five Texrope strands. There is also a sprocket on the drive shaft which, through a silent chain, transmits power to the quick-change gear-box and to the pressure-creating pump of the hydraulic system. On a shaft driven from the gear-box there is a 3-groove sheave which, through Texrope strands, drives a similar sheave that is mounted on the regulating or work-rotating wheel-spindle.

Power is transmitted by sprockets and a chain from the gear-box to a shaft along the back of the machine near the top. Power is delivered to a cam and lever mechanism at the right-hand end of this shaft, which mechanism actuates the hydraulic control valve for the regulating-wheel slide. By means of a simple valve adjustment, strokes of various lengths and speeds can be obtained for the regulating-wheel slide, and by shifting the position of the levers on the front of the gear-box, either 2, 3, 5, 9, 15 or 25 reciprocations of

the slide are obtainable per minute. There are seven regulating-wheel speeds, including the truing speed, of 8, 13, 21, 37, 62, 100, and 740, revolutions per minute. The speeds and feeds of the regulating wheel cover every grinding condition on work of any material of a size within the capacity of the machine.

In in-feed or spot grinding, the cam that controls the hydraulic system gives a slow feed to the regulating-wheel slide until the wheel reaches the position for grinding to the required diameter. The wheel dwells for a short period in this position and is finally given a quick return to the starting position. The hydraulic cylinder for the regulating-wheel slide is mounted on the inside of the base, and the piston in the cylinder acts as a feed-nut for the slide adjusting screw. This screw is fastened to the slide. An adjustable stop on the slide is set to contact with a stationary stop on the front of the bed, when the regulating wheel reaches the proper grinding position. As this occurs, the hydraulic pressure is by-passed through a relief valve. On the control valve, an adjustment regulates the volume of fluid entering the valve, permitting a variation in the speed of the slide travel. The adjustment is on the plunger that is contacted by the roll of the cam and lever mechanism previously mentioned.

When the machine is set up for through-pass grinding, the piston in the hydraulic cylinder of the regulating-wheel slide, is held in contact with one end of the cylinder so that it acts simply as a fixed feed-nut, which permits adjustments of the slide. For in-feed or spot grinding, the grinding-wheel slide is locked to the base by operating a cam-lever located on the same end of the base as the motor. This lever can be quickly released when adjustments of the grinding-wheel slide are required.

The diamond truing devices are reciprocated automatically across the faces of the wheels by the hydraulic system, as already mentioned. A movable stop on each device provides for controlling the length of traverse of the diamond and the rate of the traverse is adjustable by means of a valve. The two diamond truing valves are so arranged that the truing may be operated singly or in unison. The truing diamonds may be conveniently adjusted for taper work and may be equipped with a guide plate for special form work.

The work-holding fixture is rigidly mounted on a slide that is adjustable on the base of the machine. The proper position of the fixture for grinding work of any diameter is readily obtained by means of an adjusting screw. Settings are facilitated by a graduated scale on the fixture slide. The bar on which the work is fed is provided with

Fig. 1. Heim Centerless Grinding Machine with Hydraulic Control of the Regulating-wheel Slide and Truing Diamonds

A2-5-5-2

a surface of stellite or other wear-resisting metal. A single fixture is suitable for feeding a wide range of work. The fixture used in spot grinding is provided at the rear of the machine with an adjustable bar that is cam-actuated to eject work as it becomes finish-ground.

Means are provided for oscillating the grinding-wheel spindle slightly in spot grinding when an extremely fine finish is required. This oscillation is derived from the mechanism that also actuates the regulating-wheel slide; hence, the movements of both wheels are made in the proper sequence. The vertical lever at the front of the machine is operated to change the rotation of the regulating wheel quickly from a working speed to a truing speed. Both wheel spindles are provided at the rear end with a self-aligning end-thrust ball bearing. These bearings are completely enclosed and constantly lubricated. The positions of the grinding wheel and the regulating wheel can be closely adjusted by means of a handwheel at the outer end of their respective slides. Both handwheels are graduated on their rims to thousandths of an inch and as the graduations are about 5/16 inch apart, subdivisions of 0.001 inch can be accurately made. On previous machines, the regulating wheel and the grinding wheel were of the same diameter, but on this machine the grinding wheel is 24 inches in diameter and the regulating wheel, 14 inches. Sleeves and flanges are designed to take wheels of any face up to and including 8 inches. The grinding wheel has a bore of 12 inches.

The spindle of the regulating wheel can be swiveled from one degree above horizontal to six degrees below, to permit changing the speed at which work is fed across the grinding wheel. The slide and head of the regulating wheel are provided with a graduated scale and an index line to facilitate setting the head to the desired angle. All units, such as the quick-change gear-box and the pressure tank of the hydraulic system, are contained within the base. They are mounted on the walls of the base in such a manner that any unit can be removed intact after taking out a few bolts. Coolant is delivered to the wheels and work by a centrifugal pump mounted on a bracket fastened to the front wall of the base. All coolant is drained from the work to the tank. This tank is a separate unit, mounted on wheels and readily removed for cleaning. It is provided with three settling compartments that insure the pumping of only clean liquid back to the work. Telescopic guards on the wheel-slides, and a trough which extends across the machine, exclude all coolant from the inside of the base.

For use in grinding long bars, outboard supports are provided to supplement the work-holding fixture. These

Fig. 3. View of Centerless Grinding Machine and of Rotary Outboard Support which enables Long Bars to be ground

supports are adjustable vertically and are equipped with sets of rollers mounted on laterally adjustable brackets. These rollers are designed to overcome the resistance due to inertia of the bars being fed into the machine. Adjustments are provided on these rollers so that they may be swiveled to the angle at which the regulating wheel of the machine is inclined. Graduations enable the easy accomplishment of these settings. The weight of this machine without the motor is approximately 7500 pounds, and the floor space occupied by the machine is 5 by 9 feet.

AMERICAN "AUTO-OILED" SHAPER

One of the features of a new shaper recently brought out by the American Tool Works Co., Cincinnati, Ohio, is that all bearings and gears, all parts of the bull wheel and rocker arm, the ram guide ways, and the multiple-disk driving clutch are constantly flooded with filtered oil in such a volume as to carry away bearing heat and any foreign matter that may be deposited on them. The oil of the system is taken from a large reservoir in the base. This reservoir is provided with a settling compartment in which the heavy sediment of the oil returned to the tank, is deposited. From a second compartment into which the settled oil flows, a plunger pump forces the oil to a filtering and distributing tank at the top of the column. In this tank, the oil is forced by pump pressure through felt pads that remove all dirt or other foreign matter, after which the clean oil is distributed under pressure to the parts mentioned.

Another feature of the shaper is the centralized control. From his natural working position, the operator can lay his hands upon the primary control lever that starts and stops the machine. From the same position, he can also easily reach the speed-change lever that controls the speeds obtained through the gear-box; a lever that engages and disengages the back-gears; and the lever located at the operating end of the cross-rail, which is manipulated for engaging, disengaging, and reversing the feed.

Another important improvement is in the method of changing the length of stroke. Instead of the three operations of unclamping, setting the stroke, and reclamping, required on previous machines for changing the stroke, only one operation is necessary on the new machine. In applying a crank to the end of the stroke-adjusting shaft, a clutch lock is automatically released. When the stroke adjustment has been made, the

Fig. 2. Rear View of Centerless Grinding Machine with Coolant Tank in Place

BULLETIN No. 110

THE HEIM

CENTERLESS CYLINDRICAL GRINDER

MANUFACTURED BY

THE BALL AND ROLLER BEARING COMPANY,

DANBURY, CONNECTICUT, U. S. A.

A2-6-1-2

THE HEIM CENTERLESS CYLINDRICAL GRINDER

For Grinding Cylindrical Parts Without Centers

Such as Rollers for Roller Bearings, Wrist Pins, Cam Shafts, Valve Lifters, Valve Lifter Roller Pins, Shackle Bolts, Spring Bolts, Roller Chain Studs, Rods, Tubes,

In fact most any cylindrical piece where there is only one diameter to be ground; diameters may range from $\frac{1}{8}$" to 3" and lengths from $\frac{1}{8}$" 15" Of course much greater lengths can be ground, and also larger diamters; this, however, will, to some extent, depend on the nature of the work, and should be taken up with our engineering department.

Side and Rear Elevation
Showing wheel spindles mounted on the slides. Drive pulley and opening in rear of machine.

This machine is original in design employing exclusive features. Protected by patents in this and foreign countries, on both METHOD OF GRINDING AND MECHANISM.

The machine is self-contained and substantially built, with the highest quality of workmanship.

The grinding wheel and regulating wheel spindles are driven from a main drive shaft which is mounted on radial roller bearings and located beneath the wheel spindles in the main part of the frame.

The slides in which the grinding and regulating wheel spindles are journaled are slidably mounted on the bed of the machine, and are moved to and away from the work by hand wheels at the end of machine.

The wheel spindles are made from a special alloy steel, hardened, tempered and ground, and run in phosphor bronze bearings of liberal proportions. The bearings are designed so as to provide auomatic lubrication for the spindles and their design is such that grit and dirt is absolutely excluded, thus insuring long life. Lateral thrust on the spindles is against ball bearings.

The pump is of the centrifugal type integral with machine and with no stuffing boxes to leak or get out of order.

The design of the machine is such that one operator can take care of at least one machine, and in some instances two or more, as the work is automatically returned to the front of the machine, directly adjacent to the operator.

THE HEIM CENTERLESS CYLINDRICAL GRINDER

Front Elevation of Machine
With door removed from wheel housing exposing wheels

There are a number of features in this machine which are absolutely essential for the successful grinding of parts not ground on centers and which enables us to secure :

A high degree of precision,
Perfectly round work,
Perfectly straight work,
and
A very high degree of finish.

These features are covered by Letters Patents. The machine is provided with 16-in. grinding and regulating wheels. The diamond truing device trues both wheels parallel at the same time.

A2-6-1-4

THE HEIM CENTERLESS CYLINDRICAL GRINDER

The machine is manufactured under the following U. S. Letters Patents :

No. 1,111,254	No. 1,264,930
No. 1,210,936	No. 1,278,463
No. 1,210,937	No. 1,281,366

Other patents applied for.

There are several manufacturers, who as we understand, have started to place machines on the market, embodying many of the patented features of this machine. We give warning that we intend to protect our rights to the limit, whether the infringement is by reasons of manufacture, or use of such machines.

Accessories:

One work support to suit diameter of work to be ground. (We can furnish additional work supports at nominal charges ; price varying in accordance with the diameter of work to be supported on same.)
One dozen work wearing strips.
Water Tank.
Suspended basket in tank for receiving work.
Pump (centrifugal type) for supplying water to wheels.
Double diamond truing fixture without diamonds. (We can furnish a set of two diamonds, mounted, at an extra charge. One diamond truing fixture will answer for a half dozen machines in truing the wheels at intervals.)
Two grinding wheels, 16-in. dia., 4-in. face, 5-in. hole of proper grain and grade to suit work to be ground. (We carry a complete stock of grinding wheels suitable for rough grinding or finish grinding, of both hardened or soft work, at prices in line with those which you would receive from the wheel manufacturers.)

Specifications:

Floor space 78" x 45"	Net Weight 4,000 lbs.
Height 50"	Weight Crated 4,500 lbs.
14" diameter Drive Pulley for 6" belt	Boxed for Export 5,000 lbs.
Speed of Driving Pulley 575 R. P. M.	Dimensions boxed for export 82" x 45" x 56"

Manufactured and Sold by

The Ball and Roller Bearing Company,

Main Office and Factory :

DANBURY, CONNECTICUT, U. S. A.

333

BULLETIN No. 140

The Heim
Centerless Cylindrical Grinder

Manufactured by

THE BALL AND ROLLER BEARING COMPANY,

Danbury, Connecticut, U. S. A.

A2-6-2-2

The Heim Centerless Cylindrical Grinder

Some of the Important Features Incorporated in the Heim Centerless Cylindrical Grinder

Rigidity and Simplicity.

Both wheels mounted on heavy spindles made of special spindle steel, hardened and ground and mounted in phosphor Bronze bearings which are housed in taper pockets, these pockets are machined in the wheel slides and are provided with oil wells.

The slides in which wheels are mounted are operated from one end of the machine by graduated hand wheels, each independently toward or away from the work.

The ways in which these slides move are entirely protected from grit and water, as are all movable parts.

We provide six inch drive belts to both wheel spindles and drive from underneath; they are provided with idler pulleys mounted on anti-friction bearings for taking up slack of belts. The heavier the cut taken the greater is the pull downward on the slides, providing extreme rigidity.

A wheel diamonding fixture is mounted on each wheel slide for dressing the wheels either straight or taper; can also be arranged for dressing the wheels concave, convex, or for form grinding

Diamonding of the wheels can be effected in a few moments without disturbing the work support or work being ground

Speed of the work controlling wheel can be changed from controlling speed to diamonding speed by merely turning a hand wheel and vice-versa.

A chute is provided which facilitates feeding work into the grinder; also a chute is provided on the discharge side of the wheels over which the work can pass to the rear of the machine, or this chute can be removed and the work will return to the front of the machine.

Three speeds are provided for the controlling wheel for both feeding and diamonding purpose.

Drive shaft is mounted on radial roller bearings.

Machine will grind all diameters up to and including 4".

Certain diameters above 4" may be ground but special work fixtures must be provided.

Grinding wheels are securely enclosed in steel wheel housing with doors provided for easy access.

A centrifugal water pump is provided having no stuffing boxes and positively no leaks.

It is not necessary to take machine apart to remove or place wheels on spindles.

Wheels can be removed and replaced in a few moments.

Some of the Cylindrical Parts
That can be Ground on The Heim Centerless Cylindrical Grinder.

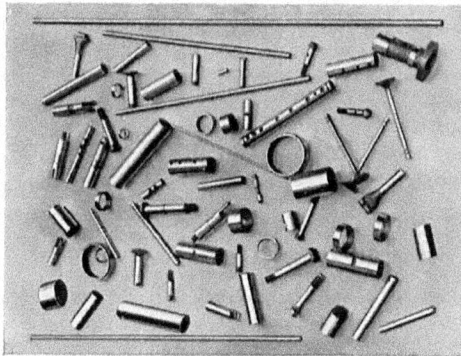

Piston Pins
Valve Lifters
Valve Lifter Rolls
Valve Lifter Roll Pins
Head Work such as
  Spring and Shackle
  Bolts, King Bolts
Valves
Pieces with two or
  more steps
Drills, Back Taper
  and straight
Taper Work
Concave or Convex
  Work
Pistons
Radial Ball Bearing
  Races
Radial Roller Bearing
  Races
Journal Roller Bearing Casings and
  Sleeves
Rollers for Roller
  Bearings

Bronze Bushings
Tubing
Roller Chain Studs
Roller Chain Rollers
Screw Driver Stock
Drill Blanks
Pipe Balls
Hardware Casters
  (Cast Iron or Pressed Steel)
Vulcanized Rubber
  Rods
Vulcanized Rubber
  Tubes
Cast Iron or Steel
  Pulleys to capacity
  of machine
Universal Joints

And many other kinds of cylindrical work where there is but one diameter to be ground.

We solicit blue prints with limits and samples of work for estimating purpose as to whether such work can be ground by the centerless process.

The Heim Centerless Cylindrical Grinder

Table Showing Approximate Lineal Feet at Which Work Will Pass Through Machine, also Approximate Amount of Stock that can be Removed from various Diameters per Pass.

Diameters	Less than ½"	½" to 1"	1" to 1½"	1½" to 2"	2" to 4"
Work advances per minute	6 feet	5½ feet	5 feet	4½ feet	4 feet
Rough Grinding stock removed per pass	.008" to .010"	.006" to .008"	.004" to .006"	.003" to .004"	.002" to .003"
Finish Grinding stock removed per pass	.0005" to .001"	.0003" to .0005"	.0002" to .0003"	.0002" to .0003"	.0002" to .0003"

Rough Grinding

Requires two to three passes to rough and straighten work.

Finish Grinding

Requires two to three passes and will finish grind work, producing a very high degree of finish and hold the work within close limits, for instance .0002" if the work requires these limits.

Accessories :

One rack.

One dozen work wearing strips.

Water Tank.

Settling tank.

Pump (centrifugal type) for supplying water to wheels.

Wheel diamonding fixtures without diamonds. (We can furnish a set of two diamonds, mounted, at an extra charge.)

Two grinding wheels, 16-in. dia. 4-in. face, 5-in. hole of proper grain and grade to suit work to be ground. (We carry a complete stock of grinding wheels suitable for rough grinding or finish grinding, of both hardened or soft work, at prices in line with those charged by wheel manufacturers.)

Can of grinding Compound.

Complete set of wrenches.

The machine is manufactured under the following U. S. Letters Patents :

Patented Sept. 22, 1914. No. 1,111,254	Patented May 7, 1918. No. 1,264,930
Patented Jan. 2, 1917. No. 1,210,936	Patented Sept. 10, 1918. No. 1,278,463
Patented Jan. 2, 1917. No. 1,210,937	Patented Oct. 15, 1918. No. 1,281,366
Reissued Jan. 25, 1921.	

Other patents pending and applied for.

There are several manufacturers, who are placing machines on the market embodying many of the patented features of this machine. We give warning that we intend to protect our rights to the limit, whether the infringement is by reasons of manufacture, or use of such machines.

A2-6-2-4

The Heim Centerless Cylindrical Grinder

Specifications:

Floor space 78" x 45"
Height 50"
14" diameter Drive Pulley for 6" belt
Speed of Driving Pulley 560 R. P. M.

Net Weight 4,500 lbs.
Weight Crated 4,750 lbs.
Boxed for Export 5,000 lbs.
Dimensions boxed for export 82" x 45" x 56"

Manufactured by

THE BALL AND ROLLER BEARING COMPANY,
DANBURY, CONNECTICUT, U. S. A

HEIM

Centerless Grinding Machine

Will grind straight cylindrical work up to 4" in diameter, 18" long, shoulder work up to 4" diameter, 8" long under the head, also taper and formed pieces.

Straight cylindrical pieces pass through the machine from 4 to 10 ft. per minute per pass, depending on diameter, amount of material removed, and accuracy required. Rubber can be ground at 70 ft. and small rods at 30 ft. per minute. Depth of cut of course varies inversely with the diameter being ground, from .032" on 1/4" to .004" on 2" pieces. Accuracy of .0001" for roundness, straightness, and size can be obtained when necessary.

Shoulder pieces are spot ground from 2 to 20 pieces per minute.

Production estimates gladly furnished from samples or blue prints.

THE HEIM GRINDER CO.
DANBURY, CONN., U. S. A.

A2-6-3-2

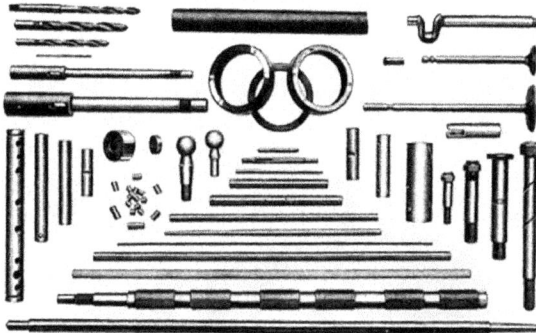

Samples of Work Ground on the Heim Centerless Grinder

SPECIFICATIONS

Wheel Spindles: Chrome nickel steel, heat treated (75 Scleroscope)
and ground.

Spindle Boxes: Phosphor Bronze supported in taper housings.

Grinding and Regulating Wheels: 16" diameter, 5" hole, 3/4" to 8" widths.
4" face wheels mounted and balanced furnished as standard equipment.

Wheel Balancing: Balancing arbor and stand with first machine.

Water Supply: Ample tank and rotary pump.

Speeds: Machine operated through driving shaft running 560 R.P.M.

Drive: We recommend chain motor drive, with 15 H.P. 1200 R.P.M. motor.
Suitable motor costs $182.00 and a hand compensator $76.00. We
will purchase and mount; or prepare machine for mounting motor,
if furnished with motor dimension sheet. Motor base plate,
sprockets, silent chain and guard included in standard equipment.

Floor Space: 6' 6" x 4' 4".

Approximate Weight: Net 5500 lbs. Crated 5750 lbs. Boxed 6150 lbs.

PRICE

(Subject to change without notice)

Heim Centerless Grinder with 16 x 4 x 5 Grinding
and Regulating Wheels, arranged for Silent Chain Motor
Drive, (without motor or starter), or arranged for
Belt Drive, including Roller Bearing Countershaft........$ 4000.00

Fixtures for different classes of work vary in
cost, but average.....................................$ 200.00

HEIM
Centerless Grinding Machine

THE HEIM GRINDER CO.
DANBURY, CONN., U. S. A.

Cable Address "HEIM"
LIEBER'S CODE

CIRCULAR HG1

A2-6-4-2

THE HEIM GRINDER CO., DANBURY, CONN.

THE HEIM CENTERLESS GRINDING MACHINE

The above cut shows the machine arranged for motor drive and set up for
spot grinding. The cut on the front cover shows the belt
drive machine set up for *through* grinding

[2]

THE HEIM GRINDER CO., DANBURY, CONN.

THE HEIM CENTERLESS GRINDING MACHINE

The Heim Centerless Grinder is the pioneer in the centerless grinding field, and the present design is the product of more than ten years' consistent study and development of this field on the part of Mr. Lewis R. Heim and his associates in the Ball & Roller Bearing Co.

The Heim Grinder Co., which was organized March 1st, 1924, will continue this development work to fully and efficiently cover every line of production to which the centerless method of grinding is applicable.

In designing grinding machinery the experienced engineer endeavors to incorporate features which shall make for maximum rigidity and in this respect the Heim Grinder is no exception.

The spindles on which the regulating and grinding wheels are mounted are of proven design and made of the best quality spindle steel obtainable. They are self-oiling and run in adjustable bronze boxes which fit taper holes in the horns of the heavy wheel slides. Flat and vee ways on the wheel slides, resting on corresponding surfaces on the well-proportioned base of the machine, provide ample support for these units.

This construction together with the weight of these units, in combination with the down-pull of the driving belts, results in a rigid, dependable machine capable of turning out a maximum production of work which will meet the most exacting specifications as to limits and quality of finish.

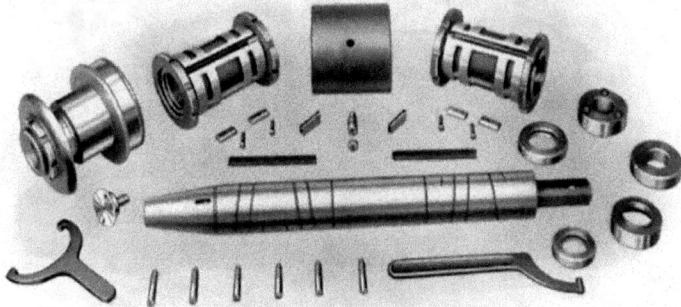

Wheel Spindle Details

While developing a machine of rigid construction, there has always been kept in mind the desirability of simplicity in the design, and the following points will serve to show how well this has been accomplished.

Wheel Housing. The wheel housing which is of ample strength to confine the wheel in case of breakage, is an independent unit and bolts solidly to the machine base. The standard housing accommodates wheels up to 6" face; a special housing can be supplied for wheels up to 8" face.

A properly designed shelf or bridge in the lower part of the housing provides a rigid support for the work fixture. Removable covers give access to the wheels, diamonds and work fixture.

Wheels. The wheels which are 16" diameter and may be any thickness up to 8", are quickly mounted or removed as they are carried on sleeves which fit the tapered ends of the spindles.

Diamond Fixtures. The diamond fixtures are mounted on the wheel slides and by using suitable guide plates the wheels can be trued for either straight, taper or form grinding.

[3]

A2-6-4-4

THE HEIM GRINDER CO., Danbury, Conn.

Wheel Slides. The regulating and grinding wheel slides are traversed independent of each other by means of feed screws operated by the hand wheels at the right hand end of the machine. Graduations in thousandths provide for the finest adjustments.

Fixed Head. In the "Fixed Head" machine the regulating wheel slide is arranged with the regulating wheel spindle set at a fixed angle of 3° in the vertical plane with relation to the grinding wheel spindle, and this fixed position of the regulating wheel has been found entirely satisfactory on the majority of work which can be ground on a centerless machine.

Swivelling Head. When a machine is required for general utility purposes, the regulating wheel slide is equipped with a swivelling head which provides for an angularity of the regulating wheel spindle with relation to the grinding wheel spindle of from 0 to 6°. This range of angularity permits the setting of the regulating wheel to produce the desired speed of work past the face of the grinding wheel; the speed of work varying from 0 to about 80 feet per minute depending on the angle employed, the speed of the regulating wheel and the diameter of the work.

Sizing Work. In the Heim machine adjustments for sizing the work are easily and quickly accomplished.

Since both the regulating and grinding wheel slides are movable, the work fixture need not be disturbed once it has been properly located on the wheel housing. By means of the hand wheels the regulating and grinding wheels are brought into proper relation with the work fixture and each other.

Once the correct position of the regulating wheel has been established, the size of work is maintained by adjustment of the grinding wheel only, to compensate for wear and truing.

No. 3 Standard Work Fixture for *Through* Grinding

Work Fixtures. All work fixtures bolt solidly to the bridge of the wheel housing and the exchange of one fixture for another is quickly accomplished.

Fixtures for *through* grinding are equipped with rests arranged to receive renewable hardened wearing strips, and the guide plates are easily adjusted to accommodate the diameter of work being ground.

No. 22 Standard Elevating Fixture for *Spot* Grinding

[4]

THE HEIM GRINDER CO., Danbury, Conn.

Fixtures for *spot* grinding are operated by a lever on the fixture, whereby the work is lowered between the wheels for the grinding operation, the wheels meanwhile being maintained in a fixed relation to each other.

All fixtures are adjustable for locating the center of the work in proper position relative to the center of the grinding wheel to produce maximum results.

Automatic Attachment. Where large production of *spot* grinding such as shackle bolts, king bolts, valves, etc., is required, we can furnish an attachment which provides for automatic traverse of the regulating wheel slide. This attachment, in combination with a fixed work rest, will materially increase the output and at the same time maintain the required accuracy. This attachment also furnishes means for oscillating the grinding wheel spindle, and in combination with the elevating fixture automatically lowers and raises the work rest and ejects the work from the fixture after grinding.

Automatic Attachment

Differential. A well designed differential unit inside the base provides the necessary range of speeds for the regulating wheel through step cone pulleys. By means of a small hand wheel at the lower left front of the machine the slow speed of the differential is cut out and the fast speed of the regulating wheel obtained for truing purposes.

Pump, Tank and Water Supply. A water tank of suitable capacity is provided and a centrifugal pump delivers an ample supply of coolant on the wheels and work. Properly designed water guards and troughs return this water to the tank which is easily accessible for cleaning and filling.

Belt Drive Machine. The machine is operated through a single speed driving shaft, running 560 R. P. M. A roller bearing countershaft can be furnished when a belt driven machine is ordered.

Motor Drive Machine. If a motor drive machine is required, a special motor base is attached to the right hand end of the machine and the motor connected to the driving shaft with sprockets and silent chain.

[5]

344

A2-6-4-6

THE HEIM GRINDER CO., Danbury, Conn.

Engineering Service. While we have already developed a considerable variety of special fixtures in equipping machines for our customers, each day opens up new fields for the centerless grinding machine and the opportunities for increasing production through continued developments in work-holding fixtures and manufacturing methods become more apparent.

This work is handled by experienced men in our Engineering Department, and we offer the services of this department to our present and prospective customers in the belief that with our experience in centerless grinding we can render a real service to the manufacturer who is interested in increasing his production on work to which the centerless grinding machine is adapted.

We solicit samples and blue prints of work under consideration, and no charge will be made for the preliminary study of these problems.

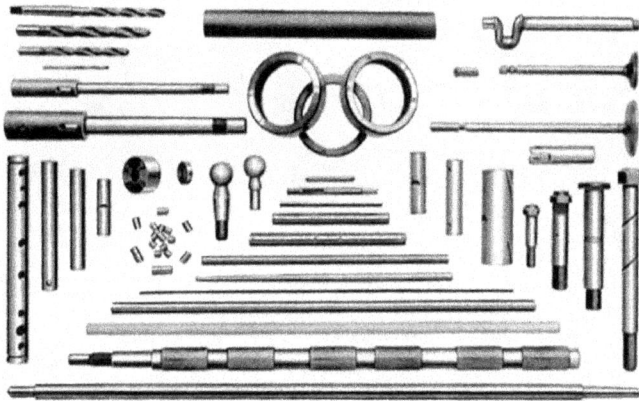

Samples of Work Ground on the Heim Centerless Grinder

The Heim Centerless Grinder is being built under the following patents issued in the name of L. R. Heim.

No. 1,210,936	No. 1,264,930	No. 1,278,463
No. 1,281,366	Reissue No. 15035	

The Heim Grinder Company has acquired the ownership of the centerless grinder patents issued and pending, formerly the property of Lewis R. Heim and The Ball & Roller Bearing Co., and is the successor to the centerless grinding business of the latter company.

Reissue Patent No. 15035 was recently sustained by the decision of the U. S. Circuit Court of Appeals, Second Circuit.

We also own the following patents covering features in centerless grinding machine design and issued in the name of M. O. Reeves.

No. 1,264,129	No. 1,410,956	No. 1,430,754
No. 1,440,796	No. 1,440,795	No. 1,456,462

We will fully enforce our rights under these patents.

[6]

THE HEIM GRINDER CO., Danbury, Conn.

The following list will be found convenient in selecting machines and equipment for various classes of work

Plain machine carrying wheels up to 6" face, but not including countershaft or attachments.

Plain machine carrying wheels up to 8" face, but not including countershaft or attachments.

Roller Bearing Countershaft.

Motor Drive (not including motor) arranged for silent chain between motor and driving shaft.

Regulating wheel slide arranged with swivelling head.

Automatic attachment for traversing the regulating wheel slide, oscillating the grinding wheel spindle, lowering and raising the elevating fixture work rest, and ejecting the work after grinding.

No. 3 Standard work fixture for *through* grinding.
Capacity, work up to 2" diameter, 7" length.

No. 3 Special work fixture for *through* grinding.
Capacity, work up to 2" diameter, 15" in length.

No. 4 Standard work fixture for *through* grinding.
Capacity, work from 2" to 4" diameter, 7" in length.

No. 4 Special work fixture for *through* grinding.
Capacity, work from 2" to 4" diameter, 15" in length.

No. 6 Special work fixture for *through* grinding.
Capacity, work up to 1" diameter, 6" in length.

No. 8 Standard work fixture for *through* grinding.
Capacity, work up to 2" diameter and over 24" in length.

No. 21 Special elevating fixture for drill grinding.

No. 22 Standard elevating fixture for *spot* grinding.
Capacity, work up to 1" diameter, $4\frac{3}{4}$" in length under head.
Capacity, with special rest, work up to 2" diameter, $4\frac{3}{4}$" in length under head.

No. 23 Standard elevating fixture for *spot* grinding.
Capacity, work up to 1" diameter, $7\frac{3}{4}$" in length under head.
Capacity, with special rest, work up to 2" diameter, $7\frac{3}{4}$" in length under head.

While the standard fixtures listed above will handle practically every class of work, it is necessary to equip these fixtures with suitable rests and wearing strips. It is therefore imperative that with the order, we receive blueprints or samples of the work for which the equipment is selected.

[7]

346

THE HEIM GRINDER CO., Danbury, Conn.

SPECIFICATIONS

Capacity: Work ranging in diameter from $\frac{1}{16}''$ to 4″ inclusive.

Wheel Spindles: Chrome nickel steel, heat treated and ground; front bearing $3\frac{1}{4}''$ diameter, $8\frac{7}{16}''$ long, rear bearing $3\frac{1}{8}''$ diameter, $8\frac{1}{4}''$ long.

Spindle Boxes: Phosphor Bronze and supported in tapered housings to provide adjustment for wear.

Grinding and Regulating Wheels: 16″ diameter, 5″ hole and can be furnished in widths from $\frac{3}{4}''$ to 8″ inclusive.

Wheel Balancing: A properly balanced grinding wheel is essential to the successful operation of a grinding machine, and with the first machine furnished a customer we include a balancing stand also an arbor which fits the standard wheel sleeve.

Wheel Truing: Each wheel slide carries an independent diamond fixture for truing the wheels for either straight, taper or form grinding. Diamonds not included.

Sizing of Work: Regulating and grinding wheel slides are operated by independent screws. Hand wheel graduated to read in thousandths.

Water Supply: Water tank of ample capacity. Centrifugal pump delivers 15 gallons per minute.

Pulleys and Belts: Regulating wheel spindle pulley 8″ diameter, $6\frac{1}{2}''$ face. Grinding wheel spindle pulley 7″ diameter, $6\frac{1}{2}''$ face. Driving belts 6″ wide, water and oil-proof.

Speeds: The machine is operated through a main driving shaft running 560 R. P. M.

Countershaft: A well designed roller bearing countershaft can be furnished with the belt drive machine.

Motor Drive: The motor should be 10 H. P. or 15 H. P., 1200 R. P. M., and is furnished by the customer. We require a diagram and the speed of the motor that we may supply suitable sprockets and silent chain which are included in the motor drive price.

Standard Equipment: One 16″ x 4″ x 5″ Regulating Wheel. One 16″ x 4″ x 5″ Grinding Wheel. 2 Wheel Sleeves. 12 Balancing Slugs. Can of Grinding Compound. Set of Wrenches. Work Chutes. Overhead work support. Belts on the machine.

Floor Space: Belt Drive Machine 6′ 6″ x 3′ 9″. Motor Drive Machine 7′ 0″ x 3′ 9″.

Weights (approximate):
Belt Drive Machine: Net **4700** lbs. Crated **5000** lbs. Boxed **6000** lbs.
Motor Drive Machine: Net **4850** lbs. Crated **5150** lbs. Boxed **6150** lbs.

Cubic Measurements: 85 x 56 x 57. 157 cubic feet.

9-24-4M

[8]

The Heim Grinder Company.   Danbury, Conn.   May 1, 1924.

TEMPORARY PRICE LIST

Heim Centerless Grinding Machines and Principal Attachments.

Plain Machine carrying wheels up to 6" face, but
not including Countershaft or Attachments--------------------$3200.00

Plain Machine carrying wheels up to 8" face, but
not including Countershaft or Attachments------------------- 3250.00

Roller Bearing Countershaft----------------------------------- 125.00

Motor Drive equipment (not including motor) arranged with
either Belt, or Silent Chain between Motor and
Driving Shaft-- 125.00

No. 3 Standard Work Fixture Complete, for work up to 2"
diameter 7" in length--------------------------------------- 80.00

No. 3 Special Work Fixture complete, for work up to 2"
diameter 15" in length-------------------------------------- 140.00

No. 4 Standard Work Fixture complete, for work ranging
from 2" to 4" diameter, 7" in length------------------------ 100.00

No. 4 Special Work Fixture complete, for work ranging
from 2" to 4" diameter, 15" in length----------------------- 150.00

No. 5 Special Work Fixture complete, for work up to 1"
diameter, 6" in length-------------------------------------- 100.00

No. 8 Standard Work Fixture complete, for work up to 2"
diameter and lengths ranging from 24" up, depending
on the work--- 225.00

No. 21 Elevating Fixture complete, for Drill grinding--------- 200.00

No. 22 Elevating Fixture complete, for small Head
work grinding--- 175.00

No. 23 Elevating Fixture complete, for large work------------ 225.00

Automatic attachment for oscillating the grinding wheel
spindle and traversing the controlling wheel slide---------- 400.00

Arranging controlling wheel slide with Swivelling Head------- 300.00

Wheel Housing for wheels up to 8" face, if ordered as
additional equipment-- 100.00

Allowance if Pump and Tank are omitted----------------------- 60.00

Hopper Feed   Quotation upon request

THE ABOVE PRICES GO INTO EFFECT MAY 1st 1924.

The Heim Grinder Company.

A2-6-6-1

LOS ANGELES OFFICE
~~~~~~~~~~~~~~
1418 SANTA FE AVE,

~~~~~~~~ SUTTER 376~~~~~~~~
PHONE DAVENPORT 4438

MACHINE TOOLS
AUTOMATIC MACHINERY
SHOP EQUIPMENT
ELECTRIC CRANES

LOUIS G. HENES

MACHINE TOOLS

PRODUCTION TOOLS
HIGH SPEED TWIST DRILLS
REAMERS, CUTTERS
HOBS AND TAPS

75 FREMONT STREET
SAN FRANCISCO, CALIF.
May 22, 1926.

SUBJECT:

Mann Mfg. Co.
9th St. & Dwight Way
Oakland, Calif.

Attention - Mr. Robert Mann.

Gentlemen:

Following up conversation I had with you regarding Centerless
Grinders, please note that I will place in your hands our Heim
photograph album together with some minute specifications
covering the machine.

I would like to give you a general resume of the Heim machine
in writing, and want to say the following:

The Cincinnati Centerless Grinder is unquestionably an excellent
machine and we have the highest opinion of it. However, the HEIM
machine is built by the holders of the basic centerless grinding
patents and as they build only one machine, Heim believe they are
able to concentrate on more details and produce a better machine
than that built by the Cincinnati people, which by the way is
built under the patents licensed from the Heim Company.

There are many minor points of difference between the two machines
but there are two prominent differences which I want to call to your
attention. First, the method of truing the wheels is far simpler
on the HEIM machine than on the Cincinnati machine. Of course, the
diamond on the grinding wheel which does not tilt is always the
same but on the slow or regulating wheel, which is tilted from 0 to
6° at will, there is a vast difference between the two machines.
On the HEIM, the diamond is always exactly opposite the point on
the wheel which contacts with the work being ground regardless of
the tilt of the wheel as compared with the grinding wheel. On the
Cincinnati machine the diamond is set approximately 45° above this
point so that as a result they have to adjust the setting of their
diamond very carefully to correspond with the angularity at which
they set their regulating wheel. This point sounds very slight
but really represents a particular operation which requires an
appreciable amount of loss of time over a period of use of the
machine.

Again, the HEIM machine is built with the two opposed wheels
set in independent slides which operate on the bed with the
conventional "V" and flat arrangement. The Cincinnati machine
is designed with the grinding wheel in the base casting itself
while the opposed regulating wheel slide and the slide on which
the work fixture rests are super-imposed. In other words, the
conventional double type of slide HEIM thinks is less satisfactory in design than theirs and they are of the opinion that
this is also indorsed by a great number of users of both makes
of machines.

Another point regarding the two machines which may or may not be
of particular interest to you is that at the present time, the
HEIM machine sells for approximately $400.00 less than the
Cincinnati machine. Cincinnati increased their price last fall
and HEIM are still selling at the price that has been in effect
since May 1924.

HEIM have amongst their users some of the largest concerns in
the country and it has only been recently that Nash Motors have
standardized on HEIM machines exclusively. Mr. Odegaard, of
the Federal Machinery Sales Co., who is agent for the HEIM
machine in Chicago, writes me that they have been very successful
in the Chicago territory, stating that there are 13 or more HEIM
grinders in the two Nash plants, 8 or 9 in the International
Harvester Co. plants and that they are also used by the Miehle
Printing Press & Mfg. Co. He also informs me that they made a
very interesting installation of HEIM machines in the shop of the
U.S. Slicing Machine Co., at LaPorte, Ind. They grind a bar
about 24" long with nothing minus and .0002 plus for roundness
and straightness. I am also informed that Mr. Odegaard sold
machines to some of the manganese steel plants on heavy duty work.

With the photographs that I will leave with you you will find
detailed specifications and comparison between the HEIM and Cincinnati
machines but please do not misunderstand the motive that prompts
me to place this data before you. It is not done with the idea
of "knocking" our competitor's machine because they make a very
good grinder and they have many of them in use but I really believe that if you study the new HEIM machine thoroughly, you will
agree with me that it is superior to the present Cincinnati machine.

                                        Yours very truly,

LGH.HF

P.S. You advised me that you probably would be buying a Universal
Milling Machine for your tool room and before you do, I wish you
would come over and inspect our new model #2-G Hendey Universal
Milling Machine. We think it is about the finest miller that is
made.

                        L.G.H.

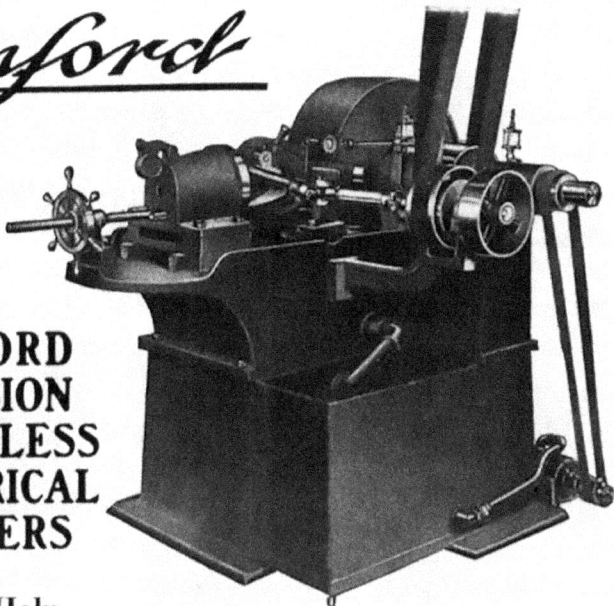
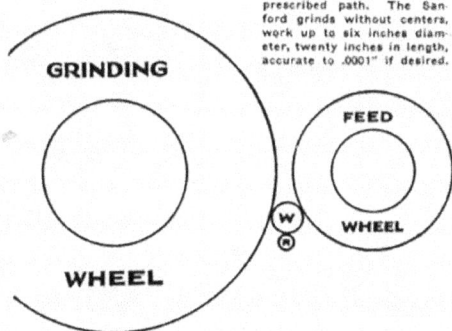

A2-7-3

A2-7-5

126   MACHINERY   August, 1921

355

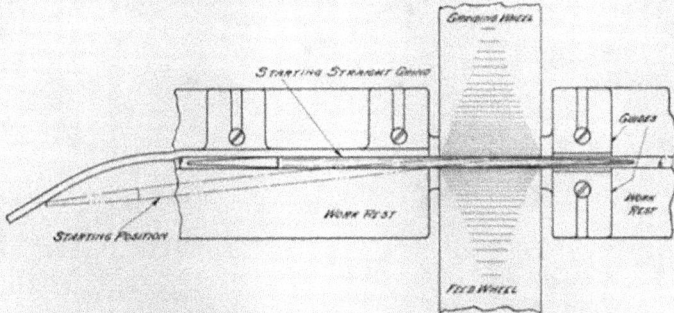

HEIM CENTERLESS GRINDERS

Finishing Difficult Work at Low Cost

These machines are grinding hardened nickel steel bearing rollers ranging in size from ½" to ⅞" diameter in various lengths to 2½". Several passes—first roughing, then finishing—are required to finish this work to limits of .0002" for roundness.

Production ranges between 300 and 600 pieces (both operations) per machine per hour, depending on the size of the work.

There are three Heim Centerless Grinders in this plant now, there was only one a year ago—a fact that speaks more strongly than words of satisfactory service.

Convenient and efficient—the simplicity of the wheel truing device makes it possible to get a close accuracy without interrupting the flow of work.

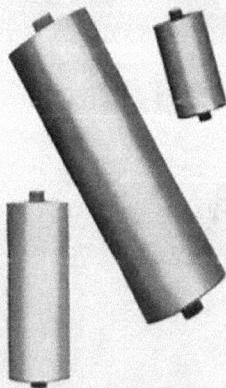

If you grind cylindrical parts, get the details of these machines

THE BALL & ROLLER BEARING COMPANY
DANBURY, CONN., U. S. A.

Reprinted from MACHINERY, March, 1920.

Why Did They Keep It?

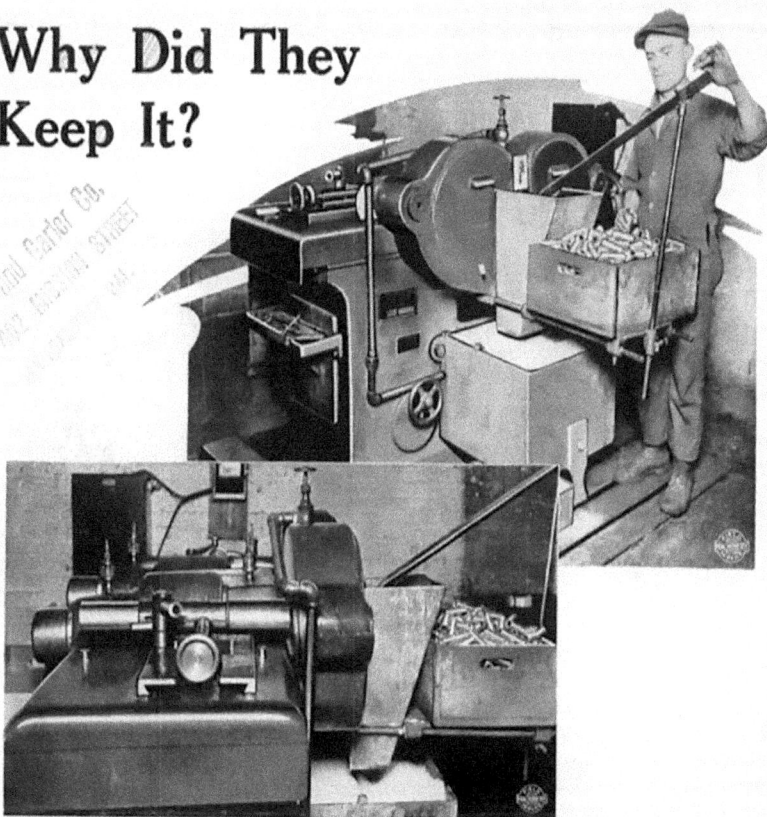

The Superintendent for W. D. Foreman, Chicago, Illinois, was frankly skeptical of our claims for Heim Grinders. This company makes automobile piston pins in tremendous quantities for the jobbing trade and he thought our cost and production estimates not only too good to be true, but quite impossible. However—he agreed to put in a Heim Centerless Grinder absolutely *at our own risk.*

That was seven months ago. Production was increased to four times the best previous records (better than we had claimed) ; the work was more uniformly accurate than before and skilled labor was no longer required.

Needless to say, the Heim Centerless Grinder is permanent equipment in this plant.

Let us tell you more about the good work Heim Centerless Grinders are doing.

THE BALL & ROLLER BEARING COMPANY
Danbury, Connecticut

Reprinted from MACHINERY, April, 1923.

The Heim Centerless Cylindrical Grinder

Greater Profits— Lower Costs

It all depends on your work. If your products are in the class served by this machine the increase in profits will be tremendous. We have records of production increases higher than six times the best figure obtainable by other grinding methods. We have records of successful application on work that is usually considered beyond the range of the centerless grinder.

So, if you do straight, taper or shoulder grinding on any class of work we can show you how to greatly increase your profits.

Tell us about your cylindrical grinding problems. Send us samples of work to be ground—we'll give you production and cost estimates that will astonish you.

The Ball & Roller Bearing Company
DANBURY, CONN., U. S. A.

362

A2-7-13

122   MACHINERY   August, 1923

NEW
DETROIT 4C
CENTERLESS GRINDER

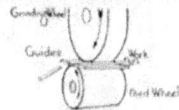

WORK REVOLVES AND TRAVELS
ALL THE TIME IN THE DETROIT
CENTERLESS GRINDER

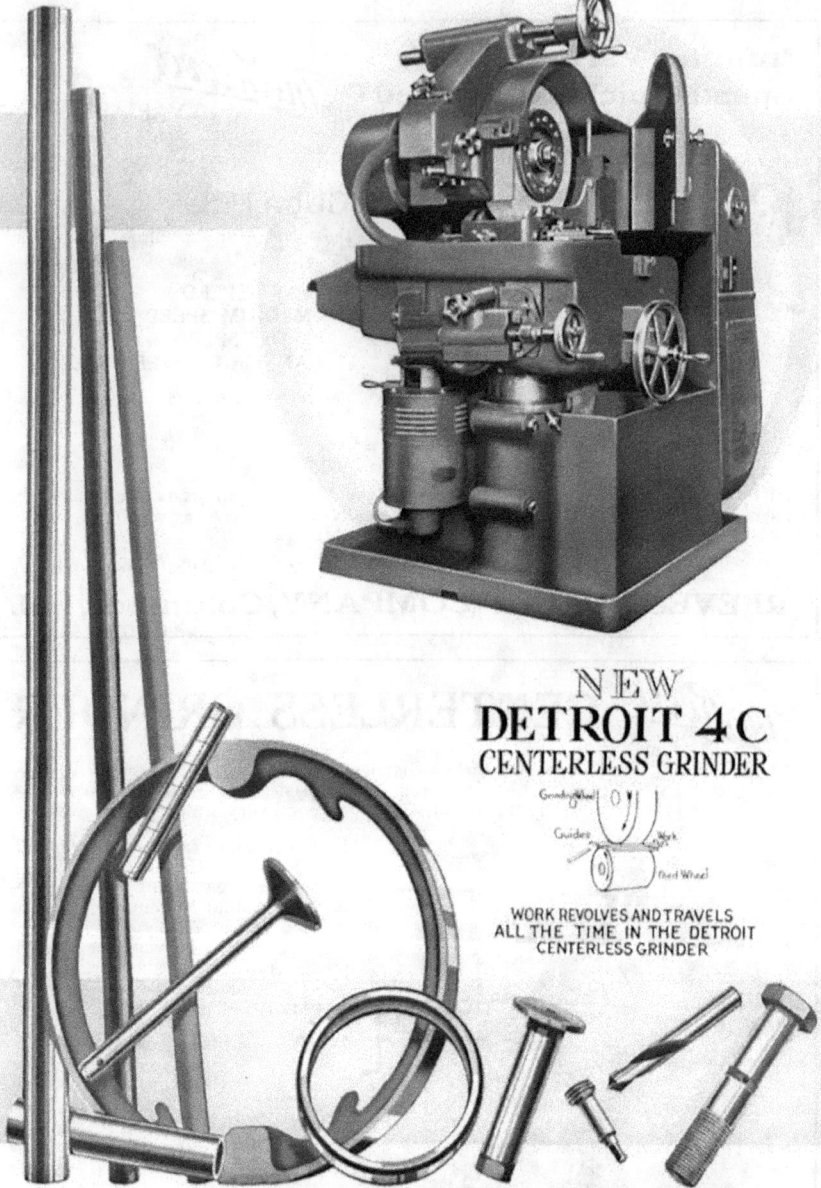

Detroit Machine Tool Company

Manufacturers and Designers of

High Production Machinery

Cable Address
"Todicom Detroit"

Centerless Cylindrical
Grinding Machinery
Semi-Automatic
Drilling Machinery

Detroit

A Message to the Production Manager
and his Grinding Superintendent

Gentlemen:

On the reverse side of this message is a photograph showing the Improved Type 4-C DETROIT Centerless Grinder. It is much heavier than any of our previous machines and has been strengthened in all ways to make it very rugged and enduring.

This machine is entirely electrically driven. A variable speed direct current motor, the power for which is obtained from a generator on the machine, drives the feed wheel. This enables us to have an absolutely flexible feed wheel drive with a range of speeds to cover every contingency.

Any desired feed speed can be obtained by merely turning the rheostat controlling the variable speed motor to the proper point. -

The grinding wheel spindle is of heavy sturdy construction. - The design of its bearings is the result of our long experience in the building of this type Grinder and is the most free from friction or belt pull of any grinder spindle on the market.

In building this Grinder all gears, shafts or other parts which could give trouble have been eliminated. There is absolutely a minimum of places on the machine where wear can occur.

The principle of the DETROIT differs from other Centerless Grinders. - It is vertical in construction with a feed wheel twice the width of the grinding wheel.

This vertical construction with the wide feed wheel guarantees you work free from waves, flats and chatter marks. - The work starts revolving before it comes in contact with the grinding wheel and continues at a uniform speed and feed past the grinding wheel.

We would appreciate a letter from you advising when our factory representative may call and discuss your grinding problems.

DETROIT MACHINE TOOL COMPANY

[signature]
Sales Manager

R.M.McGuire,
FP-

Twist Drills are speedily handled on the Heim—

The simple fool-proof full-shift performance of the Heim Centerless Grinder makes it an ideal machine for grinding all types of Twist Drills.

These production figures are not "demonstrations" but are actual figures from the job.

9/16-in. Taper Shank Twist Drills—250 to 300 per hour for grinding the flutes—200 to 250 per hour for grinding the taper Shanks—⅜-in. straight shank jobber's drills—400 to 450 per hour. ⅛-in. straight shank jobber's drills—650 to 750 per hour.

The setting and maintenance of the wheels for the grinding of any Taper Shank is as easily made and as readily maintained as the setting of a micrometer.

Through integrally mounted diamond truing devices, both the grinding wheel and the regulating wheel are kept in condition without disturbing the setting of the angle of the taper, or otherwise interfering with the steady flow of production. The same simple principle of design takes care of the longitudinal clearance on straight shank drills.

We welcome the opportunity of showing what the superior production ability of the Heim can do on this class of work.

Full information and reliable production figures gladly furnished on request.

Send for Catalog HG-1!

THE HEIM GRINDER COMPANY
DANBURY, CONN.

NOTICE

This company has acquired the ownership of the centerless grinder patents issued and pending, formerly the property of Lewis R. Heim and The Ball & Roller Bearing Co., and is the successor to the centerless grinding business of the latter company. The patents include Reissue Patent No. 15035 recently sustained by the decision of the U. S. Circuit Court of Appeals, Second Circuit.

This company is also the owner of all the centerless grinder patents formerly owned by the Reeves Pulley Co.

We will fully enforce our rights under these patents.

HEIM
CENTER LESS GRINDER

CAN YOU BEAT THIS?

Talking about return on your Investment

On the average when one Cincinnati Centerless Grinder is installed, it replaces four machines and four operators. Production is increased 50 to 300%.

And the great news about this machine is that new fields, new uses and new applications for the Cincinnati Centerless Grinder are being found daily. Steel, iron, glass, fibre, wood, raw-hide, rubber and reed are among the materials now ground.

Do not say that your job cannot be put on the Centerless Grinder. Write to us and find out what the Cincinnati Centerless Grinder will do.

PATENT NOTICE

In addition to our own patents, we are licensed under all the basic centerless grinding patents, including the Heim re-issue patent 15035, recently sustained by the decision of the U. S. Circuit Court of Appeals. In the purchase of our machines, therefore, our customers are protected against infringement of centerless grinding patents.

THE CINCINNATI MILLING MACHINE CO.,
CINCINNATI, OHIO

Investigate Centerless Grinding, the new method of grinding.

1. Form grinding
2. Straight cylindrical work
3. Two-diameter work
4. Shoulder grinding
5. Quick change over from one job to another
6. Great variety of wheel speeds providing absolute flexibility.
7. Spindles and housings mounted in rigid frames.
8. Easy to apply attachments.

CINCINNATI CENTERLESS GRINDERS

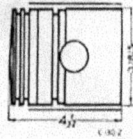

A2-8-1

Heim Original 1913 Experimental Peripheral Wheel Centerless Grinder
Reconstructed in May 1921

Heim Original 1913 Experimental Peripheral Wheel Centerless Grinder
Reconstructed in May 1921

A2-8-3

Heim Double-Ring Wheel Centerless Grinder
Used at the Ball and Roller Bearing Company: 1921

Heim Double-Ring Wheel Centerless Grinders
at the Ball and Roller Bearing Company: 1921

A2-8-5

Heim Peripheral Wheel Wheel Centerless Grinder, Front Side: 1922

Heim Peripheral Wheel Wheel Centerless Grinder, Rear Side: 1922

A2-8-7

Upper: Hem Centerless Grinder With Helical Gear System: 1923
Lower: Carrier For Spot Grinding Of Shoulder Work: 1922

A2-8-8

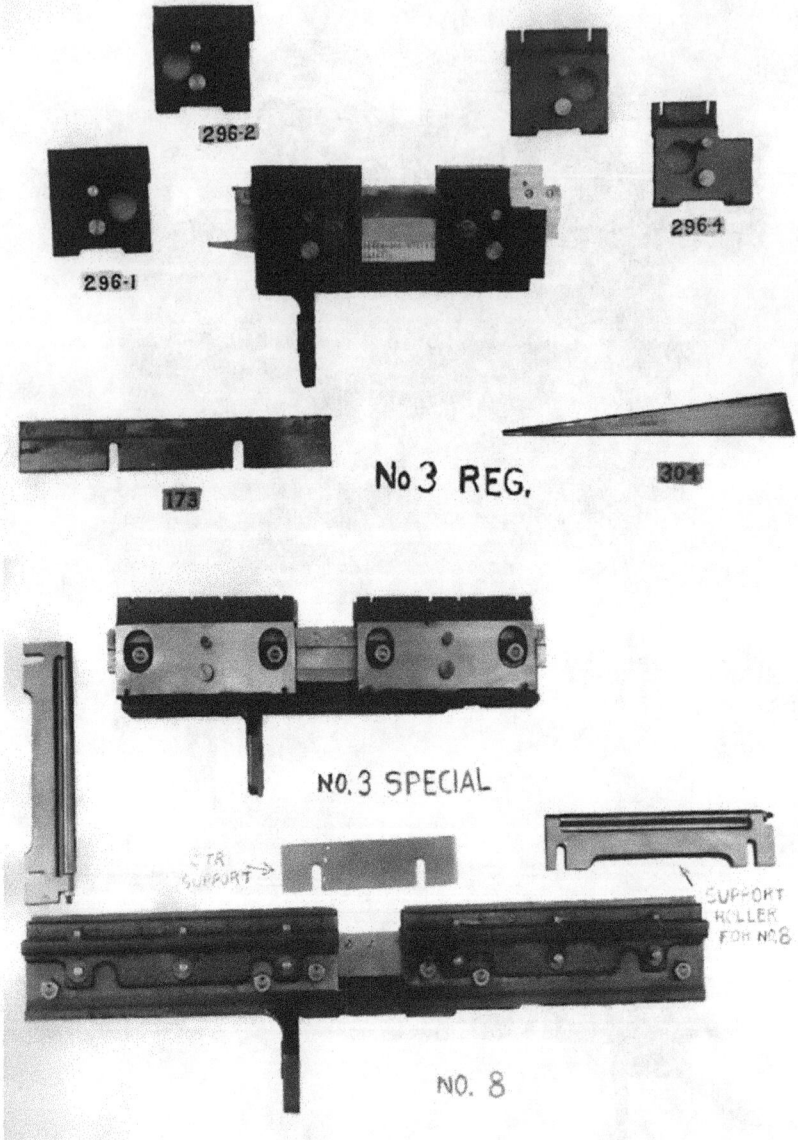

296-2

296-4

296-1

173   No 3 REG.   304

NO. 3 SPECIAL

CTR
SUPPORT →

SUPPORT
ROLLER
FOR NO 8

NO. 8

Upper: Heim No. 3 Standard Carrier For Through Grinding: 1922
Lower: Heim No. 3 Special and No. 8 Carriers For Through Grinding: 1922

A2-8-9

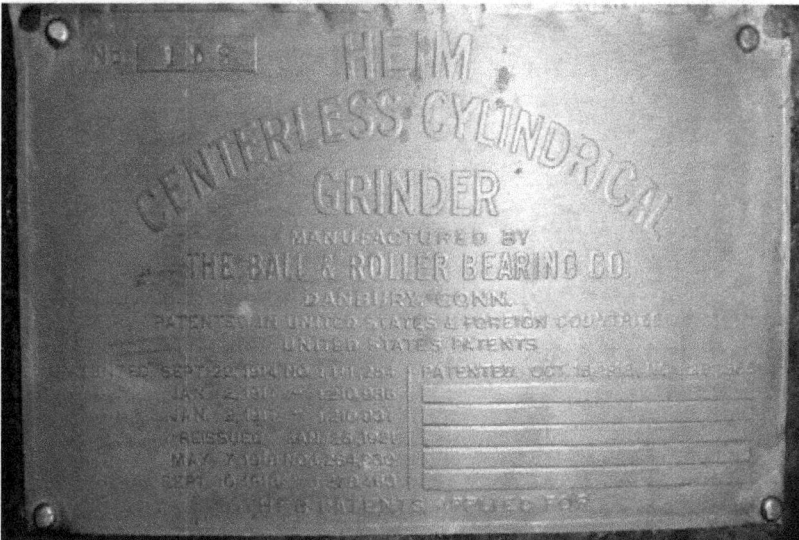

R. Jacobs (Author) With Heim Centerless Grinder No. 158 (ca. 1922)
At The American Precision Museum, Windsor, Vermont: 2014

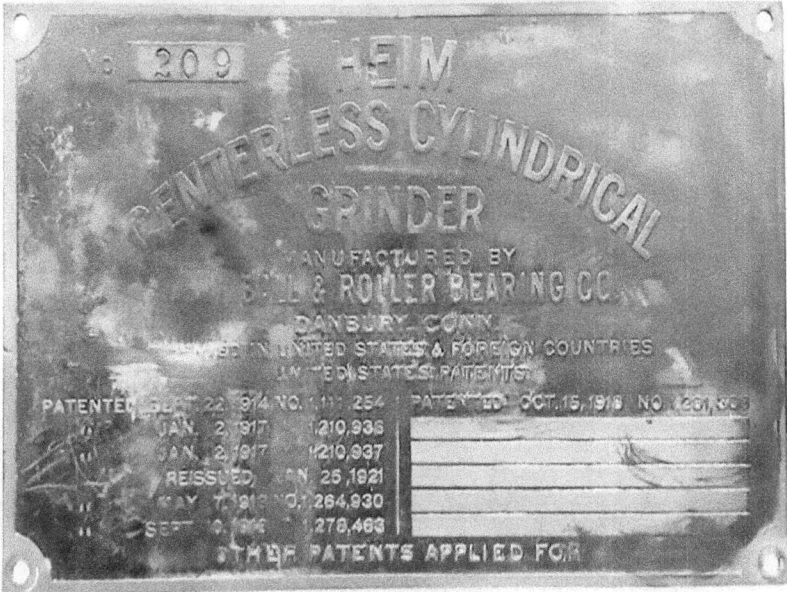

Heim Centerless Grinder No. 209 And Nameplate: ca. 1922
Courtesy of The Smithsonian Institute, Maryland: 2017

A2-9-1

Manufacture of centreless machines at LIDKÖPING - a retrospective view

by Dir. G. Ekholm

This description does not claim to be complete, but my experience of the production item in question ought to be more or less apparent.

PRIOR HISTORY

In November 1921, I received an offer to resume the position of design office manager at LMV, which position I held before I transferred one year previously to LMV's then parent company Nordiska Kullager AB (NKA) in Gothenburg.

My first task was to travel over to the USA in order to assist at Strom Bearing Co, Chicago and start ball fabrication using, among other things, machines supplied by LMV which I designed during my previous employment. At that time, in connection with my travel, I managed to spend six months studying machine tool manufacture i this country.

My focus at that time was on centreless grinding machines, which at that time had started to gain a foothold, particularly at car factories.

During the voyage, I found out in a telegram that LMV had been declared bankrupt. This was indeed an unforeseen obstacle. I decided to attempt to complete my journey according to the defined plan, despite this.

I succeeded in gaining access to Cincinnati Grinding Corporation and Detroit Machine Tool Co, the two most distinguished companies constructing centreless grinding machines in those days. However, the first of these companies was very restrictive and showed only a few finished machines. No machines were demonstrated operating. At Detroit Machine Tool Co, on the other hand, I was able to enter their demonstration room and see a number of test grinding operations for ball bearing rings and other parts. Centreless grinding of such rings, they asserted, had not happened before in the USA.

I saw Cincinnati machines in operation later on at both Ford and Noch car factories.

After these studies, supplemented with research at libraries involving journals and studies of catalogues requested from Heim Grinder Co and from a company by the name of Sanford, I started to think that it might perhaps be possible to build a better machine according to a slightly unusual design principle.

These thoughts matured still further, and on the voyage home they were made more specific with a few sketches.

ENQUIRY FROM SCHMID-ROOST, SWITZERLAND

On my return home, I received an enquiry from the above-mentioned company for a machine setup for grinding rollers for roller bearings, including circular grinding machines, flat grinding machines for the ends and edge grinding machines.

The management had decided to offer the company a machine setup, and by way of preparation for this centreless grinding had been tested in a temporary arrangement using an ordinary circular grinding machine. Contact had also been made with an engineer who was aware of a German centreless grinding machine (Hirth brand), which SKF purchased in order to test. The management at LMV had employed the engineer and decided to commence design of a machine according to the principles of this machine. However, I managed to persuade the management to abandon this decision and instead focus on a machine according to the guidelines which I proposed. Moreover, we had been made aware that the test grinding using the Hirth machine did not appear encouraging.

DELIVERY TO SCHMID-ROOST

The centreless machine was completed at the end of 1922, and grinding tests took place in the presence of representatives of the above-mentioned company, with approved results. The machine was delivered, on a sale or return basis, in early 1923 together with other machines ordered. However, once the machines had been installed in production, it turned out that it was not possible to achieve satisfactory results with the centreless machine, and this was why I travelled down to Oerlikon. I was unable to achieve any fully satisfactory results in the short time available to me. I had agreed to be in Paris one particular day in order to take part in negotiations together with the manager of LMV, and that was why we agreed that the machine should be returned to Lidköping so that we could continue grinding tests there.

- 3 -

In my view, the reason for the failure was that we did not have a suitable grinding wheel available to us, and that I was not fully conversant with the grinding technique as it related to 3-point measurement of the rollers. Only diametrical measurement took place during testing in Lidköping.

I was able to set up the machine returned from Switzerland at SKF's grinding department for test-grinding of cylindrical rollers (LMV was not part of the SKF group at that time). With their knowledge of circular grinding in general and their enormous proficiency in regard to selection of grinding wheels, grinding was splendidly successful and they took over the machine. No delivery took place to Schmid-Roost for many years. According to what we have been told, they purchased an American machine for circular grinding of the rollers.

SEPARATOR FACTORIES

In our opinion, these workshops had production which was appropriate for our centreless machines, and this was why we began to canvass them. We supplied on a sale-or-return basis a machine to AB Separator and a machine to AB Pumpseparator.

The grinding tests went very well at AB Separator, but we had problems with persuading workers to stand by the machine. They tried with six different grinders taken from ordinary circular grinding machines, but none of them stayed. The grinders were of the opinion that centreless machines would take work away from the, and so they were
 unfavourably disposed towards them. When the machine had been transferred to a different department and an untrained grinder was set to work with it, everything went well.

At AB Pumpseparator, we applied the knowledge we had gained from events at AB Separator. I trained a foreman to operate the machine. Once we had ground a few sections of various separator shafts, he was so impressed that he took me in to meet the head of the company and told him of the measurement results and the production figures
 achieved. He, too, was impressed and said that they surpassed his expectations. This company later ordered another two machines.

OTHER PRIMARY DELIVERIES

Between 1923 and 1925, we supplied machines to Atlas Diesel, which had a problem as they had to grind thin-walled cylinder pistons for air drill machines. I personally performed the test grinding and achieved satisfactory results once a suitable grinding wheel had been procured.

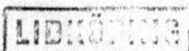

In addition, a machine was supplied to Huskvarna Vapenfabrik and, on export, to Cie d'Applications Mécaniques in France; and Fries & Höpflinger, Fichtel & Sachs and Kugelfabrik Fischer in Germany each received one machine.

MACHINES FOR RODGRINDING

In 1926, Gebr. Loewe, Düsseldorf, were appointed representatives for Germany. We began to interest ironworks in rod grinding, and Gebr. Loewe succeeded in selling a number of machines for this purpose. Rumours of these sales also spread to ironworks in Sweden, and the first deliveries to ironworks in Sandviken and Uddeholm saw very good results.

The first machines constructed were relatively simple and fitted with engines developing just 8-10 hp, but up to and including 1929 we had supplied machines developing up to 50 hp and with hydraulic transmission for the regulating wheel.

There was a lot of mistrust of the grinding method and our machines over the first few years, and the machines were often supplied on a sale-or-return basis. However, machines had to be taken back only in very few cases. By the end of 1929, around forty machines were in operation in Sweden, Germany, Great Britain, France and Switzerland.

PATENT LAWSUIT WITH CINCINNATI

At the end of 1927, we were forced to restrict our production of centreless machines, and sales to Germany had to be suspended entirely. After about ten years of preparation, Cincinnati Grinding Corp. had namely been awarded a patent for their machines. This patent was so far-reaching that it assured them of a veritable monopoly on sales of such machines. On account of this, following prior notice from Dipl.-ing. Georg Benjamin in Berlin on 2 November 1927, LMV was sued and accused of patent infringement at Landgericht Berlin on 23 February. As a countermove, LMV - via its patent attorney Dr. Moldenhauer at the Reichspatentamt - requested abrogation of the patent, reasoning that similar machines were already known. LMV's request was approved on 8 November 1928 and ratified by the Landgericht. However, Cincinnati appealed against the decision to a higher authority, the Reichsgericht in Leipzig.

The case there was rather extensive and, to a degree, dramatic. The court appointed Prof. Schwerd of the College of Technology in Hanover as its expert witness, and he ordered LMV to build a machine according to a patent which we alleged prevented innovation. The machine was to be of the same size and offer the same stability as the machine set up by Cincinnati at the college so that they could perform comparative grinding operations with the two machines However, the lawsuit and test grinding operations became something of a red herring due to the fact that Cincinnati's representatives maintained, against their better judgement, that the machine which they set up could be used for precision grind. This was stretching the truth. The rotational elements

appeared to be round when they were measured diametrically, but there was no guarantee of them being round if they were measured using a 3-point device. Neither their machine nor the machine which we set up was able to make round a previously out-of-round part if it was out-of-round when it went into the machine, provided that it was measured in the latter manner. For understandable reasons, neither of us wished to bring this fact to the attention of Prof. Schwerd as no 3-point measurement took place following test grinding.

One episode occurring during test grinding should perhaps be rescued from oblivion. During grinding in front of the professor and his assistant, Cincinnati's grinder suddenly said to our grinder: "Your rollers are ground beforehand and then blackened so that they look unground. The response of our man was not left wanting. He replied: "Yes, that is correct; and so are yours. Shall we swap? The professor overheard this exchange, and as a result he suggested that they swapped rollers. This took place; and it turned out that the other party's rollers were better in our machine than our rollers were in theirs.

I realised that after this incident, the professor's mind was made up.

During discussion of the lawsuit in Leipzig, and as late as then the parties were asked to leave the courtroom to allow the court to deliberate, we suggested to the other party that an amicable settlement should be entered into and that our nullity suit withdrawn so that the patent should remain in force. We feared that if the patent were declared invalid, additional competitors would appear on the market. However, this was rejected by the other party with the words: "We will win the lawsuit". A while later, we were asked to return to the courtroom, and the gentleman presiding explained that the case had been settled in favour of LMV.

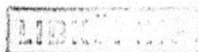

Present at the hearing were Dipl.-ing. Benjamin for Cincinnati, Prof. Riebensahm from the College of Technology in Berlin as the expert witness, and two representatives of the company in the USA. Present on behalf of LMV were patent attorney Dr. Moldenhauer, Privy Counsellor, Prof. Wallichs from the College of Technology in Aachen as the expert witness, and Engineer Loewe and myself, and also for Prof. Schwerd from the College of Technology in Hanover as the expert witness for the Reichsgericht. Also present were two higher lawyers, one for either party, who were authorised to plead their cases before the Reichsgericht. I have forgotten their names. So this was a fairly large crowd of people, more than twenty people if we include the lawyers of the Reichsgericht.

Engineer Benjamin probably behaved in this manner because Cincinnati had had another patent approved which made it impossible for LMV to build its centreless grinding machine for precision grinding without infringing this. If there was no prior warning of this before the end of the five-year deadline, there was a possibility that LMV would overlook this patent The German patent office had a five-year deadline within which it was possible to appeal against a patent decision. If this was not done within this time, the patent was considered to be unassailable. The patent related to the grinding method with the sloping supporting rail. However, we were prepared and had prepared material for use in order to attempt to have this patent declared null and void again.

Before the deadline, Engineer Loewe and I travelled to Berlin to meet Dr. Moldenhauer. This meeting was rather dramatic. After having studied our carefully compiled material, de declared that the material was too weak for us to have any prospect of success, and that he was therefore unable to take on the case. I attempted to explain just how extraordinarily important it was to have the patent in question set aside. However, Engineer Loewe was of the opinion that we should be content with what we had gained, and he left us very dejected as I explained that we had actually gained nothing as long as this patent prevented us from building centreless grinding machines for precision grinding. Good advice was now expensive. I was unable to let the matter drop as I considered it to be of vital importance to LMV's centreless production. During a stroll in Tiergarten park, I succeeded in persuading Dr. Moldenhauer to postpone his trip home to Düsseldorf, where he lived, the next day and he proposed then to show me Berlin "by night".

It was fairly late when we decided to walk back to our hotel. It was not possible to enter into any in-depth discussion of the patent issue despite the fact that I carried the papers with me all evening. On the way to the hotel, we passed the Hotel Excelsior, and I suggested - in the hope of finally having an opportunity to discuss the matter which really concerned me - that we should go to the night bar at the hotel and enjoy a bottle of wine of a brand of which I knew he was very fond and discuss the patent issue once again, which he accepted.

With a bottle of Mosel and sitting at a table strewn with papers, I explained the patent issue once again, and for more than two hours we held our discussion among dancing, surprised Berliners. Between 3 and 4 in the morning, I had managed by persuade Dr. Moldenhauer that there were certain prospects of having the patent declared null and void, and he declared himself willing to take on the case.

It was with enormous satisfaction that I was able to telephone Engineer Loewe in the morning and tell him the result, and that I had agreed with Dr. Moldenhauer that we would meet with him in the morning for a final review of the documentation. This took place, and our nullity suit was submitted to the Berlin Patent Office before it closed that same day.

As a result of our appeal, a few days after submission LMV received a telegram from Cincinnati in which they offered us free use of all their patented designs on the condition that we withdrew our nullity suit and that we did not tender centreless grinding machines to the USA for a period of ten years. Following mutual confirmation of this, we withdrew the documents from the patent office.

I have been unable to trace the time for the lawsuit at the Reichsgericht concerning the first patent and the agreement with Cincinnati regarding the latter due to the fact that the correspondence on these appears to have been lost.

PATENT DISPUTE WITH HERMINGHAUSEN

Later on, LMV also had a patent dispute initiated with a company called Herminghausen in Germany. However, this ended with an amicable settlement in February 1936.

- 8 -

While the disputes concerning the patents as related above were in progress, a certain decline in both design and sales work had of course occurred. However, intensive work on improvement of the design could now begin and new sizes designed, and sales operations were also intensified, leading to significant successes in the 1930s. The only ones that can be mentioned here are the licensing agreement with Schnellpressenfabrik Frankenthal in 1936 and the major deliveries of machine no. 5 to Bristol Aeroplan Co between 1936 and 1938 and deliveriesto the Soviet Union between 1939 and 1940.

Lidköping, 11 May 1966.

Appendix 3

Trial Documents for the Ball and Roller Bearing Co. v. F. C. Sanford Manufacturing Co., and the Heim Grinder Co. v. the Fafnir Bearing Co.

Contents

Appendix 3 contains documents with information from the patent infringement complaints filed by Heim against the F. C. Sanford Manufacturing Company in 1921 and the Fafnir Bearing Company in 1924. For the Sanford trial, documents *(A3-1 and A3-2)* include the Bill of Complaint filed by Heim and the Trial Summary developed by the author. The Bill of Complaint was copied from the original trial records obtained from the National Archives and Records Administration *(NARA)* in New York City.

For the trial against the Fafnir Bearing Company that was defended by the Detroit Machine Tool Company, documents (A3-3 and A3-4) include the Docket Entries for the trial and the Bill of Complaint that were reprinted from the original court documents obtained from the National Archives and Records Administration in Boston, Massachusetts.

A3-1: Plaintiff's Bill of Complaint: B&RB v. Sanford
The document contains both the Original and Supplemental bills dated July 23, 1920, and February 28, 1921, and summarizes the basis for the patent infringement suit against F. C. Sanford, damages sustained by Heim, and remedies demanded by Heim.

A3-2: Trial Summary: Witnesses, Dates, and Locations: B&RB v. Sanford
contains list of the witnesses plus the dates and locations of testimony and demonstrations. Included in the witness list is a short description of the person, employment history, and relationship to the trial. The trial summary is arranged to show the dates and location of testimony in chronological order.

A3-3: Summary of Docket Entries, Jan. 4, 1927: Heim Grinder v. Fafnir
Lists the dates of major events for the patent infringement trial and appeal.

A3-4:   Bill of Complaint, Nov. 11, 1924: Heim Grinder v. Fafnir
         Summarizes the basis for the patent infringement suit against Fafnir,damages
         sustained by Heim, and remedies demanded by Heim.

A3-1

Bill of Complaint and Supplemental Bill of Complaint
B&RB v. Sanford, July 23, 1920 and Feb. 28, 1921

United States District Court
District of Connecticut.

The Ball and Roller Bearing Company, Plaintiff,
vs.
F. C. Sanford Mfg. Co., Defendant.
Equity 1529

United States Letters Patents Nos. 1,210,936, 1,264,930 and Re. 15,035

Bill of Complaint.

To the Honorable the Judges of the District Court of the United States in and for the District of Connecticut in the Second Circuit:

The Ball and Roller Bearing Company, a corporation duly organized and existing under and in accordance with the laws of the State of Connecticut, a citizen of said State, having an office for the transaction of business in Danbury, in the County of Fairfield and State of Connecticut, brings this, its Bill of Complaint, against the F. C. Sanford Mfg. Co., a corporation organized and existing under and in accordance with the laws of the State of Connecticut, a citizen of said State, having an office for the transaction of business in Bridgeport, County of Fairfield and State of Connecticut, where and elsewhere it has committed the acts of infringement complained of herein.

For cause of action, plaintiff avers:

1. That the plaintiff's said cause of action arises under the patent laws of the United States and more specifically, the defendant, in the District of Connecticut and elsewhere, committed the acts of infringement herein complained of in violation of the rights of this plaintiff as sole owner of United States Letters Patents Nos. 1,210,936, granted January 2, 1917, and 1,264,930, granted May 7, 1918, in both of which Lewis R. Heim is the inventor and both of which he assigned to the plaintiff herein as hereinafter set forth; that the said Letters Patents were duly issued to plaintiff as assignee of the said Lewis R. Heim, a citizen of the United States, residing in Danbury in the State of Connecticut, the first original and sole inventor of the subject matter of the said Letters Patents, the said issue of which patents occurred January 2, 1917, and May 7, 1918, respectively, as above set forth, in due form of law under the seal of the Patent Office of the United States, duly signed by the officers thereunto authorized, bearing respectively the above dates and numbered respectively as aboveset forth, whereby there was granted and secured to plaintiff, its successors and assigns, for the term of

seventeen years from and after the respective dates of the said Letters Patents, the full and exclusive right and liberty of manufacturing, using andselling the subjects matter of the said Letters Patent, which, or duly certified copies of which, are ready in court to be produced.

2. Plaintiff further avers that the said Lewis R. Heim being as aforesaid the first, original and sole inventor of the said new and useful improvements fully set forth and described in the said Letters Patents of the United States, which said inventions and improvements had not been known or used in this country nor patented nor described in any printed publication in this or any foreign country before his said inventions and discoveries thereof, nor for more than two years prior to the respective applications for Letters Patents; had not been in public use or on sale anywhere for more than two years prior to said applications for said Letters Patents, and for which no application for foreign Letters Patents had been filed at any time by anyone and which had not been abandoned, did, on March 6, 1915, and on July 13, 1917, respectively, make due application for said United States Letters Patents, the said Heim duly complying in all respects with the statutes of the United States and the Rules of the United States Patent Office in that behalf made and provided, and after due proceedings had, the said Letters Patents did each issue in due form of law, duly signed and sealed as aforesaid unto the plaintiff herein as assignee of the said Lewis R. Heim of all his right, title and interest in both of the said Letters Patents.

3. Plaintiff further avers that the said Letters Patents issued to it as assignee of all right, title and interest in and to each of the same, by due, proper and lawful assignments in writing from the said Lewis R. Heim for valuable considerations, and that it now holds and possesses by the said assignments, all right to bring suit based upon any infringement of the said Letters Patents or either of them, and to collect profits and damages by reason of said infringements, the said assignments being respectively made prior to the dates of grant of the said patents respectively, of all of which plaintiff stands ready to make due and proper proof.

4. Plaintiff further avers that it has made extensive practical use of the inventions and improvements set forth in said Letters Patents, its business being materially built up upon the inventions of the said patents; that plaintiff has given repeated notice to the above defendant of its infringement of the said patents but that the defendant has refused to desist, from such infringement and, as hereinbefore set forth, has in the District of Connecticut and elsewhere, willfully infringed the said Letters Patents by manufacturing, using and selling in a commercial way and otherwise the inventions which form the subject matter of the said patents and in infringement of both of the same, all against plaintiff's will, notwithstanding due notice as aforesaid and without plaintiff's request or consent; that defendant has continued to infringe and is continuing to do so and threatening to do so up to the present time; that, as plaintiff is informed and believes, said infringing acts have, and have had, the effect of inducing others to so infringe and will tend to induce others to infringe; that, as plaintiff is informed and believes, defendant is conspiring and has conspired with others to infringe said

patents; that by such unlawful acts defendant has diverted to itself trade and profits which plaintiff would otherwise have received, as to the amount of which plaintiff prays recovery, whereby plaintiff has been caused great and irreparable damage, and that defendant in its said acts is devoid of equity and good conscience.

Wherefore, plaintiff prays for the issue of a permanent injunction arising out of and under the seal of this Court, enjoining and restraining the said defendant, its officers, associates, attorneys, agents, servants and employees from directly or indirectly making, using or selling in any manner whatsoever the improvements set forth and claimed in the said Letters Patents Nos. 1,210,936 and 1,264,930, or either of them, or any part thereof in violation of the rights of the plaintiff in the premises; that the defendant account to plaintiff for all profits made by it and all damages sustained by plaintiff as a direct or indirect result of said infringement; that the Court, by reason of the circumstances of said infringement, increase such damages to triple value; that plaintiff recover the costs and disbursements of this action; and plaintiff prays that a subpoena may issue out of and under the seal of this Court directed to the said defendant, requiring it upon a day certain and under certain penalty to appear and make true, full and certain answer to this Bill of Complaint, and plaintiff prays for such other and further relief as may appear proper to the Court and agreeable to equity.

The Ball and Roller Bearing Company,

By Howard I. Beard,
                    Secretary.
(Seal)

EMERY, VARNEY, BLAIR & HOGUET,
Resident Solicitors.

Robert S. Blair,
Of Counsel,
149 Broadway, New York City.

Supplemental Bill of Complaint

UNITED STATES DISTRICT COURT

DISTRICT OF CONNECTICUT.

The Ball and Roller Bearing Company, Plaintiff,

vs.

F. C. Sanford Mfg. Co., Defendant.

Equity 1529

United States Letters Patents Nos. 1,210,936, 1,264,930 and Re. 15,035

To the Honorable the Judges of the District Court of the United States, in and for the District of Connecticut in the Second Circuit:

The Ball and Roller Bearing Company, plaintiff herein, avers:

1. That on the 23rd day of July, 1920, plaintiff exhibited its original Bill of Complaint in this honorable Court against F. C. Sanford Mfg. Co., defendant, for the purpose of preventing further infringement of patents to Heim, Nos. 1,210,936 and 1,264,930; for recovery of profits and damages for past infringement and for costs and disbursements of this suit; and for such other and further relief as may appear proper to the Court and be agreeable to equity.

2. That the defendant was duly served with process, and put in its answer, and that this cause is now awaiting trial.

3. That since the filing of this suit, a patent of the United States to Lewis R. Heim, originally numbered 1,210,937 and originally granted January 2, 1917, has been reissued by the Commissioner of Patents on January 25, 1921, bearing the reissue No. 15,035; that plaintiff is the owner of all right, title and interest in and to the said original patent numbered 1,210,937 and the said reissue thereof, the said original patent having been issued to plaintiff as assignee of the entire interest of the said Heim, and plaintiff having consented to the reissue thereof.

4. That the said Heim was the first, original and sole inventor of the invention dealt with and covered in the said original patent 1,210,937 and the reissue thereof; that the said invention and improvement had not been known or used in this country nor patented

nor described in any printed publication in this or any foreign country before his date of invention and discovery thereof, nor for more than two years prior to the filing of the application for the said original patent 1,210,937, had not been in public use or on sale anywhere for more than two years prior to the said filing date and for which no application for foreign Letters Patent had been filed at any time by anyone and had not been abandoned; that the said Heim did on March 6, 1915, apply for Letters Patent of the United States and did prior to the granting of the said original patent, by due, proper and lawful assignment in writing, convey to plaintiff for valuable considerations all his right, title and interest in and to the said application and any patent or patents granted thereon and the invention covered thereby, all of which right, title and interest plaintiff now owns; that the issue of the said original patent on the said application, to plaintiff occurred in due form of law, after due proceedings had, under the seal of the Patent Office of the United States duly signed by the officers thereunto authorized, whereby there was granted and secured to plaintiff, its successors and assigns, for the term of seventeen years from and after January 2, 1917, the full and exclusive right and liberty of manufacturing, using and selling the subject matter of the said Letters Patent, which, or a duly certified copy of which, is ready in court to be produced; that the said original patent was defective or inoperative within the meaning of the statute providing for reissue thereof, and that upon due and proper application being made and due and proper proceedings thereon, the said patent was reissued to plaintiff January 25, 1921, bearing the reissue No. 15035, under the seal of the Patent Office of the United States and duly signed by the officers thereunto authorized.

5. That the invention dealt with in the said Letters Patent, Reissue No. 15035, has been extensively employed in a practical way by plaintiff, the business of plaintiff being materially built up upon the said invention; that defendant has since the issue of the said original patent on January 2, 1917, in the District of Connecticut and elsewhere, infringed the claims of the said original patent and has, since January 25, 1921, and prior thereto, in the District of Connecticut and elsewhere, infringed the said reissue patent No. 15035; that plaintiff has given repeated notice to defendant of its infringement of the said original patent and has given notice to defendant, through its counsel, of its infringement of the said reissue patent, but that defendant has refused to desist from such infringement and has, in the District of Connecticut and elsewhere, continued to willfully infringe the said reissue Letters Patent by manufacturing, using and selling in a commercial way and otherwise the inventions which form the subject matter of the said reissue patent and in infringement thereof, all against plaintiff's will and without plaintiff's request or consent; that defendant has continued to infringe and is continuing so to do and threatening to do so up to the present time; that, as plaintiff is informed and believes, the said infringing acts have, and have had, the effect of inducing others to so infringe and will tend to induce others to infringe; that, as plaintiff is informed and believes, defendant is conspiring and has conspired with others to infringe the said patent, and has contributed to the infringement thereof by others; that by such unlawful acts defendant has diverted to itself trade and profits which plaintiff would otherwise have received, as to the amount of which plaintiff prays recovery, whereby plaintiff has been caused great and irreparable damage and whereby defendant is devoid of equity and good conscience.

Wherefore, plaintiff prays for the issue of a permanent injunction issuing out of and under the seal of this Court, enjoining and restraining the said defendant, its officers, associates, attorneys, agents, servants and employees from directly or indirectly using or selling in any manner whatsoever the improvements set forth and claimed in the said reissue Letters Patent No. 15,035, or any part thereof, in violation of the rights of the plaintiff in the premises; that the defendant account to plaintiff for all profits made by it and all damages sustained by plaintiff, as a direct or indirect result of such infringement; that plaintiff recover the costs and disbursements of this action; and plaintiff prays that a subpoena may issue out of and under the seal of this Court directed to the said defendant, requiring it upon a day certain and under certain penalty, to appear and make true and full answer to this Supplemental Bill of Complaint, and plaintiff prays for such other and further relief as may appear proper to the Court and agreeable to equity.

THE BALL AND ROLLER BEARING COMPANY,

By HOWARD I. BEARD, (Seal) Secretary.

Resident Solicitor.

EMERY, VARNEY, BLAIR & HOGUET,

ROBERT S. BLAIR, of Counsel,
149 Broadway, New York, N. Y.

UNITED STATES OF AMERICA,
STATE OF
COUNTY OF

On this 28th day of February, 1921, before me, a notary public, personally appeared Howard I. Beard, who, being by me first duly sworn, deposes and says that he is Howard I. Beard, Secretary of The Ball and Roller Bearing Company, plaintiff herein; that he has read the foregoing Supplemental Bill of Complaint by him subscribed and knows the contents thereof; that the same is true of his own knowledge except as to those matters therein stated upon information and belief, and as to those matters he believes it to be true.

....
(Seal)   Notary Public

A3-2
Trial Summary

The Ball and Roller Bearing Co. v. F. C. Sanford Manufacturing Co.

This section contains two sets of information from the Heim patent infringement trial. The first is a list of the witnesses testifying at the first trial. The second part contains a summary of the trial dates, locations, and order of testimony.

The original trial testimony provided substantial information on the origins of the centerless grinder and the people involved with its development. This trial and the second trial against the Fafnir Bearing Company, which was defended by the Detroit Machine Tool Company, also illustrated the extent to which manufacturers of centerless grinders, incorporating Heim's patented features, endeavored to defeat Heim.

A: List of Witnesses

The following is a list of witnesses who testified for both the plaintiff and defendant. The list includes a short description of each witness and his/her connection to the trial.

Due to the extensive length of all testimony, only the testimony of Lewis R. Heim is provided *(see A5-10)*.

A1: List of Witnesses for the Ball and Roller Bearing Company, Plaintiff.

1: Dr. Harold Pender: Age: 42 or 43
Professor at the University of Pennsylvania at Philadelphia, Pennsylvania.
Department Head: Electrical Engineering
Expertise includes mechanical devices and machine tools.
Expert Witness for the Plaintiff.

2: Lewis R. Heim: Age: 46
Plaintiff in the trial.
President of the Ball and Roller Bearing Company at Danbury, Connecticut. Former president of the Heim Machine Company *(sold in 1915)*
Inventor of the Heim double-ring wheel centerless grinder and the Heim peripheral two-wheel centerless grinder. *(Patents: 1,210,937 and 1,264,930)*

3: James Gordon Bennett: Age: 30
Superintendent of the Ball and Roller Bearing Company at Danbury, Connecticut. Employed since 1908 *(Heim Machine Company)*
Assisted in constructing first Heim double-ring wheel centerless grinder.

4: John Henry Roth: Age: 39
Treasurer of the Ball and Roller Bearing Company and employed since 1910.
Brother-in-law of Lewis Heim.
Worked in the grinding room and witnessed development of Heim's single-wheel and peripheral wheel centerless grinders.

5: Arthur Kirkland: Age: 39
Salesman at the Ball and Roller Bearing Company at Danbury, Connecticut. Former salesman of the Sanford centerless grinder.
Former salesman of the single-wheel centerless grinder manufactured by the Detroit Machine Tool Company.

6: Charles O. Johnson: Age: 36
Foreman at Maxams Machine Company at Bridgeport, Connecticut. Former employee of the Ball and Roller Bearing Company.
Worked with and operated Heim's single-wheel, peripheral wheel, and ring wheel centerless grinders.

7: William J. Ryan:
Employed at the Precision Machine Company in New York City as a machinist and operator of roll-grinding machinery.
Precision Machine was the first company to purchase the Sanford "B" centerless grinder.

8: Clement Booth:
Consulting Engineer. Expert on grinding processes and machines.
Former employee of the Standard Roller Bearing Company of Philadelphia, Pennsylvania, from 1909 to 1918.
Witnessed use of the Grant single-wheel centerless grinder at SRB.
Witnessed demonstration of the two-wheel centerless grinder *("The Hog")* at the Ball Bearing Company at Bantam, Connecticut.
Awarded four patents for centerless grinding filed in 1925 and 1926.

9: Michael J. Dempsey: Age: 32
Mechanical Engineer at the Miamus Motor Company at Stamford, Connecticut. Former employee of the Ball Bearing Company at Bantam, Connecticut.
Testified on use of "The Hog" roll-grinding machine at Bantam.

10: Frederick L. Parker: Age: 32 Diemaker, Toolmaker, and Machinist
Former employee of the Ball Bearing Company at Bantam, Connecticut.
Testified on use of "The Hog" roll-grinding machine at Bantam.

11: Charles F. Conable: Age: 50
Foreman at Trumbell Vanderpool Company.
Former employee of the Bantam Antifriction Company that later became the Ball Bearing Company at Bantam, Connecticut.
Testified on use of "The Hog" roll-grinding machine.

12: Ernest Jeffries: Age: 38
Machinist at the Bantam Electric Manufacturing Company at Bantam, Connecticut.
Former employee of the Bantam Antifriction Company that later became the Ball Bearing Company at Bantam, Connecticut.
Testified on use of "The Hog" roll-grinding machine.

13: William H. Callahan: Age: 39 Machinist.
Former employee of the Ball Bearing Company at Boston, Massachusetts. Testified on use of the Chadwick double-disc roll-polishing machine.

14: Joseph P. Tracy: Age: 43
Electrician at the Union Manufacturing Works.
Former employee *(grinder)* at "both" the Ball Bearing Company near Boston, Massachusetts, and the Anti-Friction Company at Bantam, Connecticut.
Operated the Chadwick double-disc grinder at Boston and testified that both wheels rotated downward.
Testified that the Chadwick double-disc grinder was not used at Bantam.
Testified that the Gardner disc grinder was the machine called "The Hog" at Bantam.

15: Frederick P. Gates:
Superintendent of Hart Manufacturing Company at Hartford, Connecticut. Formerly superintendent of the Anti-Friction Company at Bantam, Connecticut. Testified on use of "The Hog" roll-grinding machine.

A2: List of Witnesses for the F. C. Sanford Manufacturing Company, Defendant.

16: Arthur Browne: Age: 60
Patent Solicitor and expert in patent litigation. Expert witness for the defendant. Testified that the Sanford centerless grinder did not infringe on the Heim centerless grinder patents.

17: Winfield S. Rogers: Age: Unknown. Estimated at 66 to 70 years old. Mechanical Engineer, Chairman of the Board of Directors, and founder of the Ball Bearing Company of Bantam, Connecticut. *(Originally the Anti-Friction Company before changing to current name.)*
Founder of the Ball Bearing Company of Boston, Massachusetts, in 1894.
Originator of the modified Chadwick two-wheel grinding machine called "The Hog" used at the Ball Bearing Company at Bantam, Connecticut.
Father-in-law to defendant witness Lee Chadwick who invented the two-wheel roll-polishing machine used at Boston.

18: William A. Weed: Age: 46
Foreman at the F. C. Sanford Company at Bridgeport, Connecticut.
Former machinist and head of the grinding department at the Ball and Roller Bearing Company.
Had thorough knowledge of construction and operation of Heim's first double-ring wheel centerless grinder.

19: James J. Daley:
Demonstrator of grinding machines for the F. C. Sanford Company.
Former employee at the Ball and Roller Bearing Company assembling roller bearings. Witnessed development of Heim's single-wheel and double-ring wheel centerless grinders.
Witnessed the building of Sanford's first two-wheel peripheral centerless grinder.

20: Francis (Frank) C. Sanford: Age: Around 49 or 50. Defendant in the trial.
President of the F. C. Sanford Manufacturing Company at Bridgeport, Connecticut.
Inventor of the Sanford Centerless Grinder, Models A & B.

21: Henry Edwards: Age: 29
Engineer at the Bantam Ball Bearing Company at Bantam, Connecticut. Demonstrated the Bantam two-wheel grinding machine *("The Hog")* during the trial.

22: Nellie Scott: Age: Unknown but at least 35.
President of the Bantam Ball Bearing Company at Bantam, Connecticut. Employed at Bantam for over 16 years, starting off as stenographer and later superintendent of the plant.
Testified on the Bantam two-wheel (Hog) grinder used as a prior art machine.

23: Samuel S. Eveland: Age: 48
Founder of the Standard Roller Bearing Company at Philadelphia, Pennsylvania. Bought the Ball Bearing Company of Boston, Massachusetts.
Inventor of the first single-wheel centerless grinder *(patents no. 747,541 and 747,542)*.

24: Lee S. Chadwick: Age: 46
Director of Manufacturing at the Cleveland Metal Products Company in Ohio. Inventor of the two-wheel centerless grinder used at the Ball Bearing Company in Boston and featured in the November 1900 issue of *Machinery* magazine.

B: Trial Summary: Testimony, Dates, and Locations

This section summarizes the dates, people, and order of testimony in the Heim patent infringement trial against Frank Sanford and his manufacturing company.

Testimony: All testimony was taken in May and June 1921.
Dates shown with an asterisk (*) are documented in the testimony transcripts.
All other dates are estimated based on order of appearance in the court documents.

Trial Locations: Unless noted, all testimony was taken at the United States District Court in South Norwalk, Connecticut.

Source Locations:
Plaintiff (P): Found in Transcript Volume 1 Defendant (D): Found in Transcript Volume 2

Witness	Type of Testimony	Page
1: Samuel S. Eveland *(D)*	Direct/Deposition	68–84
Taken out of court on May 12 in Philadelphia, Pennsylvania.		
2: Winfield S. Rogers *(D)*	Direct	173–198
Mon., May 16*		
3. Harold Pender *(P)*	Direct	48–109
Mon., May 16* and Tues., May 17*		
Tues., May 17, 10:30 a.m.:* The Norwalk Iron and Foundry Company Norwalk, Connecticut.		
4: Lewis Heim *(P)*	Direct	109–163
Tues., May 17		
5: William J. Ryan *(P)*	Direct	163–178
Tues., May 17.* Adjourned to Wed., May 28		
6: Arthur Kirkland *(P)*	Direct	178–184
Wed., May 18		
7: Charles O. Johnson *(P)*	Direct	184–191
Wed., May 18		
8: James G. Bennett *(P)*	Direct	191–195
Wed., May 18		

23: Michael J. Dempsey (P) Rebuttal 272–278
    Wed., June 29

24: Frederick L. Parker (P) Rebuttal 278–282
    Wed., June 29

25: Charles F. Conable (P) Rebuttal 282–286
    Wed., June 29

26: Ernest Jefferies (P) Rebuttal 286–291
    Wed., June 29,* Adjourned to the B&RB.

27: Harold Pender (P) Rebuttal 291–321
    Wed., June 29, 3:15 p.m.:* Demonstration of four machines
    at the Ball and Roller Bearing Company, Danbury, Connecticut.

28: William H. Callahan (P) Rebuttal 321–335
    Thurs., June 30, 8:00 a.m.*

29: Joseph P. Tracy (P) Rebuttal 335–347
    Thurs., June 30

30: Clement Booth (P) Rebuttal 347–363
    Thurs., June 30

31: Frederick P. Gates (P) Rebuttal 363–366
    Thurs., June 30

32: Harold Pender (P) Rebuttal 366–423
    Thurs. June 30

End of Testimony

Summary of Docket Entries

UNITED STATES DISTRICT COURT
DISTRICT OF CONNECTICUT.

The Heim Grinder Company

     vs.   No. 1747 Equity.

The Fafnir Bearing Company

DOCKET ENTRIES.

Nov. 11, 1924	Bill of complaint claiming infringement of Letters Patent Nos.; 1,281,366 and 15,035, etc., filed.
Nov. 11, 1924	Subpoena issued and handed to the Marshal for service.
Nov. 11, 1924	Certified copy of bill of complaint made and handed to the Marshal for service.
Nov. 19, 1924	Subpoena returned showing service under date of Nov. 15, 1924.
Dec. 8, 1924	Stipulation and order extending defendant's time for filing answer to and including December 16, 1924, filed and entered.
Dec. 16, 1924	Answer of defendant filed.
Jan. 1, 1925	Plaintiff's interrogatories filed in duplicate and duplicate mailed to Fish, Richardson & Neave.
Jan. 12, 1925	Order that defendant answer interrogatories filed and entered.
Jan. 12, 1925	Report to Commissioner of Patents.
Jan. 12, 1925	Certified copy of Plaintiff's Interrogatories made and mailed to Fish, Richardson & Neave.
Jan. 23, 1925	Objections to Plaintiff's interrogatories filed in duplicate and duplicate mailed to Bristol & White.
Feb. 16, 1925	Defendant's answers to plaintiff's interrogatories filed in duplicate and duplicate mailed to Robert S. Blair, Esq. Court copies - 29 folios.
Dec. 24, 1925	Eight subpoenas Issued and handed to the marshal for service.
Dec. 30, 1925	Subpoenas returned showing service upon Adelbert Warner and Davis Beloin.
Jan. 15, 1926	Subpoena returned showing failure to make service upon Frank A. Pardee. Also service was made upon C. P. W. Christenson, Carl E. Johnson, Allan H. Jones, John E. Hayes and C. I. Shirley.

Jan. 11, 1926	Hearing had upon merits. Attorneys Blair and Gifford ap peared for the plaintiff and Mr. Stackpole for defendant. Depositions introduced and left with stenographer - exhibits also left with stenographer. Six witnesses were sworn and testified for the plaintiff.
Jan. 12, 1926	Hearing resumes. One witness was sworn and testified for the plaintiff. Plaintiff rests. Attorney Stackpole made a statement of his case and Attorney Blair replied.

No. 1747 Eq.	-2-	Docket Entries.

Jan. 13, 1926	Hearing resumed. Three witnesses were sworn and testified on behalf of the defendant. At 10 A. M. the Judge, Attorneys and witnesses left for Hirshfield's warehouse to see demonstration of the working of the machine in question. Six witnesses were sworn and testified for the defendant.
Jan. 14, 1926	One witness was sworn and testified for the defendant. Defendant rests.
Jan. 14, 1926	The plaintiff made a motion to add Detroit Machine & Tool Co. as defendant. Decision reserved. One witness was sworn and testified for the plaintiff. Court adjourned to Jan. 18th.
Jan. 18, 1926	Hearing resumed. Six witnesses were sworn and testified for the plaintiff. Case continued to Jan. 19th.
Jan. 19, 1926	Witness Heim continued testimony in rebuttal. Other winesses recalled in rebuttal. At 6 P.M. case closed andarguments to be heard later.
Jan. 23, 1926	Copy of list of exhibits filed.
April 1, 1926	Arguments were heard upon merits of the case. Decision reserved.
April 10, 1926	Original transcript filed.
May 18, 1926	Opinion filed.
July 2, 1926	Decree filed and entered. Affidavit of Mr. Raymond R. Searles filed.
July 3, 1926	Injunction issued and with a certified copy of same handed to the Marshal for service.
July 14, 1926 1926.	Injunction returned showing service under date of July 3,
July 29, 1926	Waiver of profits and damages and order there-on filed and entered.
July 31, 1926	Defendant's petition for appeal and order allowing same entered and filed.
July 31, 1926	Defendant's assignments of error filed.
Sept. 16, 1926	Citation issued returnable thirty days from Sept. 15, 1926.

Sept. 22, 1926	Citation returned showing service accepted by Robert S. Blair.
Oct.1, 1926	Defendant's praecipe to Clerk on appeal filed.
Oct.6, 1926	Stipulation and order that depositions and evidence need not be stated in narrative form etc., filed and entered.
Oct.18, 1926	Stipulation and order extending time to file record on appeal to and including November 15, 1926, filed end entered.
Oct.26, 1926	Stipulation re transcript of record filed.
Nov.1, 1926	Stipulation and order re withdrawal of exhibits filed and entered.
Dec.30, 1926	Mandate of C. C. A. filed.
Jan.4, 1927	Consent and order modifying injunction filed and entered.
Jan.4, 1927	Satisfaction of costs filed. 1% bond.

A3-4

Bill of Complaint

UNITED STATES DISTRICT COURT
FOR THE DISTRICT OF CONNECTICUT

THE HEIM GRINDER COMPANY,
Plaintiff,

Vs.

THE FAFNIR BEARING COMPANY,
Defendant.

In Equity No.
U. S. Letters Patent No. 1,281,366 and Reissue No. 15,035,

TO THE HONORABLE THE JUDGES OF THE DISTRICT COURT OF THE UNIT-
ED STATES IN AND FOR THE DISTRICT COURT OF THE SECOND CIRCUIT:

The Heim Grinder Company, a corporation duly organized and existing under and
in accordance with the laws of the State of Connecticut, a citizen of said State, having an office for the transaction of business in Danbury, County of Fairfield, State of
Connecticut, brings this, its Bill of Complaint, against the Fafnir Bearing Company, a
corporation organized and existing under and in accordance with the laws of the State
of Connecticut, a citizen of said State, having an office for the transaction of business
in New Britain, County Of Hartford, State of Connecticut, where it has committed the
acts of infringement complained of herein.

For cause of action, plaintiff avers;

1. That the plaintiff's said cause of motion arises under the patent laws of the United
States and more specifically, the defendant, in the District of Connecticut has committed the acts of infringement herein complained of in violation of the rights of this
plaintiff as sole owner of United States Letters Patent No. 1,281,366, granted October
15, 1918, and of United States Reissue Letters Patent No. 15,035, granted January
25, 1921, in both of which Lewis R. Heim is the inventor; that on March 6, 1915, the
said Lewis R. Heim, a citizen of the United States, and then a resident of Danbury,
County of Fairfield, State of Connecticut, duly filed an application (hereinafter called
"original application") for Letters Patent of the United States, Serial No. 12,524, for
"Roll Grinding Machine"; that on July 3, 1916 the said Lewis R. Heim duly filed an
application for Letters Pate t of the United States, Serial No. 107,246 for "Method of
Grinding Hardened Rolls," said application Serial No. 107,246 being a division of
said original application Serial No. 12,524; that on October 15, 1918 Letters Patent of

the United States No. 1,281,366 did issue duly signed and sealed to The Ball & Roller Bearing Company of Danbury, Connecticut, a corporation of the State of Connecticut, as assignee of the said Lewis R. Heim, on the said divisional application, Serial No. 107,246, as by said Letters Patent or a duly certified copy thereof, which is ready in Court to be produced, will more fully appear; that on January 2, 1917 Letters Patent of the United States No. 1,210,927 did issue duly signed and sealed to the said The Ball & Roller Bearing Company, an assignee of the said Lewis R. Heim, on the said original application, Serial No. 12,524, as by a duly certified copy thereof which is ready in Court to be produced, will more fully appear; that the said Letters Patent No. 1,210,937 granted on the said original application was defective or inoperative within the meaning of the statutes providing for reissue thereof, and that on September 8, 1920 the said Lewis R. Heim, with the consent of the said The Ball & Roller Bearing Company, filed an application. Serial So. 409,011, for the reissue of the said original Letters Patent No. 1,210,937, and that on January 25, 1921 Reissue Letters Patent of the United States No. 15,035, did issue duly signed and sealed to the said The Ball & Roller Bearing Company, as assignee of the said Lewis R. Heim, as by said Letters Patent or a duly certified copy thereof, which is ready in Court to be produced, will more fully appear.

2. That the said Lewis R. Heim was the first, original and sole inventor of the inventions dealt with and covered in the said Letters Patent; that the said inventions and improvements had not been known or used by others in this country before his invention or discovery thereof, and was not patented or described in any printed publication in this or any foreign country before his invention thereof, nor more than two years prior to his original application for Letters Patent therefor and upon which said Letters Patent No. 1,281,366 and Reissue Letters Patent No. 15,035 were ultimately issued, and was not in public use or on sale in this country for more than two years prior to his original application for Letters Patent therefor in the United States; that no application for patent for said invention had ever been filed by said Lewis R. Heim or his legal representatives or assigns in any country foreign to the United States prior to his filing said original application for Letters Patent therefor in the United States; that his said applications for Letters Patent therefor in the United States and the inventions were never abandoned; and that, in the making of the hereinbefore mentioned applications for Letters Patent, the said Lewis R. Heim duly complied in all respects with all conditions and requirements of the United States Statutes in such cases made and provided and then in force.

3. Plaintiff further avers that it now holds and possesses the entire right, title and interest in and to the said United States Letters Patent No. 1,281,366 and Reissue No. 15,035, and the inventions thereof, by due, proper and lawful assignment in writing, from the said The Ball & Roller Bearing Company for a valuable consideration, together with all right to bring suit based upon any infringement of the said Letters Patents or either of them, and to collect profits and damages by reason of said infringements, whereby the entire legal and equitable title to the said Letters Patents and inventions and all rights thereunder, are now vested in plaintiff, of all of which plaintiff

stands ready to make due and proper proof.

4. Plaintiff further avers that, in 1921, and shortly after the issue of the said Reissue Letters Patent No. 15,035, the said The Ball & Roller Bearing Company, then the owner of the entire right, title and interest in and to the said Reissue Letters Patent, brought suit in the United States District Court for the District of Connecticut to restrain infringement of the said Letters Patent by the F. C. Sanford Manufacturing Company; that the trial of the said suit was held in the said United States District Court for the said District; that upon full proofs and upon appeal to the Court of Appeals for the Second Circuit, the said Court of Appeals held the said Reissue Letters Patent No. 15,035, to be valid and to be infringed by said The F. C. Sanford Manufacturing Company; that the plaintiff, having acquired prior to the entry of the decree of the said District Court on mandate of the said Court of Appeals, the entire right, title and interest in and to the said Reissue Letters Patent as aforesaid, brought its Supplemental Bill of Complaint in the said United States District Court for the District of Connecticut against the said The F. C. Sanford Manufacturing Company; and that the said Court thereupon, in conformity with the mandate of the said Court of Appeals, decreed the said Reissue Letters Patent No. 15,035 to be valid and to be infringed by the said The F. C. Sanford Manufacturing Company, and issued, in behalf of the plaintiff herein, a writ of injunction restraining such infringement, that the said injunction is still in force, and that heavy damages were received by plaintiff arising out of such infringement.

5. Plaintiff further avers that the said The Ball & Roller Bearing Company, its predecessor in right, title and interest in and to the said Letters Patents had made extensive practical use of the inventions and improvements therein set forth, and that plaintiff, since It has acquired the entire right, title and interest in and to said Letters Patents and inventions, as above set forth, has made extensive practical use of the inventions and improvements therein set forth, plaintiff's business being built up upon the said inventions; that, with the exception of the manufacturers of the machine used by the defendant, no one is at present, in so far as plaintiff is aware, manufacturing infringing apparatus, other manufacturers of infringing machines having been enjoined, gone out of the business of making infringing apparatus, or taken a license under the said Letters Patent; that plaintiff has marked the apparatus which it has put out in accordance with said invention as set forth in the said Reissue Letters Patent No. 15,035, and the apparatus which it has put out for carrying out the invention set forth in Letters Patent No. 1,281,366, as patented, together with the date of the respective patents. In compliance with the statutes, hereby it has given due notice to the public; and that since the issuance of the said Letters Patent and prior to this time, the defendant has, in the District of Connecticut, wrongfully infringed the said Reissue Letters Patent No. 15,035 by using apparatus made in accordance with, and in infringement of the said Letters Patent and of the invention therein set out and claimed, and has wrongfully infringed the said Letters Patent No. 1,281,366 by practicing the invention therein set out and claimed, and in infringement of the said Letters Patent, all against plaintiff's will, notwithstanding due notice of said Letters Patents as above, as well as by advertising, and without plaintiff's license of consent; that,

upon information and belief, defendant has continued to infringe and is continuing and threatening so to do up to the present time; that by said wrongful acts, defendant has diverted to itself trade and profits which plaintiff would otherwise have received, as to the amount of which plaintiff prays recovery, whereby plaintiff has been caused great and irreparable damage.

Wherefore, plaintiff prays for the immediate issue of an injunction Pendente Lite, and for the issue of a permanent injunction, issuing out of and under the seal of this Court enjoining and restraining the said defendant, its attorneys, agents, servants and employees, from directly or indirectly or in any way or manner whatsoever, infringing the said Reissue Letters Patent No. 15,035, and Letters Patent No. 1,281,366 in violation of the rights of the plaintiff in the premises; that defendant account to the plaintiff for the profits made by it and the damage sustained by the plaintiff; the plaintiff recover the costs and disbursements of this action, as a direct or indirect result of such infringement; and plaintiff prays that a subpoena may issue forthwith out of and under the seal of this Court directed to the said defendant, requiring it upon a day certain and under certain penalty, to appear and make full, true and certain answer to this Bill of Complaint, and plaintiff prays for such other and further relief as may appear proper to the Court and agreeable to equity.

THE HEIM GRINDER COMPANY

By _Henry N. Flynt_ ← Signed
President

Resident Solicitors

ROBERT S. BLAIR
Of Counsel,
50 East 42nd Street,
New York, New York

UNITED STATES OF AMERICA:
STATE OF NEW YORK: ss.
COUNTY OF NEW YORK:

On this 10th day of November, 1924, before me, a Notary Public, appeared HENRY N. FLYNT, who being by me first duly sworn, deposes and says: That he is Henry N. Flynt, President of The Heim Grinder Company, plaintiff herein; that he has read the foregoing Bill of Complaint by him subscribed and knows the contents thereof; that the same is true of his own knowledge except as to those matters herein stated upon information and belief, and as to those matters he believe it to be true.

_____
Notary Public

The Heim Company

Contents

This section contains various documents to support the history of the Heim Company and the bearings invented by Lewis Heim.

A4-1:   Article: "Fairfield War Plant Beats Nazis at Own Game,"
Bridgeport Sunday Post, April 23, 1944
Reprint of the article published in 1944 on the origins of the rod end bearing invented by Lewis Heim. Included are four pictures from the original article showing Heim, his son Charles, and other employees working at the plant.

A4-2:   Heim Company Catalog #14, December 1950.
Selected pages from the catalog that describe Heim's history and a summary of applications of Unibal rod end and spherical bearings. The catalog contains a list of factory representatives and distributors, one of which is the company of Edward Maltby, the bearings broker who contacted Heim to duplicatethe Messerschmitt slotted bearing.

A4-3:   Heim Company Catalog #23, ca. 1950
Selected pages from the catalog showing ball and roller bearings constructed from stamped sheet metal construction.

A4-4:   Advertisements: Examples of Heim Rod End and Spherical Bearing Applications: ca. 1960
Contains ten advertisements published by the Heim Company to promote rod end and spherical bearings. The advertisements illustrate the many applications of Heim's bearings, including their primary features for handling misalignment in linkages and transmitting motion at varying angles. The last advertisement includes as statement that rod end bearings were developed during World War II for exclusive use in aircraft.

Bridgeport Post Article on the Heim Company

Reprinted from the *Bridgeport Sunday Post*: Feature Section The Bridgeport Sunday Post, April 23, 1944

Fairfield War Plant Beats Nazis At Own Game

AAF Materiel Center Aids Heim Co.

Better Nazi Plane Secret At Plant Here

Fairfield Plant Outblitzes Nazis On Plane Jobs

At 46 Sanford street, Fairfield, there's a little war plant which has some of the topkicks in American military aircraft manufacture shaking their heads and wondering how they do it—turn out the dependable, highly-efficient Heim Company rod-end bearing in quantities thought impossible a mere 18 months ago.

Probably 999 people out of a thousand have never heard of a rod-end bearing and probably aren't too eager to, and yet there's a romance back of the Heim product which smacks of the best efforts of a Hollywood scenario writer.

"Enemy secrets," said Charlie Heim, telling the story. "Actually we're indebted to the Nazis for the basic idea back of our spherical rod-end bearing. To the Nazis and British Intelligence and the U.S. Army Air Forces Materiel Command."

Former Yale Grid Star

Although at the outset Charlie Heim's yarn sounds like something E. Phillips Oppenheim might have written, the youthful treasurer and general manager of the Heim Co. looks far different from anybody's conception of a sinister Oppenheim character. Charlie is stocky, smiling and affable. A dozen years ago he played football for Yale, a teammate of the never-to-be-forgotten Albie Booth. Today he's demonstrating some of his quarterback pluck and ingenuity making goals for Uncle Sam in the shape of vital aeroplane parts.

Back in 1942 the British and Field Marshal Rommel were clawing at one another in North Africa. Of serious annoyance to the British was a new type Messerschmitt fighter, ducking in, strafing, and getting away fast because of its easy maneuverability. Finally the British brought down a specimen and intelligence officers ordered it sent to this country—to Vultee Field, California, for dismantling.

Experts from our AAF Materiel did the job, ending up with the aeronautical equivalent of an autopsy report. The maneuverability of the downed Messerschmitt was attributed to bearing at the end of a rod or cable running from the controls of the plane to its rudder, wingflaps and ailerons.

The Materiel Command officers rushed the devices to Edward D. Maltby, Los Angeles bearing broker. "Can you find somebody to duplicate this?" they asked him, and straightaway a Maltby representative flew to Connecticut, the center of the bearing industry.

Shop of Coincidence

"Funny," says Charlie Heim, going on with the story. "This whole shop of ours is full of coincidence. Take Eric Hildes-Heim, our sales manager, for instance. If it hadn't been for Eric, the Maltby man would never have heard of us, and if Eric's name wasn't Hildes-Heim, we wouldn't have heard of him."

Hildes-Heim is a Dane, slight, intent. One of the first to fly the grandfather of the present-day helicopter, he was chosen because he was light and had a better chance of surviving ascent in the fragile, embryonic craft than a heavy man. Even so, he wore a crash helmet.

During World War I, Hildes-Heim flew the Danish neutrality patrol and bears the distinction of having seen absolutely no action. From 1914 through 1918 the Danes came to blows with but one German plane, shooting it down. Hildes-Heim was grounded that memorable day with a belly-ache.

At the end of the war, he came to this country with a Danish aeronautical purchasing commission. In 1928, he returned as a representative of Danish banking interests, this time to stay. He joined the Heim Co. in 1940 when they were making roller bearings, taper pins and dowel pins.

"The fact that his name is Hildes-Heim doesn't mean he's a relative," Charlie Heim explained. "Before he showed up here at the plant one day, we'd never seen him. He came out of curiosity. I guess he wanted to find out what people looked like who had part of his own name."

Because of his aviation background, Hildes-Heim knew Maltby's man from California. The two got together, Hildes-Heim took the Messerschmitt part to Lewis R. Heim, president and founder of the Heim company. "That's my dad," Charlie Heim said, a note of pride in his voice. "He's a mechanical genius; there's no doubt about it. There isn't any problem involving machinery he can't solve. Just try him."

Lewis R. Heim will be 70 his next birthday but looks 10 or 15 years younger. He's stocky like his son, even stockier. Despite his age, he works long hours in the shop and office, and when he isn't to be found in either place, he's still working at his Fairfield Beach home, bent over a drawing board in a glassed-in study overlooking the sound.

Lewis Heim was born on a farm at New Hamburg, N.Y. near Poughkeepsie. He attended a back-road school through the Sixth grade and that's all the formal education he's ever had. At 16 he left the farm; agriculture didn't appeal to him; already a fondness for things mechanical was beginning to shape his career.

He got a job in a Danbury hat factory. Two years later he'd invented and patented a felt-stiffening machine still in use by hatters all over the world. Unfortunately, his mechanical ability exceeded his youthful business acumen. He sold his patent rights for the proverbial song, missing early riches.

Next he turned his attention to laundry machinery, filing a series of patents, at least one of which proved a real boon to the male of the species. The Heim Collar Roller takes the Gay Ninety razor edge off stiff collars, allowing a man to attend a dinner party without having his throat cut.

Invented Centerless Grinder

In 1913 an invention of Lewis Heim's revolutionized machine shop practice throughoutthe globe. Skeptics said a centerless grinder couldn't be built, but Lewis Heim built it. As a result, such important industrial items as cylindrical shafting, roller bearings, bullet cores and all cylindrical motor and machine parts are ground today in less than two percent of the time it used to take and with a close tolerance and smooth finish previously believed impossible.

"Of all his inventions," Charlie Heim said, "that's the one my dad wants to be remembered for. The centerless grinder conserves manpower and saves long hours of backbreaking labor."

As in World War II, the talents and facilities of Lewis Heim were devoted to war work in World War I—100 percent. Operating under the style of the Ball and Roller Bearing Co., he made submarine bearings in Danbury. In 1932 the present Heim Co. was formed, moving to Fairfield three years later.

In June, 1941, they experienced a fire destroying their assembly and inspection departments and had not yet recovered completely from the disastrous affects of the flames when called upon by the AAF Materiel Command to duplicate the Messerschmitt rod-end bearing. At that time they had less than one-third their present shop space, one-fourth as many employees.

Withal, Lewis Heim faced the problems with characteristic fortitude, realizing its urgency. Up to now the rod-end bearings used in American aircraft were for the most part dependent on ball bearings, a circlet of balls contained in each housing.

414

New Design in Five Minutes

"My dad studied the Messerschmitt part for five minutes," Charlie Heim went on, "and then began sketching with his pencil. He came up with an answer that same day. The rod-end bearing he designed was something like Messerschmitt's, to be sure, but simpler to manufacture—more efficient—and made of less critical material!"

The Messerschmitt bearing consists of a shank ending in a slotted housing, with a drilled-through spherical slice inserted and turned at a 90-degree angle. The spherical slice is all; there are no other balls. Necessarily, the entire part is made of high-alloy steel.

Heim's bearing omits the slotting and slicing. He inserts a drilled-through perfect sphere, thus allowing 360 degree freedom as opposed to the Messerschmitt's 90. The sphere is secured in the housing by stamped-in brass inserts. Steel against brass creates less friction than steel against steel. Hence Heim is able to use a steel low in carbon and manganese and a low-alloy brass—a type no more precious than that in ordinary kitchen faucets.

Passed Rigid Tests

AAF Materiel Command officers took Heim's bearing to Wright Field and put it through rigid engineering tests. It stood up—better than any rod-end bearing ever tested. On September 4, 1942, the Materiel Command approved the first Heim Co. rod-end bearing contract. Just 67 days later the Heim Company began shipping finished products to aircraft manufacturers.

"During those two months we were busier'n all get-out," Charlie Heim continued, "Thank goodness Carl had come with us then. If it hadn't been for Carl—well, I don't know what we would've done. Carl was a big help."

Carl C. Van Etten is Lewis Heim's son-in-law and secretary of the Heim Co., joining the concern earlier that same year. A graduate of Lafayette, he formerly taught American History in a Scarsdale, N.Y. High School, and laughs over his pedagogical experience as adequate background for an industrial executive. Van Etten handles office management, correspondence, accounting supervision and a myriad other details. His flair for top-speed business might amaze some of his old associates.

Former school teacher, former Danish aviator, former Yale football player—these three, led by the native genius of a farm-bred mechanic, and backed by a patriotic collection of both skilled and unskilled workers—have done a truly remarkable war-winning job.

New Heim company buildings went up almost overnight and additional floors in the old buildings suddenly found themselves occupied by Heim lathes, screw machines, grinders and drill presses. Withal, little or no new machinery was acquired. Grey-haired Lewis Heim got into an old-stained suit of coveralls and began on-the-spot direction of one of the fastest and most ingenious machine-converting and re-tooling program ever heard about.

A "Help Wanted" sign went up outside and they hired anybody who came along, a practice in which they still persist. They took on grandmothers of 65 and boys and girls with their first working papers, farmers, butlers, hairdressers, housewives, soda-jerkers, sales clerks and hash-slingers. Even today they haven't enough help, despite the fact that representatives of almost every nationality under the sun are on their payroll.

They paid wages as high as the law allowed—and they still pay them. Worker morale is high. Most Heim employees have a brother, son, husband or sweetheart in some branch of the armed services and they're in there pitching—backing up their fighting men.

Aid U.S. and Allies

As a result, Heim rod-end bearings aid in the flight of nearly every American warplane and thousands are shipped to our Allies under lend-lease. AAF Materiel Command experts from the Area Office at 109 Church street, New Haven, visit the plant constantly, ironing out production snags, testing and inspecting finished work, helping to solve raw material procurement and transportation problems.

Production mounts and continues to mount. The second month the Heim company began rod-end bearing manufacture, they turned out three times as many as they did the first—in March, 1944, more than 20 times as many as in November, 1942.

"We're doing our best," Charlie Heim concluded his story. "Every worker in this plant is giving all he's got. But we aren't satisfied yet—and we won't be satisfied until every Allied plane is fully equipped with spherical rod-end bearings both inside and out."

Lewis R. Heim, president and founder of the Heim company in Fairfield, now busy on important war work.

Charles Heim, left, former Yale football star, and now treasurer and general manager of the company, confers with Edward Dardani, general foreman of the Fairfield plant, while they inspect a tapping machine for finishing threads in shanks.

Eric Hildes Heim, sales manager of the Heim company, former Danish aviator, who joined the firm after he had made a visit out of curiosity to see if a concern with the name Heim was run by any relative of his.

Women workers form a large percentage of the Heim company. Above are; Ann Konduk of Westport, Mrs. Helen Nassef of Fairfield and Mildred Hanson of Bridgeport.

HEIM
UNIBAL
SPHERICAL BEARINGS
AND
SPHERICAL BEARING
ROD ENDS

Catalogue No. 14, December 1950

THE HEIM COMPANY, FAIRFIELD, CONN.

A4-2-2

2

LEWIS R. HEIM

HISTORY

With forty years of experience in the anti-friction bearing industry and with a long list of inventions to his credit on hatting machinery, steam laundry machinery, ball bearings, roller bearings and the centerless grinder and most of its refinements, Lewis R. Heim was a natural source from whom the aviation industry might expect something out of the ordinary.

When aircraft deliveries were held up because of a critical shortage in rod ends and self-aligning bearings, this fact was brought to the attention of Mr. Heim and the result is this new and revolutionary Heim Unibal spherical bearing.

It is with pardonable pride that we introduce to American manufacturers generally this new type of bearing which has been accepted so enthusiastically by the aircraft industry.

THE HEIM COMPANY

THE HEIM COMPANY ... FAIRFIELD, CONN.

FOREWORD

Heim Unibal Spherical Bearings and Spherical Bearing Rod Ends have a greater carrying capacity than the conventional type since they have a greater surface supporting area. Because of this area, they are not subject to false brinelling. By their very design, maximum correction of misalignment is obtained. Because of their revolutionary construction, they will take a greater radial and axial thrust load with resulting longer life.

Heim Unibal Spherical Bearings are manufactured in many sizes to cover a variety of applications. Heim Unibal Spherical Bearings and Rod Ends are produced in standard and light series. Male and female rod ends can be furnished in both standard and light series.

The ball is made from SAE 52100 steel, hardened, precision ground and lapped. The bearing inserts, commonly called the race, are made of bearing bronze. The outer ring of the spherical bearings and the outer member of the rod end bearings can be made of almost any suitable material. Standard units are made from carbon steel. Units to be used on aircraft are made from approved aircraft steel. All bearings are prelubricated and grease fittings will be furnished on rod ends when requested.

All parts of the Heim Unibal Bearings and Rod Ends are produced in our own modern plant by highly trained and experienced personnel. Every bearing must pass the most rigid inspection after which they are thoroughly cleaned, rust proofed and packed in moisture and dust-proof wrappers and containers. All aircraft rod ends will be magnetically tested if required under Army or Navy contracts.

We strongly recommend the use of our standard bearings and rod ends wherever possible. However, we are prepared to produce special bearings and rod end bearings for any given application. Tell us your bearing problem and our Engineering Department will be pleased to submit their recommendations.

GUARANTEE — Heim Unibal Spherical Bearings and Spherical Bearing Rod Ends are guaranteed against defects in material and workmanship. Any bearings proving defective may be returned, charges prepaid, and we will replace, without charge, F.O.B., Fairfield, Conn. Under no circumstances will The Heim Co. assume responsibility for any contingent charges.

UNIBAL SPHERICAL BEARINGS AND SPHERICAL BEARING ROD ENDS

A4-2-4

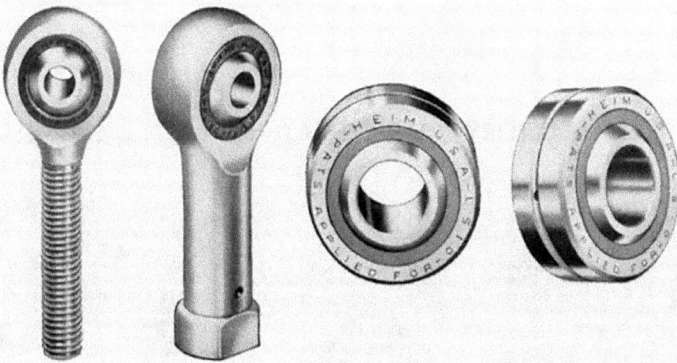

PARTIAL LIST OF APPLICATIONS WHERE HEIM UNIBAL
SPHERICAL BEARINGS AND SPHERICAL BEARING
ROD ENDS ARE USED—

ELECTRIC FANS
SHOE MACHINERY
TEXTILE MACHINERY
CARPET LOOMS
ENGINE CONTROLS FOR TRUCKS AND MOTORS
MARINE APPLICATIONS
FLEXIBLE COUPLINGS
CAM FOLLOWERS
AIRCRAFT
OSCILLATING FANS
ELECTRICAL EQUIPMENT
SEWING MACHINES
ROAD BUILDING MACHINES
PRINTING PRESSES
OIL FIELD EQUIPMENT
ELECTRIC SWITCHES
WASHING MACHINES AND LAUNDRY EQUIPMENT
HYDRAULIC EQUIPMENT
REFRIGERATORS
CANNING MACHINERY
MACHINE TOOLS
GOVERNOR CONTROLS

UNIBAL SPHERICAL BEARINGS AND SPHERICAL BEARING ROD ENDS

DIRECT FACTORY REPRESENTATIVES AND DISTRIBUTORS

ALLIED BEARINGS SUPPLY CO.
822 S. Boulder Avenue, Tulsa 5, Okla.

ANDERSON BEARING CO.
314 Circle Ave., Forest Park, Ill.

BEARING ENGINEERING COMPANY
1545 Mission Street, San Francisco, Cal.

BEARING ENGINEERING COMPANY
1019 Harrison Street, Oakland, Cal.

BEARING INDUSTRIES
6112 E. 14th Street, Oakland 3, Cal.

BEARING SALES & SERVICE, INC.
919 East Pine Street, Seattle, Wash.

BEARING SALES & SERVICE, INC.
1718 Pacific Avenue, Tacoma 9, Wash.

BEARING SALES & SERVICE, INC.
1109 N. W. Glisan Street, Portland, Oregon

BEARING SALES & SERVICE, INC.
Eugene, Oregon

BEARING SALES & SERVICE, INC.
Roseburg, Oregon

BEARING SERVICE CO.
340 St. Francis Street, Wichita, Kans.

BEARINGS SERVICE COMPANY, INC.
412 E. Sample St., South Bend, Ind.

BEARINGS SERVICE COMPANY, INC.
9 N. W. First Street, Evansville, Ind.

BEARINGS SERVICE COMPANY, INC.
111 South Market Street, Marion, Ill.

BEARINGS SERVICE COMPANY, INC.
211 West Center Street, Madisonville, Ky.

BEARING SERVICE & SUPPLY CO.
1850 Market St., Denver 2, Colorado

BEARING SERVICE & SUPPLY CO.
1207 S. Main Street
P.O. Box 1166, Salt Lake City 11, Utah

BEARINGS SUPPLY CO
2412 First Avenue N., Billings, Mont.

BEARING & TRANSMISSION COMPANY
P.O. Box 667
214 North Market Street, Shreveport, La.

BEHRING'S BEARING SERVICE
P.O. Box 57, Houston 3, Texas

BENSON ENGINEERING COMPANY
30½ Highland Park Village, Dallas, Texas

BENSON ENGINEERING COMPANY
519 S. Broadway, Wichita, Kan.

BERRY BEARING COMPANY
2633 Michigan Avenue, Chicago, Ill.

BERRY BEARING COMPANY
4828 Calumet Avenue, Hammond, Ind.

ROGER BROWN BEARING CO.
111 East Missouri St., El Paso, Texas

CARPENTER BEARING COMPANY
P. O. Box 2336, Abilene, Texas

CARTER ENGINEERING & SALES
20365 Mack Avenue, Grosse Pointe Woods,
30, Michigan

COTTINGHAM BEARING CO.
401 Exposition Avenue, Dallas, Texas

DETROIT BALL BEARING CO. of Michigan
110 West Alexandrine Avenue, Detroit 1, Mich

DETROIT BALL BEARING CO. OF OHIO
325-327 Tenth Street, Toledo 2, Ohio

DIXIE BEARING & SUPPLY CO., INC.
733 N. 21st Street, Baton Rouge, La.

DIXIE BEARING AND SUPPLY CO., INC.
807 St. Charles Street, New Orleans, La.

ILLINOIS BEARING COMPANY
827 N. Broadway, Decatur, Ill.

ILLINOIS BEARING COMPANY
513 Franklin Street, Peoria, Ill.

INDIANA BEARINGS, INC.
801 N. Capitol Avenue, Indianapolis, Ind.

INDIANA BEARINGS, INC.
301 N. Madison Street, Muncie, Indiana

INDUSTRIAL BEARINGS & TRANSMISSION CO.
1535 Broadway, Kansas City, Mo.

INDUSTRIAL SUPPLY CO., INC.
1100 Third Avenue S., Minneapolis 4, Minn.

IOWA BEARING COMPANY, INC.
321 E. 2nd St., Des Moines, Iowa

KENTUCKY BEARINGS SERVICE, INC.
413-15 South Second Street, Louisville 2, Ky.

LIBERT BEARING COMPANY
Green Bay, Wisconsin

EDWARD D. MALTBY COMPANY, INC.
1718 S. Flower Street, Los Angeles, Cal.

EDWARD D. MALTBY COMPANY, INC.
120 South Fourth Avenue, Phoenix, Arizona

EDWARD D. MALTBY COMPANY, INC.
745 15th St., San Diego 2, Cal.

EDWARD D. MALTBY COMPANY, INC.
1358 Kapiolani Blvd.
Waterhouse Bldg.
Honolulu, Hawaii

MIDCAP BEARING SERVICE
605 Main Street, San Antonio, Texas

MINNESOTA BEARING COMPANY
1619 Hennepin Avenue, Minneapolis, Minn.

MINNESOTA BEARING COMPANY
409 E. Superior Street, Duluth, Minn.

MOFFATT BEARINGS COMPANY
1640 Fairmount Avenue, Philadelphia, Pa.

MOFFATT BEARINGS COMPANY
1128 Cathedral Street, Baltimore, Md.

MOFFATT BEARINGS COMPANY
304 West Morehead Street, Charlotte, N. C.

MOFFATT BEARINGS CO.
605 Hogan St., Jacksonville 2, Florida

MOFFATT BEARINGS COMPANY
432 Luckie St. N.W., Atlanta, Georgia

MOFFATT BEARINGS COMPANY
700 Third Ave., Birmingham, Ala.

MOFFATT BEARINGS COMPANY
816 W. Broad Street, Richmond, Va.

MOFFATT BEARINGS COMPANY
1501 N. Olden Ave. Ext., Trenton, N. J.

THE OHIO BALL BEARING CO.
3634 Euclid Avenue, Cleveland 3, Ohio

PENNSYLVANIA BEARINGS, INC.
5536 Baum Blvd., Pittsburgh, Pennsylvania

POWER TRANSMISSION SALES, INC.
1276 West Third St., Cleveland 13, Ohio

POWER TRANSMISSION SALES, INC.
626 Broadway, 212 Building Industries Bldg.
Cincinnati 2, Ohio

PRECISION BEARING & SUPPLY CO.
2852 Farnum Street, Omaha, Neb.

R-J BEARINGS CORP.
3300 Lindell Blvd., St. Louis, Mo.

R & M BEARINGS CANADA, LTD.
1006 Mountain Street, Montreal, Canada

R & M BEARINGS CANADA, LTD.
50 Edward Street, Toronto, Canada

R & M BEARINGS CANADA, LTD.
1006 Seymour Street, Vancouver, B.C., Canada

R & M BEARINGS CANADA LTD.
130 Ferguson Ave., Hamilton, Ont., Canada

R & M BEARINGS CANADA LTD.
1024 Oxford St., London, Ont., Canada

R & M BEARINGS CANADA LTD.
5 Bourlamaque Ave., Quebec, Canada

SYRACUSE BEARING CO.
428 E. Jefferson Street, Syracuse, N. Y.

SYRACUSE BEARING CORP.
246 St. Paul St., Rochester, N. Y.

SYRACUSE BEARING CORP.
1541 Jefferson Avenue, Buffalo, N. Y.

SYRACUSE BEARING CORP.
2903 Pine Avenue, Niagara Falls, N. Y.

SYRACUSE BEARING UTICA CORP.
State & Cooper Sts., Utica, N. Y.

TEK BEARING COMPANY, INC.
924 Lafayette Street, Bridgeport, Conn.

TEK BEARING CO., INC.
2500 East Main Street, Waterbury, Conn.

TEK BEARING COMPANY, INC.
177 Lafayette Street, N. Y., N. Y.

TEK BEARING COMPANY, INC.
100 Washington Street, Newark, N. J.

TEK BEARING COMPANY, INC.
510 Cambridge Street, Boston, Mass.

TEK BEARING COMPANY, INC.
33 Central Avenue, Albany, N. Y.

TEK BEARING COMPANY, INC.
849 N. Main Street, Providence, R. I.

TEK BEARING COMPANY, INC.
1733 Park St., Hartford, Conn.

TEK BEARING COMPANY, INC.
50 Portland St., Worcester, Mass.

TENNESSEE BEARINGS COMPANY
Kingsport & Knoxville, Tennessee

WISCONSIN BEARING COMPANY
915 N. Market Street, Milwaukee, Wisconsin

WISCONSIN BEARING COMPANY
743 W. College Ave., Appleton, Wisc.

WYOMING AUTOMOTIVE CO.
500 E. Yellowstone, Casper, Wyoming

HEIM

BALL BEARINGS

ROLLER BEARINGS

PILLOW BLOCKS

FLANGE UNITS

THE HEIM CO.

FAIRFIELD CONN., U.S.A.

MOFFATT BEARINGS COMPANY

PHILADELPHIA   TRENTON   BALTIMORE   RICHMOND
ATLANTA   BIRMINGHAM   CHARLOTTE   JACKSONVILLE

HEIM

BALL BEARINGS

ROLLER BEARINGS

PILLOW BLOCKS

FLANGE UNITS

CATALOGUE NO. 23

THE HEIM CO.
FAIRFIELD CONN., U.S.A.

H E I M
FLANGED ROLLER BEARINGS

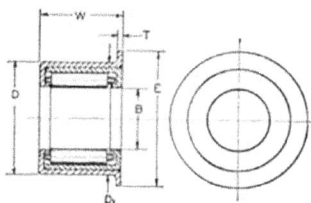

Heim Flanged Roller Bearings are self-contained. They consist of an outer casing, an inner roller race, and a full complement of rollers. The outer casing is flanged at both ends, one flange projecting outwardly and the other toward its axis, while the flange of the inner race also extends towards its axis (see sectional cut).

The casing and roller race are formed from sheet metal by the use of accurate forming dies resulting in a uniformity of size. The flanges also assure roundness of casing and roller race.

Their construction allow for the use of a full quota of rollers which results in a greater load carrying capacity for length of bearing.

The flanges at each end of the bearing retain the lubricant and effectively exclude foreign matter.

The end flanges are square with the shaft, thus providing a better surface for taking a limited amount of end thrust for location of shaft and parts.

Roller race and rollers are heat treated and rollers are ground.

RECOMMENDED MOUNTING LIMITS AND LOAD RATINGS

*Bearing No.	Shaft Limits		Recommended Housing Limits For Unground O. D.		Recommended Housing Limits For Ground O. D.		No. of Rolls	Load Ratings in Pounds Based on a Min. Shaft Hardness of 58 Rockwell C				
			HIGH	LOW	HIGH	LOW		50 RPM	100 RPM	300 RPM	500 RPM	1000 RPM
FB6-8	.375"	.374"	1.385"	1.383"	1.377"	1.375"	6	1695	1383	960	810	642
FB8-8	.500"	.499"	1.385"	1.383"	1.377"	1.375"	8	1880	1532	1062	895	710
FB10-9	.625"	.624"	1.385"	1.383"	1.377"	1.375"	10	1988	1623	1125	950	754
FB12-9	.750"	.749"	1.385"	1.383"	1.377"	1.375"	14	2130	1739	1205	1017	806
FB14-8	.875"	.874"	1.385"	1.383"	1.377"	1.375"	18	2155	1759	1219	1030	815
FB15-8	.9375"	.9365"	1.385"	1.383"	1.377"	1.375"	22	2243	1831	1270	1070	850
FB16-8	1.000"	.999"	1.385"	1.383"	1.377"	1.375"	29	2365	1931	1338	1130	895
FB16-8D	1.000"	.999"	1.8325"	1.8300"	1.8205"	1.8195"	13	3010	2460	1690	1430	1130
FB17-8	1.0625"	1.0615"	1.8325"	1.8300"	1.8205"	1.8195"	15	2510	2050	1420	1200	950
FB18-8	1.1250"	1.1240"	1.8325"	1.8300"	1.8205"	1.8195"	17	2540	2075	1435	1210	960
FB19-8	1.1875"	1.1865"	1.8325"	1.8300"	1.8205"	1.8195"	20	2620	2140	1480	1250	990
FB20-8	1.250"	1.249"	1.8325"	1.8300"	1.8205"	1.8195"	24	2700	2200	1520	1285	1020
FB21-8	1.3125"	1.3115"	2.2075"	2.2050"	2.1945"	2.1935"	15	3075	2520	1735	1465	1165
FB22-8	1.375"	1.374"	2.2075"	2.2050"	2.1945"	2.1935"	17	3150	2580	1790	1510	1200
FB23-8	1.4375"	1.4365"	2.2075"	2.2050"	2.1945"	2.1935"	19	3185	2610	1810	1530	1210
FB24-8	1.500"	1.499"	2.2075"	2.2050"	2.1945"	2.1935"	22	3275	2675	1850	1565	1240

*Letter "G" in bearing number indicates ground outside diameter. Example: FBG6-8.
Load ratings are based on 2500 hour life expectancy under normal conditions.

THE HEIM COMPANY • FAIRFIELD, CONN., U. S. A.
ANTIFRICTION BEARINGS AND PILLOW BLOCKS

8

HEIM
ROLLER BEARING PILLOW BLOCK

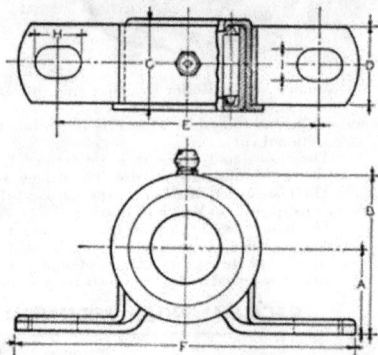

This Roller Bearing Pillow Block has a steel housing and is suited for medium and light loads with speeds up to 1000 R.P.M.

PART NO.	SHAFT DIA.	A	B	C	D	E	F	I	H
P6	3/8 "	15/16 "	1 25/32 "	1 "	7/8 "	2 7/8 "	3 3/4 "	11/32 "	1/2 "
P8	1/2 "	15/16 "	1 25/32 "	1 "	7/8 "	2 7/8 "	3 3/4 "	11/32 "	1/2 "
P10	5/8 "	15/16 "	1 25/32 "	1 1/8 "	1 "	2 7/8 "	3 3/4 "	11/32 "	1/2 "
P12	3/4 "	15/16 "	1 25/32 "	1 1/8 "	1 "	2 7/8 "	3 3/4 "	11/32 "	1/2 "
P14	7/8 "	15/16 "	1 25/32 "	1 "	7/8 "	2 7/8 "	3 3/4 "	11/32 "	1/2 "
P15	15/16 "	15/16 "	1 25/32 "	1 "	7/8 "	2 7/8 "	3 3/4 "	11/32 "	1/2 "
P16	1 "	15/16 "	1 25/32 "	1 "	7/8 "	2 7/8 "	3 3/4 "	11/32 "	1/2 "

The P series is adoptable for applications where loads are light and speeds moderate.

THE HEIM COMPANY • FAIRFIELD, CONN., U.S.A.
ANTIFRICTION BEARINGS AND PILLOW BLOCKS

H E I M

UNGROUND

BALL BEARING PILLOW BLOCK

SELF-ALIGNING

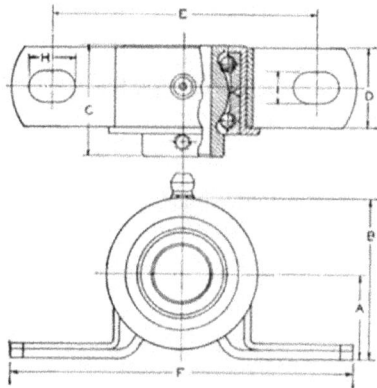

Recommended for use on commercial cold rolled shafts of fractional dimensions.

PART NO.	SHAFT DIA.	A	B	C	D	E	F	I	H
PB6	3/8 "	15/16"	1 25/32"	1 1/4 "	7/8 "	2 7/8 "	3 3/4 "	11/32"	1/2 "
PB7	7/16"	15/16"	1 25/32"	1 1/4 "	7/8 "	2 7/8 "	3 3/4 "	11/32"	1/2 "
PB8	1/2 "	15/16"	1 25/32"	1 1/4 "	7/8 "	2 7/8 "	3 3/4 "	11/32"	1/2 "
PB9	9/16"	15/16"	1 25/32"	1 1/4 "	7/8 "	2 7/8 "	3 3/4 "	11/32"	1/2 "
PB10	5/8 "	15/16"	1 25/32"	1 1/4 "	7/8 "	2 7/8 "	3 3/4 "	11/32"	1/2 "

Load ratings for PB series same as shown on page 6 for FBB series.

THE HEIM COMPANY • FAIRFIELD, CONN., U.S.A.

ANTIFRICTION BEARINGS AND PILLOW BLOCKS

Applications of Heim "Unibal" Rod End and Spherical Plain Bearings

This section contains advertisements published by the Heim Company for Spherical Rod End and Spherical Plain Bearings. The advertisements focus on Heim Rod Ends that illustrate the features and benefits of this class of plain bearing. The advertisements also illustrate the many configurations and applications of Heim Spherical Bearings across a wide range of industries.

The ten advertisements listed below were obtained from a catalog of fifty "Examples of Heim Unibal Rod End Applications" published by the Heim Company of Fairfield, Connecticut around 1960.

1. Rod End Bearings In Linkage Mechanisms on Gas-Diesel Engine made by the Cooper-Bessemer Corporation.

2. Rod End Bearings Compensate Shaft Misalignment on a Cigar Wrapping Bander made by Package Machinery Company.

3. Rod End Bearings Take Greater Radial and Thrust Loads and Correct Misalignment Conditions in Navy's Sea Master Attack Seaplane.

4. Double End Offset Rod End Bearings on "Speed-Trol" Vertical Mounted Motors made by Sterling Electric Motor, Inc.

5. Male and Female Rod End Bearings Used in Linkages on Pineapple Coning Machine made by Foster Machine Company.

6. Rod End Bearings Transmit Motion at Odd and Varying Angles on Multi-Engine Synchronizer made by Gemsco Syncontrol.

7. Rod End Bearings Provide Smoother Linkage Operation on High Speed Looms made by Draper Corporation.

8. Rod End Bearings Easily Handle Misalignment to 25 Degrees in Linkages on Portable Sewing Machines made by the Minneapolis Sewing Machine Company.

9. Unibal Spherical Bearings and Rod Ends Transmit Motion at Varying Angles in Turret Trucks made by the Hyster Company.

10. Unibal Spherical Bearings Support Heavier Loads and Correct Misalignment in Food Packing Machines made by the Anchor Hocking Glass Corporation. This advertisement contains the statement that Heim spherical bearings were developed during World War II for the exclusive use in aircraft.

A4-4-4

advanced thinking at

MARTIN...

uses the efficient

HEIM

Unibal

ROD ENDS

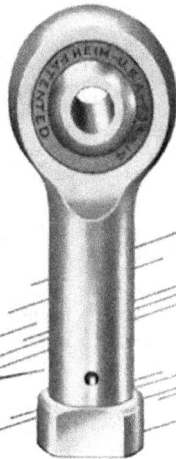

in the new *SeaMaster* XP6M-1 multi-jet attack seaplane.
Built for the Navy at Martin's Middle River plant, the big, swept-wing
flying boat is powered by four jet engines, and is designated in the over 600
mile-per-hour class. Four Allison J-71 jet engines are equipped
with afterburners to give the craft additional speed and power.

The Heim Unibal consists of a single ball, revolving in bronze bearing inserts, in an outer member of carbon steel, approved aircraft steel, or any other suitable material. They are used extensively by the aircraft industry, and were chosen by Martin for use in the fabulous, new Sea Master for their strength, efficiency, and dependability.

Heim Unibal Rod Ends are manufactured in many sizes to cover a variety of applications, other than in the aircraft industry, and fill a long-felt need for a bearing that corrects misalignment conditions, and has maximum carrying capacity because of their greater surface supporting area. They will take a greater radial and axial thrust load.

Tell us your bearing problem, and our Engineering Department
will be pleased to submit their recommendations for any application.

THE HEIM COMPANY
FAIRFIELD, CONNECTICUT

431

432

A4-4-6

HEIM
SPHERICAL BEARING
ROD ENDS

AT WORK

Foster Machine Co., Westfield, Mass.
Model 75 Pineapple Coning Machine

The Heim Unibal Spherical Bearing, as its name indicates, consists of a single, hardened steel ball through which the shaft passes. This ball revolves in bronze bearing inserts which in turn are housed in the bearing body. A really amazing bearing principle, for which machinery manufacturers are finding more and more applications every day.

One such application of this bearing in the Heim Unibal Spherical Bearing Rod End is shown above in the Foster model 75. This machine is used for winding nylon, and rayon yarns to packages for full fashion knitting machines. The linkage for which the Heim rod ends are used, must be universally flexible, free-turning and ruggedly built to withstand considerable vibration.

Wherever motion is to be transmitted at varying angles, you too will find that Heim Unibal Spherical Bearings will prove to be more dependable, more efficient, and less costly.

Complete catalogs are
available upon request

THE HEIM  COMPANY
FAIRFIELD · CONNECTICUT

HEIM *Spherical Bearing* ROD ENDS

are used wherever motion is to be transmitted at odd and varying angles.

Here is a typical application of **HEIM ROD END BEARINGS**
on the giant Le Roi 600 h.p. L-3460 engine for tough pipe line power applications.

When two or more engines are operated as a unit, the Gemsco Syncontrol is the vacuum synchronizer which efficiently synchronizes multi-engine installations. Here again is illustrated the application of the Heim Spherical Bearing Rod End to an instrument extremely sensitive in its response to the slightest change in manifold pressures.

These applications are shown through the courtesy of the General Machine & Supply Company of Odessa, Texas.

The **HEIM SPHERICAL BEARING** *embodies a simple but revolutionary principle.*

A single ball revolves in spherical bronze bearing inserts which are shaped around the ball and expanded into interlocking relationship with the outer member to form the complete rod end or bearing unit.

To illustrate this construction, we show here an exploded view of this single ball principle — the longer lasting, more efficient HEIM Spherical Bearing.

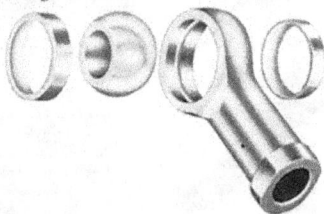

Whether your linkage mechanism is on delicate, small parts, or on heavy, giant machinery, you will find a HEIM Spherical Bearing Rod End to meet your requirements, for they are available from stock in sizes from 1/8" to 1" bore, or made to order in larger sizes.

Please write for catalog or any special engineering data to

THE HEIM COMPANY
FAIRFIELD, CONNECTICUT

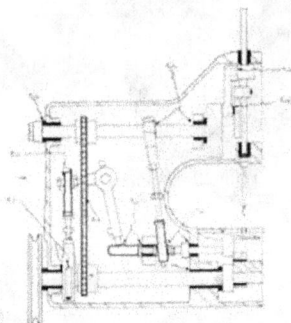

436

Throttle Control...

Simplified with

HEIM *Unibal* BEARINGS

When the new improved Hyster Turret Truck was introduced, some of the improvements included simplification and strengthening of the throttle control mechanism.

More durability, longer life, and trouble-free service were accomplished through the wider use of Heim Unibal Bearings — often referred to as spherical self-aligning bearings.

THE SINGLE BALL CORRECTS MISALIGNMENT IN EVERY DIRECTION.

Heim Unibal Spherical Bearings and Rod Ends are the solution to most problems of transmitting motion at varying angles. Rod Ends are available in a complete range of bore and thread sizes in both male and female types.

Write for Complete Catalog and a free sample of Unibal.

THE HEIM COMPANY
FAIRFIELD, CONNECTICUT

437

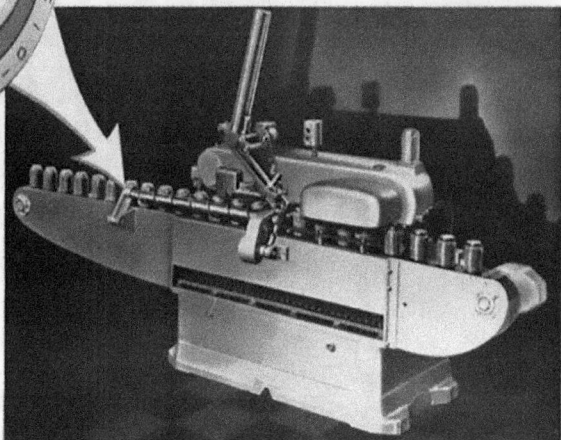 

References

Chapter 1: The Late 1800s

1. "Estimated Population of American Colonies: 1610 to 1780," U.S. Census Bureau, *Historical Statistics of the United States, Colonial Times to 1970*, Chapter Z, Colonial and Pre-Federal Statistics, 1970, p. 1,168.
2. "Population and Area: 1790 to 2010," *Statistical Abstract of the United States*, 2012, U.S. Census Bureau, 2012, Table 1, p. 8.
3. "German Immigration to America," *Siteseen Limited* (UK, 2017), http://www.emmigration.info/german-immigration-to-america.htm.
4. *Ibid.*
5. "Where and When Was the First German Immigration? German Immigrants and Their Impact in America," (Prezi, 2017), https://prezi.com/_yufapi0tgl0/where-andwhen-was-the-first-german-immigration.
6. Stanley Lebergott, "Labor Force and Employment 1800–1960," Tables 1 and 2, *National Bureau of Economic Research*, 1966.
7. "Dutchess Country, NY," Wikipedia, https://en.wikipedia.org/wiki/Dutchess_County,_New_York#History.
8. "Lewis R. Heim," *Encyclopedia of American Biography*, vol. 35 (The American Historical Company, 1966), 622–624.
9. *Ibid.*

Chapter 2: Danbury and Hatting Machines

1. "Danbury Hatting History," Danbury Historical Society (Danbury, CT), Feb. 2012, https://danburymuseum.org/history.
2. Wikipedia, "Hat Making in Danbury; Danbury, Connecticut," Feb. 2012, https://en.wikipedia.org/wiki/Danbury,_Connecticut.
3. "How a Felt Hat is Made," Nine Monthly Series of Articles: *The American Hatter*, Jan.–Sept. 1921.
4. "Lewis R. Heim," *Encyclopedia of American Biography*, Vol. 35 (The American Historical Company, 1966), 622–624.
5. *Ibid.*
6. *Ibid.*

Chapter 3: The Collar Roller and the Heim Machine Company

1. "Collars: Detachable Collars," Encyclopedia.com, 2012.
2. List of Company Directors, "Targett and Seimon Company," Danbury Land and Incorporation Records (Danbury, CT, 1915).
3. Beth Savage, "Registration Form for the Ball and Roller Bearing Company," August 25, 1989, Section 8:1, National Register of Historic Places.
4. "Bill of Sale from Prentiss Tool and Supply Company, NYC," Danbury Land and Incorporation Re-

cords, Vol. 126, p. 100 (Danbury, CT, March 31, 1905).

5. "Bond Deed: Buyers: L. R. and A. H. Heim. Seller: Frederick M. Thompson,"
Danbury Land and Incorporation Records, Vol. 126, pp. 108–110 (Danbury, CT, April 13, 1905).

6. "Bill of Sale from Jones and Lamson Machine Company," Danbury Land and Incorporation Records, Vol. 126, p. 304 (Danbury, CT, Oct. 5, 1905).

7. "Certification of Incorporation, Danbury Troy Laundry," Danbury Land and Incorporation Records, p. 309 (Danbury, CT, Recorded: October 5, 1906).

8. "Lewis R. Heim," Encyclopedia of American Biography, Vol. 35 (The American Historical Company, 1966), pp. 622–624.

9. "Ball and Roller Bearing v. Commissioner of Internal Revenue," Board of Tax Appeals (15 B.T.A. 862, 1929), March 14, 1929. (see Appendix 2, A2-4).

10. Diane Hassan, "Ball and Roller Bearing Company Expansion," Danbury Museum Historical Society (Danbury, CT, August 29, 2011).

11. "List of Officers and Directors; Danbury Troy Laundry," Danbury Incorporation Records (Danbury, CT, July 1912 and July 1913).

Chapter 4: The Ball and Roller Bearing Company

1. Robert Woodbury, "The Bicycle and Ball Bearings," History of the Grinding Machine (Cambridge, MA: MIT Press, 1959), 109–114.

2. "Dodge Brothers," Dodge Brothers Motor Car Company History, 2014, http://dodgemotorcar.com.

3. "FAG Bearings," Schaeffler Technologies AG and Co. KG, Germany; Company History from 1883 to 1945 (2016), www.fag.com/content.fag.de/en/index.jsp.

4. Ibid.

5. Testimony of W. S. Rogers, The Ball and Roller Bearing Company v F. C. Sanford Company, No. 1529, 280 F 415 (D.Conn. 1922), 173–198.

6. "Among The Shops," Machinery, Nov. 1900, 68–70.

7. Testimony of S. S. Eveland, The Ball and Roller Bearing Company v F. C. Sanford Company, No. 1529, 280 F 415 (D.Conn. 1922), 68–84.

8. "MRC Bearings, History," MRC Bearings, Wayne, PA (2016), www.mrosupply.com/popular_products/mrc-bearings.

9. Ibid.

10. "Ball and Roller Bearing v Commissioner of Internal Revenue," Board of Tax Appeals (15 B.T.A. 862, 1929), March 14, 1929. (see Appendix 2, A2-4).

11. "History of Ball and Roller Bearing Company," Ball and Roller Bearing Company website (July 2010).

12. Personal communication with Mike Smith (sales manager) and David Nohe (CEO, the Ball and Roller Bearing Company), New Milford, CT, April 11, 2013.

13. "Certificate of Incorporation of the Ball and Roller Bearing Company," Danbury Land and Incorporation Records (Danbury, CT; May 19, 1914), p. 457.

14. "Ball and Roller Bearing v. Commissioner of Internal Revenue." (see Appendix 2, A2-4).

Chapter 5: A Short History of Grinding Machines

1. Robert Woodbury, "The Grinding Machine Comes of Age," History of the Grinding Machine (Cambridge, MA: MIT Press, 1959), 51–71.

2. *Ibid*. Henry Leland Statement on Joseph Brown, 66–67.

3. Robert Woodbury, "Wheel Dimensions, About 1800"; *History of the Grinding Machine,* table inside front cover.

4. Robert Woodbury, "Charles Norton and Heavy Production Grinding," *History of the Grinding Machine,* 97–102.

5. "Heald Machine Co. History," *Vintage Machinery,* accessed April 3, 2014, www.vintagemachinery.org.

6. Robert Woodbury, "Specialized Grinding Machines," *History of the Grinding Machine,* 129.

7. "The Heald Cylinder Grinding Machine," Heald Machine Company, *Worcester Magazine,* May 1915, p. IV.

8. Robert Woodbury, "Knoop Hardness Scale Values," *History of the Grinding Machine,* 91.

9. Robert Woodbury, "Artificial Abrasives Become Widely Available," *History of the Grinding Machine,* 95.

Chapter 6: The Centerless Grinder

1. Testimony of Lewis R. Heim, The Ball and Roller Bearing Company v. The F. C. Sanford Manufacturing Company. No. 1529, 280 F 415 (D.Conn. 1922), Plaintiff's Transcript of Record), Question Q.10, 109–163 and 215–272.

2. *Ibid.,* Question Q.12.

3. *Ibid.,* Question Q.34.

4. *Ibid.,* Question Q.30.

5. *Ibid.,* Question Q.47.

6. *Ibid.,* Question Q.51.

7. *Ibid.,* "Testimony of James G. Bennett," 191–195.

8. Testimony of Lewis R. Heim, Questions Q.9, Q.60, RDQ.241, and RDQ.246.

9. Product Data Sheet Titled "Heim Centerless Grinding Machine," The Heim Grinder Company, ca. June 1924. (see Appendix 2, A2-6-3).

10. "Ball and Roller Bearing v. Commissioner of Internal Revenue" (Board of Tax Appeals, 15 B.T.A. 862, 1929), March 14, 1929. (see Appendix 2, A2-4).

11. *Ibid.*

12. "Ball and Roller Bearing v. Commissioner of Internal Revenue."

13. Beth Savage, Registration Form.

14. L. G. Henes, salesman at L. G. Henes Machine Tools, "Letter to Mann Mfg. Company Regarding Heim Centerless Grinder," May 22, 1926, (see Appendix 2, A2-6-6).

Chapter 7: Centerless Grinder Competitors

1. Harrison Gilmer, "Birth of the American Crucible Steel Industry," *The Western Pennsylvania Historical Magazine,* 36:17–36, March 1, accessed July 2017.

2. "Crucible Industries History," Crucible Industries LLC, accessed July 2018, www.crucible.com.

3. Harrison Gilmer, "Birth of the American Crucible Steel Industry, " *The Western Pennsylvania Historical Magazine,* Vo. 36, Number 1, March 1953, https://journals.psu.edu/wph/article/view/2428/2261.

4. "History of Crucible Steel," Lipsitz and Ponterio LLC, accessed July 2014, www.lipsitzponterio.com/jobsites-Crucible_Steel_History.html.

5. "Catalog of Products of the Sanderson Bros. Steel Works," Crucible Steel Company of America, Sanderson Works: 1913 edition.

6. *Ibid.*

7. Testimony of Lewis R. Heim; The Ball and Roller Bearing Company v. The F. C. Sanford Manufacturing Company. No. 1529, 280 F 415 (D.Conn. 1922), Plaintiff's Transcript of Record), 1:109–163 and 215–272.

8. *Ibid.*

9. Liberty L-12 Aircraft Engine, "History and Specifications," Wikipedia, accessed 2015, https://en.wikipedia.org/wiki/Liberty_L-12.

10. "History of Ball and Roller Bearing Company," Ball and Roller Bearing Company website (accessed July 2010).

11. Testimony of Samuel S. Eveland, The Ball and Roller Bearing Company v. The F. C. Sanford Manufacturing Company, No. 1529, 280 F 415 (D.Conn. 1922), Defendant's Transcript of Record, 68–84.

12. Testimony of Arthur Kirkland, The Ball and Roller Bearing Company v. The F. C. Sanford Manufacturing Company. No. 1529, 280 F 415 (D.Conn. 1922), Plaintiff's Transcript of Record), 178–184 and 202–206.

13. Testimony of Lewis R. Heim.

14. *Ibid.*

15. "Ball and Roller Bearing v. Commissioner of Internal Revenue" (Board of Tax Appeals, 15 B.T.A. 862, March 14, 1929), (see Appendix 2, A2-4).

16. "Reeves Pulley Company Collection, 1896–1969," Nicholson and Smith, Indiana Historical Society, Indianapolis, IN.

17. F. C. Sanford Manufacturing Company, "Automobile parts and supplies, mfr.: 45 Dewey Ct.," First listing in Bridgeport City Directory (1907, p. 489).

18. "Detroit No. 4 Heavy-Duty Centerless Grinding Machine," *American Machinist,* Jan.–June 1921, 54: 574–575.

19. "Detroit Centerless Grinder Improved Type No. 4C," *American Machinist*, March 19, 1925, 62: 479–480.

20. Edward Zdrojewski, "Frederick Geier and the Cincinnati Mill," *Cutting Tool Engineering*, June 1993, accessed Nov. 2014, www.libraries.uc.edu/business/research/bios/frederick-geier.html.

21. Frederic B. Jacobs, "Cincinnati Centerless Grinders," *Production Grinding* (Cleveland, OH: The Penton Publishing Company, 1922), 107–108.

22. Fukuo Hashimoto, Ivan Gallego, Joao F. G. Oliveira, David Barrenetxea, Mitsuaki Takahashi, Kenji Sakakibara, Hans-Olof Stalfelt, Gerd Staadt, and Koji Ogawa, "Advances in Centerless Grinding Technology," *CIRP Annals–Manufacturing Technology,* Vol. 61, issue 2, 2012, 748.

23. Carl G. Ekholm, "Manufacture of Centerless Machines at Lidkoping," letter dated May 11, 1966, (see Appendix 2, A2-9).

Chapter 8: Ball and Roller Bearing Company v. F. C. Sanford

1. Frederic B. Jacobs, "Cincinnati Centerless Grinders," *Production Grinding*, (Cleveland, OH: The Penton Publishing Company, 1922), Preface.

2. Testimony of Lewis R. Heim; The Ball and Roller Bearing Company v. The F. C. Sanford Manufacturing Company. No. 1529, 280 F 415 (D.Conn. 1922), Plaintiff's Transcript of Record, 109–163 and 215–272.

3. *Ibid.*

4. Testimony of Francis C. Sanford, The Ball and Roller Bearing Company v. The F. C. Sanford Manufacturing Company, No. 1529, 280 F 415 (D.Conn. 1922), Defendant's Transcript of Record, 230–269 and

307–309.

5. *Ibid.*

6. Testimony of Lewis R. Heim.

7. Testimony of Arthur Kirkland, The Ball and Roller Bearing Company v. The F. C. Sanford Manufacturing Company. No. 1529, 280 F 415 (D.Conn. 1922), Plaintiff's Transcript of Record, 178–184 and 202–206.

8. Bill of Complaint, The Ball and Roller Bearing Company v. The F. C. Sanford Manufacturing Company. No. 1529, 280 F 415 (D.Conn. 1922), (see Appendix 3, A3-1).

9. Testimony of Lee Chadwick, The Ball and Roller Bearing Company v. The F. C. Sanford Manufacturing Company. No. 1529, 280 F 415 (D.Conn. 1922), Plaintiff's Transcript of Record, 99–121.

10. Testimony of W. S. Rogers, The Ball and Roller Bearing Company v. The F. C. Sanford Manufacturing Company. No. 1529, 280 F 415 (D.Conn. 1922), Plaintiff's Transcript of Record, 173–198.

11. Opinion, Ball and Roller Bearing Co. v. F. C. Sanford Mfg. Co., No. 1529, 280 F 415 (D.Conn. 1922), May 6, 1922, 417.

12. Testimony of Harold Pender, The Ball and Roller Bearing Company v. The F. C. Sanford anufacturing Company. No. 1529, 280 F 415 (D.Conn. 1922), Plaintiff's Transcript of Record, 291–320.

13. *Ibid*, p. 366–423.

14. Opinion, Ball and Roller Bearing Co. v. F. C. Sanford Mfg. Co. No. 1529, 280 F 442, D.Conn. June 9, 1923, 442.

15. Opinion, Ball and Roller Bearing Co. v. F. C. Sanford Mfg. Co. No. 215, 297 F 163, 2nd Cir. Feb. 18, 1924, 163.

Chapter 9: Heim Grinder v. Fafnir

1. Frederic B. Jacobs, "Cincinnati Centerless Grinders," *Production Grinding*, (Cleveland, OH: The Penton Publishing Company, 1922), 101–102.

2. Edward Zdrojewski, "Frederick Geier and the Cincinnati Mill," *Cutting Tool Engineering*, June 1993, accessed Nov. 2014, www.libraries.uc.edu/business/research/bios/frederick-geier.html.

3. "Heim Grinder Company Incorporation," City of Danbury, Connecticut Incorporation Records, Feb. 25, 1924, 13–14.

4. "Assignment of Patents: The Reeves Pulley to The Cincinnati Milling Machine Company:6 Patents for Centerless Grinding," April 9, 1924, U.S. Patent and Trademark Office, Digest of Assignments, Vol. R-31: 205.

5. "Assignment of Patents: L. R. Heim and The Ball and Roller Bearing Company to The Heim Grinder Company, 28 Patents for Centerless Grinding," April 10, 1924, U.S. Patent and Trademark Office, Digest of Assignments, Vol. H-53: 194; and Transfer of Patents: Liber T-121, 378.

6. "Assignment of Patents: The Cincinnati Milling Machine Company to The HeimGrinder Company, 6 Patents for Centerless Grinding," May 17, 1924, U.S. Patent and Trademark Office, Digest of Assignments, vol. R-31: 215.

7. "Assignment of Patents: The Heim Grinder Company to The Cincinnati Milling Machine Company, 42 Patents for Centerless Grinding," December 30, 1926 and March 9, 1927, U.S. Patent and Trademark Office, Digest of Assignments, Vol. H-58, p. 114. Transfer No. 1: Liber G130, p. 228. Transfer No. 2: Liber V129, p. 158.

8. Plaintiff Statement Regarding Companies Manufacturing Centerless Grinders in 1924, Heim Grinder v. Fafnir Bearing (Equity No. 1747, Item 5), November 11, 1924, (see Appendix 3, A3-4).

9. Affidavit by Raymond R. Searles, VP at Fafnir Corporation, "Grinding of Outer Rings at Fafnir," Heim Grinder v. Fafnir Bearing (Equity No. 1747), May 25, 1925.

10. Testimony of Lewis R. Heim; The Ball and Roller Bearing Company v. The F. C. Sanford Manufacturing Company. No. 1529, 280 F 415 (D.Conn. 1922), Plaintiff's Transcript of Record, 109–163 and 215–272.

11. Opinion, Heim Grinder Co. v. Fafnir Bearing Co. (No. 1747, 13 F2d 408, D.Conn. May 17, 1926), 408.

12. "Assignment of Patents: The Heim Grinder Company to the Cincinnati Milling Machine Company," December 30, 1926 and March 9, 1927.

13. "Cincinnati and Heim in Consolidation," *American Machinist*, September 16, 1926, 510a, (see Appendix 2, A2-5-6).

14. "Lewis Rasmus Heim," Obituary, *Danbury News Times*, April 1, 1964.

15. Edward Zdrojewski, "Frederick Geier and the Cincinnati Mill," *Cutting Tool Engineering*, June 1993. Source: University of Cincinnati website, accessed 2014, www.libraries.uc.edu/business/research/bios/frederick-geier.html, Gold Coin Payment.

16. Carl G. Ekholm, "Manufacture of Centerless Machines at Lidkoping," letter dated May 11, 1966. (see Appendix 2, A2-9).

17. Robert Woodbury, "Heim Automatic Correction Action," *History of the Grinding Machine*, 151–155.

18. "Lewis R. Heim," Encyclopedia of American Biography, vol. 35 (The American Historical Company, 1966), 622–624.

19. Fukuo Hashimoto et al, "Advances in Centerless Grinding Technology," *CIRP Annals–Manufacturing Technology*, Vol. 61, issue 2, 2012, 747–770.

20. Beth Savage, "Registration Form for the Ball and Roller Bearing Company," *National Register of Historic Places*, August 25, 1989.

Chapter 10: The Florida Years

1. James M. Laux, *Mount Dora, Florida: A Short History* (Dollar Bill Books, Ann Arbor, MI., 2003), 28–30.

2. *Ibid.*

3. "Patent Assignment: L. R. Heim of Mount Dora, Florida to The Cincinnati Grinder Company of Cincinnati, Ohio," July 30, 1927, One Patent for Grinding Machinery (granted no. 1,958,001 on May 8, 1934, US Patent and Trademark Office: Digest of Assignments) H-58: 293.

4. "Patent Assignment: The Cincinnati Grinder Company of Cincinnati, Ohio to The Heald Machine Company of Worcester, MA," May 6, 1932, 26 patents assigned. Included One Patent for Grinding Machinery (no. 1,958,001, granted May 8, 1934, U.S. Patent and Trademark Office: Digest of Assignments) K-31:104.

5. Beth Savage, "Registration Form for the Ball and Roller Bearing Company," *National Register of Historic Places*, August 25, 1989, Section 8, p. 3.

6. *Ibid.* Chapter 11: The Heim Company

Chapter 11: The Heim Company

1. "Fairfield War Plant Beats Nazis at Own Game," *Bridgeport Sunday Post*, April 23, 1944, *(see Appendix 4, A4-1)*.

2. "Heim-Made Bearings Fly the Skies in 747 and All Other U.S. Planes," *Bridgeport Sunday Post*, March 1, 1970.

3. "Lewis R. Heim," *Encyclopedia of American Biography,* Vol. 35 (The American Historical Company, 1966), 622–624.

4. "Fairfield War Plant Beats Nazis at Own Game," *Bridgeport Sunday Post,* April 23, 1944.

5. "Heim Hardened Dowel Pins," advertisement in *American Machinist,* June 1, 1938, 488.

Chapter 12: The Rod End Bearing

1. "Fairfield War Plant Beats Nazis at Own Game," *Bridgeport Sunday Post*, April 23, 1944. *(see Appendix 4, A4-1).*

2. "Heim Rod End Bearings Shipped to AAFMC at Wright Field," personal communication with Carl H. Van Etten, grandson of Lewis Heim and former Heim Company employee, April 1, 2017.

3. "Fairfield War Plant Beats Nazis at Own Game," *Bridgeport Sunday Post.*

4. "Rose Joint," personal communication with Carl H. Van Etten, grandson of Lewis Heim and former Heim Company employee, April 1, 2017.

5. "Rose Joint," personal communication with Andy Henn, Heim Company machinist and mechanical engineer from 1961–2015, December 14, 2016.

6. "Lewis R. Heim," *Encyclopedia of American Biography*, Vol. 35 (The American Historical Company, 1966), 622–624.

Chapter 13: Unibal and the Art of Making Industrial Bearings

1. Personal communication with Carl H. Van Etten, grandson and former Heim Company employee, April 1, 2017.

2. *Ibid.*

3. *Ibid.*

4. Personal communication with Andy Henn, Heim Company machinist and mechanical engineer from 1961–2015, December 14, 2016.

5. Personal communication with Carl H. Van Etten.

6. Personal communication with Andy Henn.

7. "The History of Van Kampen Investments," Milestones Historical Consultants *(accessed 2017)*, http://milestonespast.com/vankampen.htm, p. 1.

Chapter 14: Accomplishments

1. "Lewis R. Heim," *Encyclopedia of American Biography*, Vol. 35 (The American Historical Company, 1966), 622–624.

2. Personal communication with Carl H. Van Etten, grandson and former Heim Company employee, April 1, 2017.

3. Personal communication with Andy Henn, Heim Company machinist and mechanical engineer from 1961–2015, December 14, 2016.

4. "Heim Universal Headed by Albert R. McCloskey," *Bridgeport Post*, March 29, 1970, p. 60.

5. Personal communication with Carl H. Van Etten.

6. *Ibid.*

7. *Ibid.*

8. Personal communication with Charles W. Heim, grandson, January 22, 2017, and August 26, 2017.

9. "Heim-Made Bearings Fly the Skies in 747 and All Other US Planes," *Bridgeport Sunday Post,* March 1, 1970.

10. Heim family records of Bonnie Heim Perkins, granddaughter of Alfred Heim, 2010–2017.

11. Personal communication with Carl H. Van Etten.

12. Heim family records of Bonnie Heim Perkins.

13. Personal communication with Carl H. Van Etten.

14. Personal communication with Charles W. Heim.

15. Personal communication with Carl H. Van Etten.

16. Personal communication with Charles W. Heim.

17. Personal communication with Carl H. Van Etten.

18. "Draper Corp. Acquires Heim, Universal Bearing," Bridgeport Post, April 3, 1963.

19. *Ibid.*.

20. "Heim-Made Bearings Fly the Skies in 747 and All Other US Planes," *Bridgeport Sunday Post*, March 1, 1970.

21. "Draper Corp. Acquires Heim, Universal Bearing," *Bridgeport Post*, April 3, 1963.

22. "Heim Among 11 Rockwell Companies Acquired by New Parent Corporation," *Bridgeport Post*, January 23, 1978.

Index

spherical bearing: See "bearings"

spot grinding: xv, 85-86, 175, 286

stamped sheet metal bearing: xvi, xvii, 187-191, 195, 198-201, 205, 210, 226, 228, 259, 411

Standard Oil Company: 2

Standard Roller Bearing Company: 37, 41, 42, 105-107, 146, 398, 400

Standard Steel Bearing Company: 42

steel: 1, 2, 26, 33, 34, 35, 38, 39, 41, 42, 45, 46, 48, 55, 57, 60-63, 65, 67-69, 74, 96, 99-101, 105, 131, 141, 157, 173, 187, 192, 205, 209, 212, 214, 225, 238, 242, 258, 283, 415

Stephenson, George: 101. See also French and Stephenson (F & S)

Suriray, Jules: 35

sweepstick (definition): 232 - 234

Sylvan Shores and Lake Gertrude, Florida: xvi, 184 – 186, 260 -261

T

Taft, Howard: 174

Targett, Edwin: 22, 24

T. Brothwell and Company: 11

TEK Bearing Company (Bridgeport, CT): 258

telescopically interlocked construction (bearings): xvii, 191–193, 198 – 199, 210

Thomas, Edwin S.(Judge) : 139, 142-143, 150, 154-160, 162, 172-174

Thompson, Frederick M.: 26-27

through grinding: xv, 85, 86

Timken Roller Bearing Axle Company: 37, 45, 50

Torrington Company: xiv, 38, 42, 48, 50, 97, 195

Troy, New York: 21

truing devices (for grinding and regulating wheels): xv, 59, 61, 81, 84, 87, 88-89, 92, 94, 119, 124, 168, 175

TRW Bearing Company: 42

Turner Machine Company: xiii, 10, 12, 15, 17-19, 263

U

Unibal (definition): 210

Unibal ball bearing (deep groove): xix, 259–260

Unibal rod end bearing: xviii, 227-228, 411

Unibal spherical bearing: xviii, 227-228, 235, 411

United Motors Parts and Accessories Company: 39, 46

United States Court of Appeals; 2nd Circuit (including U. S. Federal Judges Manton, Hough, and Mayer): 159-162, 165-167, 169, 174, 315, 409

United States District Court of Connecticut: 139-141, 147, 150, 152, 154-156, 159, 161-162, 165-166, 172-174

United States Patent and Trademark Office (USPTO): vii, 72, 82, 103, 113, 138, 160, 168, 188, 250, 391-392, 395

Universal Bearing Corporation: 259, 262

U.S. Ball Bearing Manufacturing Company: 50

V

Van Etten: See Heim

W

Waters, Charles: 131, 193, 267, 270,

Watt, James: 53–54

Weed, William: 72, 82, 103, 113, 138, 160, 168, 188, 250, 391-392, 395

White, Bob: 183

White Star Laundry (company): 260

Whitin Machine Works (Whitinville, MA): 232

Wilkie, Leighton: 195

Wilkinson, John: xiv, 53-55, 59

Woodbury, Robert: 53, 83-84, 178

Wooster, A. M.: xvi, 75, 137-139, 198

Wright Field: 210, 415